CRUSH MECHANICS
of Thin-Walled Tubes

CRUSH
MECHANICS
of Thin-Walled
Tubes

Dai-heng Chen

CRC Press
Taylor & Francis Group
Boca Raton London New York

CRC Press is an imprint of the
Taylor & Francis Group, an **informa** business

CRC Press
Taylor & Francis Group
6000 Broken Sound Parkway NW, Suite 300
Boca Raton, FL 33487-2742

First issued in paperback 2017

© 2016 by Taylor & Francis Group, LLC
CRC Press is an imprint of Taylor & Francis Group, an Informa business

No claim to original U.S. Government works

ISBN-13: 978-1-4987-5517-7 (hbk)
ISBN-13: 978-1-138-74858-3 (pbk)

Library of Congress Cataloging-in-Publication Data

Names: Chen, Dai-heng, author.
Title: Crush mechanics of thin-walled tubes / Dai-heng Chen.
Description: Boca Raton : Taylor & Francis, 2016. | Includes bibliographical references and index.
Identifiers: LCCN 2015037872 | ISBN 9781498755177 (alk. paper)
Subjects: LCSH: Tubes. | Deformations (Mechanics) | Buckling (Mechanics)
Classification: LCC TA492.T8 .C45 2016 | DDC 671.8/3--dc23
LC record available at http://lccn.loc.gov/2015037872

Visit the Taylor & Francis Web site at
http://www.taylorandfrancis.com

and the CRC Press Web site at
http://www.crcpress.com

Contents

Preface

Many thin-walled structures are used in vehicles as impact-energy absorption components, and it is quite important to correctly evaluate crushing phenomena exhibited by such structures during impact. In 2002, the author began a new research project aimed at development of novel collision devices for use in vehicles. These devices were developed within a consortium including a number of Japanese automotive parts suppliers. Since that time, there has been extensive investigation in the literature on the crushing behaviors of thin-walled structures, and many research results have been written and published. These describe the current understanding of how such crushing deformations happen and which factors determine the crushing behaviors. Initially, these reports were used as seminar documents for graduate students and partners within the consortium. These reports proved popular among the initial recipients, and it was decided that commercial printing would be of interest to the broader scientific and engineering community.

The purpose of this book is to provide a basic understanding of the fundamental concepts and mechanisms of crushing deformations in thin-walled structures. The hope is that interested readers, particularly beginners, can easily and systematically grasp the knowledge required in the analysis and design of energy absorption components. Although there are many textbooks on buckling and post-buckling of thin-walled structures, there are few books of a similar nature aimed at those new to the field. In this book, the crushing deformations of circular and square tubes—the simplest hollow thin-walled structures—under axial compression, bending, and torsion are explained in detail. The book focuses on these simple and basic problems and attempts in this way to provide readers with a clear physical understanding and useful analytical techniques.

When examining the deformation behaviors of thin-walled structures under dynamic loading, such as in a collision, it is necessary to account for various effects, such as inertial force and the strain-rate dependence of the dynamical characteristics of specific materials. However, in order to grasp such dynamic behavior, it is illuminating to first investigate the deformation behaviors of thin-walled structures under a quasi-static load, for which the influences of inertial force, strain-rate dependence, and other such factors can be ignored. Moreover, the impact velocity during a vehicle crushing accident is about 40 m/s at most. The influence of the stress-wave speed on deformation characteristics during impact weakens as the impact velocity decreases. This was shown in a study that the author conducted on the effect of impact velocity on the peak load of circular tubes subjected to axial compression. Taking into consideration the aims mentioned above, this book discusses mainly quasi-static crushing behaviors.

The structure of this book is as follows. Chapter 1 addresses axial compression of circular tubes, and Chapter 2 addresses axial compression of square tubes. Through these two chapters, the underlying concepts, such as features of buckling and post-buckling behavior of thin-walled hollow members, are shown in detail. In particular, the mechanism of collapse and associated calculations for the initial peak force and the average compressive force are summarized.

The geometry of a tube is an important factor in its energy absorption capacity. One well-known way, described by many researchers, to create a high-performance energy absorption device is to introduce corrugations along the axis or circumference of a tube. Chapter 3 discusses how introducing axial and circumferential corrugations affects the force-displacement and energy-absorption characteristics of tubes.

In general, both an axial compressive load and a bending moment are applied to all members of a vehicle when a collision occurs. It is therefore important to understand the collapse behavior of members undergoing bending deformation when trying to evaluate strength and energy-absorption characteristics. Chapter 4 discusses the bending deformation of circular and square tubes. In particular, the characteristic flattening phenomenon, the maximum moment in bending deformation, and the moment-rotation relation during bending collapse are discussed in this chapter.

Thin-walled members with an open cross section are commonly used as structural components. Chapter 5 discusses the collapse behavior of thin-walled structures with an open cross section during axial crushing and bending deformation.

Pure torsional load is seldom applied to thin-walled members in practice, but both axial and bending deformations are frequently accompanied by torsional load. To understand deformation behavior under such combined loads, we must first clarify the properties of collapse deformation of thin-walled members under torsional loads. Chapter 6 discusses torsion of tubes.

Data related to the topics of this book that have been made public are selected and summarized throughout the book. However, in order to obtain a more precise understanding, the author conducted a series of investigations about these issues. The book therefore contains many new results obtained by the author and members of the research consortium to which the author belongs. The investigations are carried out mainly by means of numerical simulation using the finite element method (so-called FEM analysis) because modern computers have advanced greatly, and nowadays, computer analysis is on equal footing with laboratory experiments. Therefore, to demonstrate the maximum load in collapse and the load-displacement relation after collapse for the circular and square tubes subjected to axial compression, bending, or torsion, the author also shows a comparison with numerical results at the same time that the existing theoretical approaches are introduced. In addition, some other original research results, beyond the comparisons, are included in the book. These results are summarized as follows.

In Chapter 1, "Axial compression of circular tubes," the effect of flanges in cylinder ends on the deformation mode of tubes, as well as conditions of changes in deformation modes, were investigated. Additionally, formulas are developed for evaluating the curvature radius ρ_f at the wrinkle tip, the eccentricity ratio m and the maximum strain $\varepsilon_\theta|_{max}$.

In Chapter 2, "Axial compression of square tubes," it is found that the deformation of square tubes is characterized by two stages, which is not the case with circular tubes. First, shallow buckles are created over the entire length of the square tube, after which they are crushed one by one. Additionally, for the collapse behavior of a plate subjected to axial compression, it is shown that the axial collapse of the plate is dominantly controlled by two factors: the stress limitations of the material and the limitations on in-plane lateral deformation at the side edges of the plate. Further, a simplified method is proposed for predicting such collapse stresses as a function of the two factors.

In Chapter 3, "Crushing of corrugated tubes," the deformation characteristics of tubes with a surface corrugated in the axial or radial direction under axial loading are investigated systematically on the basis of the author's research results. In particular, it is shown that the axisymmetric deformations corresponding to the introduced axial corrugations can be classified into two modes: S-mode deformation and P-mode deformation. The crushing force in S-mode deformation continues increasing without fluctuation. Additionally, a simple method is proposed for predicting the average crushing force of a corrugated tube.

In Chapter 4 "Bending collapse of tubes," the flattening phenomenon and the maximum bending moment are investigated in detail. This includes the proposition of a new method for evaluating the maximum bending moment of square tubes with consideration of sidewall buckling.

In Chapter 5, "Thin-walled structure with an open cross section," the mechanism of collapse due to buckling is analyzed by using U-shaped and V-shaped beams (with compressive stress applied to the flange) as examples, and the mechanism of collapse due to cross-sectional flattening is analyzed by using V-shaped and W-shaped beams (with tensile stress applied to the flange) as examples. For flattening-induced collapse, because it is typically difficult to evaluate the correspondence between cross-sectional flattening and rotation angle, the author proposes a new technique that can be used to determine the relation between the bending moment M and the rotation angle θ.

In Chapter 6, "Torsion," it is found that there are two possible cases for torsion-induced collapse of square tubes, as for bending collapse: collapse due to buckling and collapse due to flattening, and analysis methods are proposed for predicting the maximum torsion moment in each case.

The author believes that this book will be helpful for deepening readers' understanding of the basic concepts and analysis techniques of crushing deformation of thin-walled structures. The target audience for this book is engineering students, practicing mechanical and structural engineers, and researchers interested in analyzing energy absorption and designing structures that may undergo impacts. The book can be used for both teaching and self-study.

The author would like to thank Dr. Kuniharu Ushijima for his cooperation in research, valuable advice and support in manuscript preparation.

Jiangsu University, Professor
Tokyo University of Science, Professor Emeritus
Chen Daiheng

Notation

Major notation used in this book is given below.

Notation applied common to each chapter

Symbol	Definition
	—— Geometry ——
R	Radius of a circular cross section
$C_i \;\; (i = 1, \cdots, n)$	Side length of a n-gonal cross section
C	Side length of a regular polygonal cross section
b	Width of a plate
t	Thickness of a plate, or wall thickness of a tube
L	Length of a plate, or length of a tube
A	Area of cross section for a tube
A_s	Area surrounded by the cross section for a tube
V	Volume of a tube
A'	Equivalent area of cross section for a tube with stiffeners $(= V/L)$
	—— Material ——
E	Young's modulus
G	Shearing modulus
ν	Poisson's ratio
σ_s	Initial yield stress
σ_s^d	Dynamic plastic yield stress
σ_u	Tensile strength, namely ultimate tensile stress
σ_0	Energy equivalent flow stress
E_t	Tangent modulus $(= \partial\sigma/\partial\varepsilon)$
E_s	Secant modulus $(= \sigma/\varepsilon)$
$\left. \begin{array}{ll} E_{xx}, & E_{x\theta} \\ E_{\theta\theta}, & G_{xz} \end{array} \right\}$	Coefficients in the relation between stress increment and strain increment, as shown in Eq. (1.26)
H_T, η_T	Coefficients used in J_2 incremental theory, as shown in Eq. (1.28) and Eq. (2.22)
H_s, η_s	Coefficients used in J_2 deformation theory, as shown in Eq. (1.29) and Eq. (2.23)
ν_p	Plastic Poisson's ratio
D	Flexural rigidity of plate $(= Et^3/12(1 - \nu^2))$

Symbol	Definition
M_p	Fully plastic bending moment considering effect of plane strain $(= \sigma_s t^2 / (2\sqrt{3}))$
M_0	Fully plastic bending moment based on the flow stress σ_0 $(= \sigma_0 t^2 / 4)$
$f_\sigma(\varepsilon)$	Stress-strain formula of material
μ_f	Friction coefficient

—— Mechanics ——

Symbol	Definition
x, y, z	Cartesian coordinates, as shown in Fig. 1.3, Fig. 2.9, etc.
u, v, w	Displacements in x, y and z directions
N_x, N_y, N_{xy}, N_{yx}	Force components defined in Figs. 1.3, 2.9
M_x, M_y, M_{xy}, M_{yx}	Moment components defined in Figs. 1.3, 2.9
σ	Stress
ε	Strain
τ	Shear stress
γ	Shear strain
ε^p	Plastic strain
ε_θ	Circumferential strain
$\varepsilon_\theta\vert_{max}$	The maximum value of ε_θ
ε_f	Average final strain, see Eq. (1.92)
P	Crushing force in axial compression
δ	Crushing distance in axial compression
δ_e	Effective crushing distance
P_x	Compression force in x direction for compressed plate
U_x	Compression displacement in x direction for compressed plate
P_{buc}	Buckling force
σ_{buc}	Buckling stress
σ_{buc}^e	Elastic buckling stress
σ_{buc}^p	Plastic buckling stress
σ_{buc}^d	Dynamic buckling stress
k	Buckling coefficient
P_{col}	Collapse force (the first peak crushing force in axial compression)
σ_{col}	Collapse stress $(= P_{col}/A)$
P_{ave}	Average crushing force
σ_{ave}	Average crushing stress $(= P_{ave}/A)$
E_{Ab}	Absorbed energy in crushing
M	Moment
θ	Rotation angle

Notation used for a specific chapter

Symbol	Definition	Chapter
a_t, b_t	Radiuses of toroidal shell element, as shown in Fig. 2.33	2
a_{el}, b_{el}	Semi-major and minor axis of an ellipse approximating the flattened cross section of a tube	4
B	Coefficient in Eq. (2.68)	2
B_1, B_2	Coefficient in Eq. (2.29)	2
b_{eff}	Effective width	2
$b_{eff}|_{cri}$	Critical effective width	2
b_{e1}, b_{e2}	Effective widths for a plate under stress gradient shown in Fig. 4.36	4
b_f	Width of flange for U-Shaped or V-Shaped thin-wall member	5
b_w	Width of web for U-Shaped thin-wall member	5
C	Elastic wave speed for a one-dimensional rod	1
C_μ	Coefficient of flattening ratio, see Eq. (4.17)	4
e	Coefficient defined by Eq. (2.25)	2
e	Location at which force P is applied, as shown in Fig. 5.4	5
I_1, I_3	Integrals defined in Eq. (2.84), (2.92)	2
m	Eccentricity ratio $(= (R_{out} - R)/(R_{out} - R_{in}))$	1
m	Number of half-wave in the length direction	1, 2, 6
M_{buc}	Moment at buckling of a tube subjected to bending or torsion	4, 6
M_{max}	Maximum moment $(= M_{col})$	4, 6
M_{col}	Collapse moment $(= \min(M_{cri}, M_{lim}))$	4, 6
M_{cri}	Critical moment caused by the stress limit of the material	4
M_{lim}	Limit moment caused by flattening	4, 6
M_{el}	Maximum elastic bending moments of a square tube	4
M_{pl}	Cross-sectional fully plastic bending moments of a square tube	4
M_{pm}, M'_{pm}, M''_{pm}	Fully plastic moments, as shown in Fig. 5.2	5
M_s	Cross-sectional fully plastic torsional moments of a square tube	6
n	Number of sides for a n-gonal cross section	2
n	Number of circumferential waves in buckling of cylinder	1, 6
n	Number of ribs for a tube with stiffeners, as shown in Fig. 1.67(b)	1

Symbol	Definition	Chapter
n	Number of waves for a radially corrugated tube, as shown in Fig. 3.44	3
n	Strain hardening exponent	1, 2, 4
R_{out}, R_{in}	Outer and inner radiuses of a wrinkle, as shown in Figs. 1.26 and 1.27	1
r_s	Radius at the corner of square cross section, see Fig. 3.19	3
t_f	Thickness of flange for U-Shaped or V-Shaped thin-wall member	5
t_f	Thickness of flange for a tube, as shown in Fig. 1.59	1
t_w	Thickness of web for U-Shaped thin-wall member	5
t'	Thickness of partition wall in a polygonal tube	2, 4
V_t	Movement velocity of a point in the toroidal area, as shown in Fig. 2.33	2
α, β	Inclination angles, as shown in Figs. 1.31 and 1.32	1
α, β	Angles used in folding mechanism, as shown in Fig. 2.32	2
γ	Coefficient in Eq. (2.65)	2
Δ	Out-of-plane displacement, as shown in Fig. 5.3	5
$\Delta - \tau$	Flattening quantity in square tube under torsion, as shown in Fig. 6.19	6
δ_Q	Deflection at point Q, as shown in Fig. 5.24(a)	5
δ_c	Deflection at the midpoint of a beam, as shown in Figs. 5.8, 5.10 and 5.19	5
$\dot{\varepsilon}$	Equivalent strain rate	1
ε_{cri}	Critical compressive strain in side area at collapse of a plate subjected to compression	2
ε_{edge}	Strain at the side edge of a plate subjected to compression	2
ε_{lim}	Compressive strain in side area at collapse due to stress limit for a plate subjected to compression	2
ε_{lim}	Maximum compressive strain at the bottom surface when the bending moment reaches the maximum value M_{lim}	4
θ	Angle in polar coordinates as shown in Figs. 1.4, 4.3, etc.	1, 4
θ_2	$= \theta/2$	4, 5
θ, ϕ	Angle coordinate used in toroidal curved surface, as shown in Fig. 2.33	2
η	Structural effectiveness $(= \sigma_{ave}/\sigma_0)$	1, 2

Symbol	Definition	Chapter
η_p	Plastic reduction factor $(= \sigma_{buc}^p / \sigma_{buc}^e)$	2, 6
κ_{buc}	Curvature corresponding to M_{buc}	4
κ_g	Bending curvature	4
κ_{lim}	Curvature corresponding to M_{lim}	4
k_m	Mindlin's shear correction factor	1
λ	Half-wavelength for wrinkle formed in crushing	1
λ	Plate slenderness $\left(= \sqrt{\dfrac{\sigma_s}{\sigma_{buc}^e}} \right)$	2, 4, 6
λ_0	Half-wavelength of fold for non-corrugated tube to distinguish from that for corrugated tube	3
λ_c, a_c	Half-wavelength and amplitude of corrugation for a corrugated tube	3
λ_g, w_g, d_g	Interval, width and depth of grooves for an axially grooved tube, as shown in Fig. 3.1, or for a radially corrugated tube, as shown in Fig. 3.48	3
λ_r	Half-interval of rings for a circular tube with ring stiffeners, see Fig. 1.67	1
μ	Flattening ratio for tube bending	4
ρ	Density	1
ρ	Knock-down factor $(= \sigma_{col}|_{exp} / \sigma_{buc}^e)$	1
ρ	Curvature radius at a wrinkle vertex	1, 3
ρ_b	Circumferential curvature radius at the bottom surface of a tube under bending	4
ρ_c	Curvature radius at a wave vertex for corrugation, see Fig. 3.17	3
ρ_f	Final curvature radius at the crest of a fold	1, 3
ρ_s	Radius of a circular arc used to approximate the flattened cross section of a square tube as shown in Figs. 4.33 and 6.19	4, 6
σ_c	Constant term in Eq. (2.35)	2
σ_{cri}	Maximum average stress of a compressed plate caused by in-plane lateral deformation	2
σ_{lim}	Maximum average stress of a compressed plate caused by the stress limit in the material	2
τ_{buc}	Buckling stress of a tube subjected to torsion	6
τ_{buc}^e	Elastic buckling stress of a tube subjected to torsion	6
τ_{buc}^p	Plastic buckling stress of a tube subjected to torsion	6
ϕ	Solidity ratio $(= A/A_s)$	1,2
ϕ	Taper angle for a frustum, as shown in Fig. 3.33	3
ϕ_0, ϕ_0'	Angle shown in Fig. 4.23(c), Fig. 4.26	4

Symbol	Definition	Chapter
ϕ_s	Central angle of a circular arc used to approximate the flattened cross section of a square tube as shown in Figs. 4.33 and 6.19	4, 6
ψ_0	Half the external angle for a regular polygonal cross section $(= \pi/n)$	2
ψ_k	Half the external angle for the kth corner of a polygonal cross section	2

As for the strain-hardening characteristic of material, the following five kinds of approximate expressions are mainly used in this book.

(1) Bilinear hardening rule:

$$\sigma = \sigma_s + E_t \left(\varepsilon - \sigma_s / E \right)$$

(2) n power hardening rule:

$$\sigma = E_a \left(\varepsilon^p \right)^n$$

(3) Swift's n power hardening rule:

$$\sigma = E_a \left(\varepsilon_0 + \varepsilon^p \right)^n$$

(4) Ludwik's n power hardening rule:

$$\sigma = \sigma_s + E_a \left(\varepsilon^p \right)^n$$

(5) Ramberg-Osgood equation:

$$\varepsilon = \frac{\sigma}{E} \left[1 + \frac{3}{7} \left(\frac{\sigma}{\sigma_s} \right)^{N-1} \right]$$

In addition, to avoid repetitive explanations of the parameters used in this book, unless otherwise noted, the material is assumed to obey bilinear hardening rule.

1

Axial compression of circular tubes

For efficiently absorbing the impact energy of collisions involving various types of transport machinery including automobiles, lightweight crash elements with thin-walled cross sections are now widely used in the front and rear parts of vehicles. To accurately understand the deformation behavior of these elements under compression, it is first necessary to clarify the deformation behavior of simply shaped thin-walled circular and square tubes under axial compression. This chapter addresses axial compression of circular tubes, and the next chapter addresses axial compression of square tubes.

After Fairbairn and Hodgkinson's experimental study on the quasi-static axial compression of thin-walled circular tubes in 1846 [201], many theoretical and experimental studies have been conducted on the axial compression and buckling of circular tubes (e.g., Lilly [126], Lorenz[128], Timoshenko[199], Southwell[191], Dean[62]).

In a thin-walled circular tube made of aluminum alloy A5052 with one end supported and the other end under axial compression (Fig. 1.1), the deformation mode depends on parameters such as the thickness/radius ratio t/R and the length/radius ratio L/R of the tube. The deformation modes are roughly classified into two types: (1) a type called **progressive folding collapse**, where wrinkles are formed by local buckling of the tube wall and then the circular tube collapses while standing straight up as local buckles are formed progressively one after another (Fig. 1.1(a) and (b)); and (2) a global bending type called **Euler buckling**, where the circular tube flexes and collapses by buckling (Fig. 1.1(c)). For efficient impact energy absorption, it is preferable to design progressive buckling compression of circular tubes.

In discussing the progressive folding collapse of circular tubes, it is first necessary to get to know the main features of the compressive force variation. Fig. 1.2 shows a schematic of a typical relation between axial compressive force and compression displacement of a circular tube. Generally, as shown in the figure, the force monotonically increases until the first buckling occurs (this usually gives the maximum force throughout the compression process); after the **collapse force**, the force repeatedly increases and decreases, showing some number of intermediary force peaks until total collapse occurs. The shaded area in the force-displacement diagram shows the amount of energy absorbed during axial compression of the circular tube, as evaluated by the product of average force and maximum displacement (called **effective crushing distance**) in the compression process. In designing a circular tube as an energy absorber, it is desired that absorbed energy be maximized while the

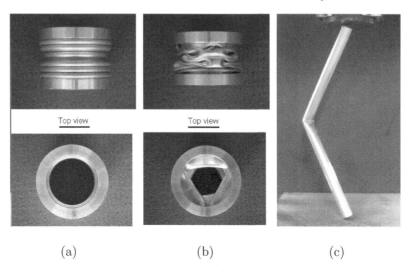

(a)   (b)   (c)

FIGURE 1.1
Examples of circular tubes under axial compression (aluminum alloy A5052):
(a) axisymmetric compressive deformation ($R = 26$ mm, $t = 2$ mm, $L = 150$ mm); (b) non-axisymmetric compressive deformation ($R = 25$ mm, $t = 1$ mm, $L = 150$ mm); (c) global bending ($R = 24$ mm, $t = 2$ mm, $L = 1000$ mm).

maximum force, transmitted as impulsive force, is minimized. From this viewpoint, in regard to the axial compressive force applied to a circular tube, the following four parameters are widely used: (1) **collapse force** (P_{col} in the figure), (2) **average crushing force** (P_{ave} in the figure) or **force efficiency** defined as the ratio of average force to maximum force, (3) **absorbed energy** per unit mass and (4) **effective crushing distance** (δ_e in the figure).

The **progressive folding collapse** of circular tubes is further divided into two groups: the **axisymmetric deformation mode** (Fig. 1.1(a)), and the **non-axisymmetric deformation mode** (Fig. 1.1(b)) with waveforms in the circumferential direction. The non-axisymmetric progressive folding collapse is normally found in thin-walled circular tubes, while axisymmetric progressive folding collapse is often found in relatively thick-walled circular tubes (ordinarily $R/t \leq 40$ in aluminum alloys). Note that, as discussed later in Section 1.4, depending on factors such as the tube wall thickness, the constraint at the end part, and the flange thickness, wrinkles of the axisymmetric deformation mode are observed prior to the transition to non-axisymmetric deformation. However, the reverse course, the transition from non-axisymmetric to axisymmetric deformation, is not observed under any circumstances.

In this chapter, the axisymmetric deformation is first discussed in detail, and the non-axisymmetric deformation is then discussed. Then, the deformation modes found in an axially compressed circular tube are surveyed.

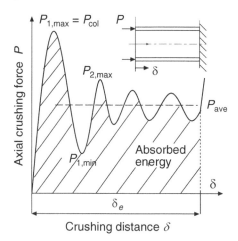

FIGURE 1.2
Relationship between crushing force and crushing distance in progressive folding collapse of a circular tube.

1.1 Axisymmetric buckling of circular tubes

In the progressive folding collapse of a circular tube, the basic phenomenon is the formation of buckles, which is divided into two steps: local buckling deformation due to the compressive force, and then the folding of the tube wall to form a wrinkle.

Buckling is an instability that occurs when an external force causes the deformation of a structure to reach a certain threshold where a new deformation mode different from the prior mode emerges and the deformation switches to this new mode. If the compressive force reaches a certain value in a circular tube under axial compression, not only compressive deformation but also buckling deformation appears; there are two possible types of buckling: **global buckling**, called **Euler buckling** (Fig. 1.1(c)), and **local buckling** (Fig. 1.1(a) and (b)), in which wrinkles appear on a part of the cylindrical surface. To absorb impact energy through the axial compressive deformation of a circular tube, local buckling is used.

Local buckling that occurs during the compression of a circular tube is classified into two groups: **axisymmetric buckling** and **non-axisymmetric buckling**. In this section, the axisymmetric buckling is discussed.

Axisymmetric buckling of a circular tube is studied either as **elastic buckling** or **plastic buckling** depending on whether the buckling stress exceeds the yield stress of the material. If the wall thickness of a circular tube is thin, the buckling stress is less than the yield stress of the material, and the phenomenon is elastic buckling. In regard to the elastic buckling stress of a circular tube, Lorenz [128], Timoshenko [199] and Southwell [191] derived theoretical solutions, which are today used to find exact solutions for the elastic buckling of a circular tube. If the wall of the circular tube is moderately thick, however, buckling occurs after plastic yielding, and the result is plastic buckling. In the analysis of the plastic buckling stress of a circular tube, because the problem is complex, a unified solution as that for elastic buckling has not been obtained. Among the many studies to date, Gerard [80] and Batterman [20] have proposed representative analytic solutions based on the J_2 incremental theory and the J_2 deformation theory of plasticity, respectively.

In the following, the elastic buckling of thin-walled circular tubes, which is the simplest mode to handle, will be discussed in detail.

1.1.1   Elastic buckling

Theoretical analysis of the **axisymmetric elastic buckling** of a thin-walled circular tube under axial compressive force was presented in a book by Timoshenko and Gere [203]. In that work, for a circular tube of wall thickness t and radius R, the critical stress σ_{buc}^e for axisymmetric elastic buckling is given by

$$\sigma_{buc}^e = \frac{Et}{R\sqrt{3(1-\nu^2)}} \tag{1.1}$$

where E and ν are the Young's modulus and Poisson's ratio of the material. Here, a positive value means a compressive stress for buckling stress.

As one of the simplest shell theories, the **Donnell's equations** [65] will be employed here to analyze the elastic buckling of a circular tube. For the sign convention shown in Fig. 1.3, the equilibrium equations are derived by employing the stationary potential energy criterion for isotropic circular cylindrical shells.

$$\begin{cases} N_{x,x} + N_{xy,y} = 0 \\ N_{xy,x} + N_{y,y} = 0 \\ M_{x,xx} + (M_{xy} + M_{yx})_{,xy} + M_{y,yy} - \dfrac{1}{R}N_y + N_x w_{,xx} \\ \quad + 2N_{xy}w_{,xy} + N_y w_{,yy} = -p \end{cases} \tag{1.2}$$

where N_x, N_y, N_{xy} are the stress resultants, M_x, M_y, M_{xy} are the moment resultants, p is the internal pressure and w is the radial displacement. Partial differentiation is here denoted by a comma.

Using an Airy stress function f such that

$$f_{,xx} = N_y \quad f_{,xy} = -N_{xy} \quad f_{,yy} = N_x \tag{1.3}$$

a simpler set of two equations in two variables w and f can be derived as

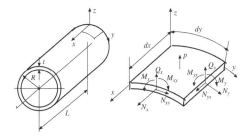

FIGURE 1.3
Circular cylinder shell.

follows:

$$\begin{cases} D\nabla^4 w + \dfrac{1}{R}f_{,xx} - \left(f_{,xx}w_{,yy} - 2f_{,xy}w_{,xy} + f_{,yy}w_{,xx}\right) = p \\ \nabla^4 f - Et\left(w_{,xy}^2 - w_{,xx}w_{yy} + \dfrac{1}{R}w_{,xx}\right) = 0 \end{cases} \tag{1.4}$$

where

$$D = \frac{Et^3}{12(1-\nu^2)}$$

$$\nabla^4(\ ) = (\ )_{,xxxx} + 2(\ )_{,xxyy} + (\ )_{,yyyy}.$$

To investigate the buckling force at the bifurcation point one assumes that the two variables w and f are given by

$$w = w_0 + \hat{w}, \quad f = f_0 + \hat{f} \tag{1.5}$$

where w_0, f_0 are the pre-buckling solutions and $\hat{w}$, $\hat{f}$ are small perturbations at buckling. By substituting Eq. (1.5) into Eq. (1.4) and neglecting products not containing the w_0 or f_0, a set of linearized stability equations can be obtained as follows.

$$\begin{cases} D\nabla^4 \hat{w} + \dfrac{1}{R}\hat{f}_{,xx} - \left(f_{0,xx}\hat{w}_{,yy} - 2f_{0,xy}\hat{w}_{,xy} + f_{0,yy}\hat{w}_{,xx}\right) \\ \qquad - \left(\hat{f}_{,xx}w_{0,yy} - 2\hat{f}_{,xy}w_{0,xy} + \hat{f}_{,yy}w_{0,xx}\right) = 0 \\ \nabla^4 \hat{f} - \dfrac{Et}{R}\hat{w}_{,xx} + Et\left(\hat{w}_{,xx}w_{0,yy} - 2\hat{w}_{,xy}w_{0,xy} + \hat{w}_{,yy}w_{0,xx}\right) = 0 \end{cases} \tag{1.6}$$

Now consider the stability of a cylindrical shell shown in Fig. 1.4 that is simply supported at its ends and subjected only to a uniformly distributed axial compressive force N_{x0}. Under this loading the pre-buckling state of the shell is axisymmetric and can be expressed as follows.

$$N_{y0} = N_{xy0} = 0, \quad w_0 = \text{constant} \tag{1.7}$$

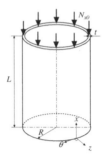

FIGURE 1.4
Cylindrical shell subjected to a uniformly distributed axial compressive force.

Thus, Eq. (1.6) is reduced to the following equations:

$$\begin{cases} D\nabla^4\hat{w} + \dfrac{1}{R}\hat{f}_{,xx} - N_{x0}\hat{w}_{,xx} = 0 \\ \nabla^4\hat{f} - \dfrac{Et}{R}\hat{w}_{,xx} = 0 \end{cases} \tag{1.8}$$

In the analysis of axisymmetric buckling of a circular tube which is simply supported at its ends, $\hat{w}$ and $\hat{f}$ are assumed as follows:

$$\hat{w} = A_w \sin\frac{m\pi x}{L}, \quad \hat{f} = A_f \sin\frac{m\pi x}{L} \tag{1.9}$$

Substituting Eq. (1.9) into Eq. (1.8) yields

$$A_w \left[D\left(\frac{m\pi}{L}\right)^4 + N_{x0}\left(\frac{m\pi}{L}\right)^2 + \frac{Et}{R^2} \right] \sin\frac{m\pi x}{L} = 0 \tag{1.10}$$

The term $[\cdots]$ should be zero on the left-hand side of the above equation so that a nontrivial solution for the coefficient A_w exists. Thus,

$$D\left(\frac{m\pi}{L}\right)^4 + N_{x0}\left(\frac{m\pi}{L}\right)^2 + \frac{Et}{R^2} = 0 \tag{1.11}$$

Therefore, from Eq. (1.11), the **elastic buckling stress** σ_{buc}^e is given by

$$\sigma_{buc}^e = -\frac{N_{x0}}{t} = D\left(\frac{m^2\pi^2}{tL^2} + \frac{E}{R^2D}\frac{L^2}{m^2\pi^2}\right) = \frac{1}{2} \times \frac{Et}{\sqrt{3(1-\nu^2)}R}\left(\alpha_m^2 + \frac{1}{\alpha_m^2}\right) \tag{1.12}$$

where

$$\alpha_m^2 = \frac{Rt}{2\sqrt{3(1-\nu^2)}}\left(\frac{m\pi}{L}\right)^2 \tag{1.13}$$

The actual buckling stress σ_{buc}^e is given by the minimum value of Eq. (1.12) when m is varied.

If the circular tube of length L is sufficiently long and m is sufficiently large, Eq. (1.12) takes the minimum value when m satisfies $\alpha_m = 1$, namely

$$\frac{L}{m} = \pi\sqrt{R\sqrt{\frac{D}{Et}}} \tag{1.14}$$

and the minimum value is given by

$$\sigma^e_{buc} = -\frac{N_{x0}}{t} = \frac{Et}{R\sqrt{3(1-\nu^2)}} \tag{1.15}$$

From Eq. (1.14), the **half-wavelength** λ $(= L/m)$ of buckles is given by

$$\lambda = \frac{L}{m} = \frac{\pi}{[12(1-\nu^2)]^{1/4}}\sqrt{Rt} \cong 1.73\sqrt{Rt} \tag{1.16}$$

which is of the order of $\sqrt{Rt}$.

1.1.2 Plastic buckling

Eq. (1.1) for elastic buckling stress can be applied to only the elastic buckling of circular tubes; in other words, it can be applied when the buckling stress σ^e_{buc} is smaller than the plastic yield strength of the material. For example, in the case of aluminum 6063-T5, the ratio between plastic yield stress σ_s and Young's modulus E is approximately given by $\sigma_s/E \cong 0.002$. Therefore the wall thickness t of a circular tube to which Eq. (1.1) is applicable is

$$t/R < 0.0033.$$

Since $t/R > 0.01$ usually holds for the thickness of circular tubes used as collision energy absorbers, buckling stress generally exceeds the yield stress of the material. To evaluate the buckling stress after plastic yielding, it is necessary to consider the plastic deformation of the material.

Among the many analyses proposed for the **axisymmetric plastic buckling stress** of a circular tube under axial compressive force, representative simple theoretical solutions are summarized below.

(1) Buckling stress is derived by replacing the Young's modulus E in Eq. (1.1) with the tangent modulus E_t in the stress-strain relation after yielding [203]:

$$\sigma^p_{buc} = \frac{E_t t}{R\sqrt{3(1-\nu^2)}} \tag{1.17}$$

(2) Assuming that, if the mechanical properties of the material may differ in the axial and circumferential directions after yielding, the stress-strain

relation in the circumferential direction is still a function of Young's modulus E, Timoshenko's solution [203] can be used, in which the tangent modulus E_t is simply substituted into the flexural rigidity:

$$\sigma^p_{buc} = \frac{\sqrt{E_t E t}}{R\sqrt{3(1-\nu^2)}} \tag{1.18}$$

(3) The solution of Gerard [80], which was based on the J_2 **deformation theory of plasticity**, can also be used:[1]

$$\sigma^p_{buc} = \sqrt{\frac{E_t E_s}{3\left(1-\nu^2_p\right)}} \cdot \frac{t}{R} \tag{1.19}$$

where E_s is secant modulus in the stress-strain relation, ν_p is the Poisson's ratio in the plastic deformation calculated from the elastic Poisson's ratio by using

$$\nu_p = \frac{1}{2} - \left(\frac{1}{2}-\nu\right)\frac{E_s}{E} \tag{1.20}$$

Batterman [20] conducted a more detailed analysis based on the J_2 deformation theory and derived the following theoretical solution for plastic buckling stress, which is different from the solution proposed by Gerard [80]:

$$\sigma^p_{buc} = \frac{2Et}{R} \cdot \frac{1}{\sqrt{(3\eta_s + 2 - 4\nu)\eta_T - (1-2\nu)^2}} \tag{1.21}$$

where

$$\eta_s = \frac{E}{E_s}, \qquad \eta_T = \frac{E}{E_t} \tag{1.22}$$

Note that by using ν_p from Eq. (1.20), Eq. (1.21) can be rewritten as

$$\sigma^p_{buc} = \sqrt{\frac{E_t E_s}{\left(1-\nu^2_p\right) + \left(\frac{1}{2}-\nu\right)^2\left(\frac{E_s}{E}\right)\left(\frac{E_s}{E}-\frac{E_t}{E}\right)}} \cdot \frac{t}{R} \tag{1.23}$$

(4) Also available is the solution of Batterman [20] based on the J_2 **incremental theory of plasticity**:

$$\sigma^p_{buc} = \frac{2Et}{\sqrt{3}R} \cdot \frac{1}{\sqrt{(5-4\nu)\eta_T - (1-2\nu)^2}} \tag{1.24}$$

Here,

$$\eta_T = \frac{E}{E_t} \tag{1.25}$$

---

[1]Bijlaard [23] also evaluated the plastic buckling stress based on the J_2 deformation theory of plasticity, but his result was not as simple as Eq. (1.19).

In the following, the analysis of plastic buckling stress is briefly discussed with consideration given to the plastic constitutive equation of the material; this will explain the solution, given by Eq. (1.24), based on the J_2 incremental theory and the solution, given by Eq. (1.21), based on the J_2 deformation theory. Then, the improvement of the solution method is discussed by considering the thickness of a thick-walled circular tube. Although the utility of the improved analysis accuracy is limited, as discussed later, and doubt remains in regard to practicality, the discussion will nonetheless facilitate understanding of the analytic techniques for the buckling of thick-walled circular tubes.

1.1.2.1 Theoretical analysis considering the plastic constitutive equation of materials

To evaluate the effects of plastic deformation of a material, it is necessary to use the following relation between the stress increment and the strain increment after yielding:

$$\begin{cases} \dot{\sigma}_x &= E_{xx}\dot{\varepsilon}_x + E_{x\theta}\dot{\varepsilon}_\theta \\ \dot{\sigma}_\theta &= E_{x\theta}\dot{\varepsilon}_x + E_{\theta\theta}\dot{\varepsilon}_\theta \\ \dot{\tau}_{x\theta} &= G_{x\theta}\dot{\gamma}_{x\theta} \end{cases} \tag{1.26}$$

If an analysis similar to that in Section 1.1.1 is performed using Eq. (1.2), the following equation for the buckling stress can be obtained [159]:

$$\sigma^p_{buc} = \frac{t}{\sqrt{3}R} \cdot \sqrt{E_{xx}E_{\theta\theta} - E^2_{x\theta}} \tag{1.27}$$

The coefficients E_{xx}, $E_{x\theta}$, $E_{\theta\theta}$ and G_{xz} in Eq. (1.26) depend on both the stress and the plastic deformation theory being applied. In the theoretical analysis to derive buckling stress, the stress in the circular tube before buckling is usually assumed to be uniaxial, existing only in the axial direction. Thus, the coefficients of Eq. (1.26) are given by

$$E_{xx} = H_T \frac{\eta_T + 3}{4}, \qquad\qquad E_{\theta\theta} = H_T \eta_T$$
$$E_{x\theta} = H_T \frac{\eta_T + 2\nu - 1}{2}, \qquad\qquad G_{x\theta} = G \tag{1.28}$$
$$H_T = \frac{4E}{(5 - 4\nu)\eta_T - (1 - 2\nu)^2}, \qquad \eta_T = \frac{E}{E_t}$$

in the case using J_2 incremental theory, and by

$$E_{xx} = H_s \frac{\eta_T + 3\eta_s}{4}, \qquad\qquad E_{\theta\theta} = H_s \eta_T$$
$$E_{x\theta} = H_s \frac{\eta_T + 2\nu - 1}{2}, \qquad\qquad G_{x\theta} = G_s$$
$$H_s = \frac{4E}{(3\eta_s + 2 - 4\nu)\eta_T - (1 - 2\nu)^2}, \qquad \eta_s = \frac{E}{E_s} \tag{1.29}$$
$$G_s = \frac{E}{3\eta_s - (1 - 2\nu)}$$

in the case using J_2 deformation theory. Here, G is the shearing modulus of the material.

By substituting Eq. (1.28) into Eq. (1.27), Batterman's equation, Eq. (1.24) for the J_2 incremental theory can be obtained. Also, by substituting Eq. (1.29) into Eq. (1.27), Batterman's equation, Eq. (1.21) for the J_2 deformation theory can be obtained [20].

1.1.2.2 Theoretical analysis considering the effect of wall thickness

The expressions for the moment in the above equations are based on **Kirchhoff-Love's theory** for a thin-walled plate. Under this plate theory, it is assumed that the shear strain γ_{xz} and γ_{yz} in the thickness direction of the plate can be neglected (the plate is in the x, y plane and z is the thickness direction) and that the displacement functions u, v, w in the x, y, z directions satisfy **Navier's hypothesis** and can be expressed by the following equations:

$$\begin{cases} u & = & u(x,y) - z\dfrac{\partial w(x,y)}{\partial x} \\ v & = & v(x,y) - z\dfrac{\partial w(x,y)}{\partial y} \\ w & = & w(x,y) \end{cases} \tag{1.30}$$

Eq. (1.30) will suffice if the plate is thin; if the plate is thick, however, a better result should be obtained by considering the shear strains γ_{xz} and γ_{yz} in the thickness direction.

Mindlin's plate theory is often used for thick plates. Under this plate theory, the following displacement function is assumed in order to consider the shear deformation [150, 27]:

$$\begin{cases} u & = & u(x,y) + zu_1(x,y) \\ v & = & v(x,y) + zv_1(x,y) \\ w & = & w(x,y) \end{cases} \tag{1.31}$$

Applying Mindlin's theory to the axisymmetric buckling of a circular tube, the displacement function is given by

$$\begin{cases} u & = & u_0(x) + z\psi(x) \\ v & = & 0 \\ w & = & w_0(x) \end{cases} \tag{1.32}$$

where x is the axial direction and z is the radial direction of the circular tube. Therefore, the strain is given by

$$\begin{cases} \varepsilon_x & = & \dfrac{du_0}{dx} + z\dfrac{d\psi}{dx} + \dfrac{1}{2}\left(\dfrac{dw_0}{dx}\right)^2 \\ \varepsilon_\theta & = & \dfrac{w_0}{R} \\ \gamma_{xz} & = & \dfrac{dw_0}{dx} + \psi \end{cases} \tag{1.33}$$

On the basis of Eq. (1.33), the plastic buckling stress in the axisymmetric case of a thick-walled circular tube can also be obtained from equilibrium equation, Eq. (1.2), through an analysis similar to the one discussed in Section 1.1.1. As a result, the buckling stress is given by [49]

$$\sigma_{buc}^{p} = \frac{t}{\sqrt{3}R} \sqrt{E_{xx}E_{\theta\theta} - E_{x\theta}^{2}} \left(1 - \frac{\sqrt{3}t}{12k_{m}GR} \sqrt{E_{xx}E_{\theta\theta} - E_{x\theta}^{2}} \right) \quad (1.34)$$

where **Mindlin's shear correction factor**, k_{m}, was assumed to be

$$k_{m} = \frac{\pi^{2}}{12}$$

in Mindlin's study [150], and

$$k_{m} = \frac{5}{6}$$

in Reissner's study [169].[2]

Furthermore, Eq. (1.34) considers the shear deformation in the thickness direction using Mindlin's plate theory, but does not consider the variation of the radius of the circular tube in the thickness direction. To consider this radial variation, the formula [141]

$$\varepsilon_{\theta} = \frac{w_{0}}{R+z} \quad (1.35)$$

should be used for ε_{θ} in Eq. (1.33); using the same analytic procedure as in the derivation of Eq. (1.34), the plastic buckling stress of a thick-walled circular tube is obtained as follows:

$$\sigma_{buc}^{p} = \frac{t}{\sqrt{3}R} \sqrt{C_{1}F_{1}} \left[C_{2} - \frac{\sqrt{3}t}{12k_{m}GR} \sqrt{C_{1}F_{1}} \right] - \frac{1}{12} \left(\frac{t}{R} \right)^{2} E_{x\theta}(1+C_{2}) \quad (1.36)$$

considering both the shear strain in the thickness direction and the radial variation of the circular tube in the thickness direction [49]. In Eq. (1.36),

$$
\begin{aligned}
F_{1} &= E_{xx}E_{\theta\theta}\frac{R}{t}\log\left(\frac{R+t/2}{R-t/2}\right) - E_{x\theta}^{2} \\
C_{1} &= 1 - \frac{t^{2}}{12R^{2}} \\
C_{2} &= 1 + \frac{E_{x\theta}t^{2}}{12k_{m}GR^{2}}
\end{aligned}
\quad (1.37)
$$

---

[2]A study by the author [49] showed that, in deriving the buckling stress of a circular tube, the calculation result is not highly dependent on Mindlin's shear correction factor k_{m}. Even if shear correction is taken into consideration by setting $k_{m} = \pi^{2}/12$ or $k_{m} = 5/6$, the result is the same to three digits as that without such a consideration, indicating that the influence of k_{m} is negligibly small.

1.1.3   Buckling stress in axial compression experiments

This section examines whether the buckling stress $\sigma_{buc}|_{exp}$ observed in axial compression experiments on circular tubes can be evaluated by using the formulas for buckling stress as discussed in Sections 1.1.1 and 1.1.2. Since the systematic evaluation of the effects of strain hardening and yielding stress of a material is experimentally difficult and computer-based finite element method (FEM) simulation is now advanced, the problem is divided into the following two subjects: (1) The possibility of using FEM simulation to evaluate the buckling stress $\sigma_{buc}|_{exp}$ observed in axial compression experiments on circular tubes; (2) the possibility of using the buckling stress obtained from the theoretical formulae discussed in Sections 1.1.1 and 1.1.2 to evaluate the buckling stress $\sigma_{buc}|_{FEM}$ obtained from FEM simulation.

1.1.3.1   Elastic buckling

The experimental data on buckling stress in elastic buckling are widely scattered, but many researchers have reported that, compared with theory, experiment gives a much smaller buckling stress (e.g., Robertson [176], Lundquist [129], Wilson and Olson [220]). The ratio $\rho = \sigma_{buc}|_{exp}/\sigma_{buc}^{e}|_{theory}$ of the experimental buckling stress $\sigma_{buc}|_{exp}$ in the elastic buckling and the theoretical value $\sigma_{buc}^{e}|_{theory}$ for elastic buckling stress is called the "**knock-down factor**". The ratio $\rho = \sigma_{buc}|_{exp}/\sigma_{buc}^{e}|_{theory}$ decreases as the circular tube becomes thinner, and is usually in the range 0.10–0.65. The value $\sigma_{buc}|_{exp}$ is much smaller than $\sigma_{buc}^{e}|_{theory}$ mainly due to the initial geometric imperfections, which may be comparable with the wall thickness of the circular tube [114, 225]. For example, Koiter [115] showed that the buckling stress decreases to about 24% if the imperfection is close to the thickness. The wall of a circular tube in which elastic buckling occurs is usually very thin, with a wall thickness/radius ratio of $t/R < 1/500$. In this type of circular tube, the initial geometric imperfection Δ_0 close to the thickness falls in the range $\Delta_0/R < 1/500$ compared with the radius; this is a conceivable range of values in manufacturing. Therefore, the buckling stress decreases to about 0.10–0.65 mainly because of the initial geometric imperfections in the test pieces of circular tubes.

However, as shown in Fig. 1.5, the buckling stress $\sigma_{buc}|_{FEM}$ in the compressed circular tube obtained from FEM simulation agrees well with the elastic buckling stress obtained theoretically from Eq. (1.1). In the analysis of Fig. 1.5, the friction coefficient between the rigid plate, which applies compressive force to the tube, and the end part of the tube is assumed to be 0.3.

1.1.3.2   Plastic buckling

Many researchers (e.g., Gerard [81], Batterman [20]) have shown that the **initial geometric imperfections** of a circular test tube do not appreciably affect the buckling stress in plastic buckling. This is because a circular tube

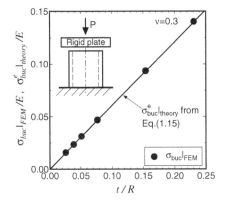

FIGURE 1.5
Comparison between $\sigma_{buc}|_{FEM}$ and $\sigma_{buc}^{e}|_{theory}$ in the elastic buckling of circular tubes [49].

that undergoes elastic buckling and one that undergoes plastic buckling have generally different wall thicknesses. Specifically, $t/R > 1/50$ normally holds in the thickness/radius ratio for the circular tube showing the plastic buckling. In a circular tube of this thickness, initial imperfection close to the thickness can usually be avoided. This explains why the scatter is small in the experimental buckling stress $\sigma_{buc}|_{exp}$ in plastic buckling and why the experimental buckling stress agrees very well with the FEM simulation, as shown, for instance, in Fig. 1.6. Fig. 1.6 shows a force-displacement curve for the compression of a circular tube of aluminum alloy A5052 as evaluated by a quasi-static axial compression test and FEM analysis. Fig. 1.7 shows the deformation behavior of circular tubes corresponding to the data of the force-displacement curve. The comparison between the experiment and the FEM analysis shows that they agree very well with each other with respect to the initial collapse force, the variation of compressive force, and the deformation behavior.

Note here that, in the cross-sectional deformation of a circular tube as shown in Fig. 1.7, local bending is observed at the end part of the tube when a buckling stress (point A in Fig. 1.6) is observed in both experiment and numerical analysis, because there is a frictional force in the radial direction at the upper and bottom end sections of the tube, and the growth of the radius of the circular tube due to the compressive force in the axial direction is interrupted. In other words, in the end part of a tube, not only uniform compression but also local bending occurs from the start. The buckling stress shown in the figure appears when deflection increases rapidly in the radial direction due to the local bending. Buckling occurs due to a similar mechanism in both experiment and numerical analysis, and similar values of buckling stress are obtained in both of them.

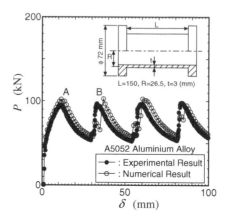

FIGURE 1.6
$P - \delta$ curve obtained by quasi-static axial compression tests and FEM analysis
for the axial compression of circular tubes of aluminum alloy A5052 [49].

In the following, in Fig. 1.8, the buckling stress $\sigma_{buc}|_{FEM}$ obtained from
the FEM simulation and the predicted values $\sigma^p_{buc}|_{theory}$ from the various
theoretical formulas discussed in Section 1.1.2 are compared.

(1) Computation indicates that the use of Mindlin's plate theory, namely
Eq. (1.34), instead of Kirchhoff-Love's plate theory, namely Eq. (1.27),
or considering the variation of radius in the thickness direction, namely
Eq. (1.36), has only a small effect; in other words, the accuracy of analysis
is not improved by considering them.

(2) By using the J_2 deformation theory, namely Eq. (1.29), and the J_2 incremental theory, namely Eq. (1.28), the resulting theoretical curves for
buckling stress are found to be roughly equal to the predictions of the
simplest formulas, Eqs. (1.17) and (1.18).

(3) $\sigma^p_{buc}|_{theory}$ obtained from the J_2 deformation theory, namely Eq. (1.29), is
smaller than $\sigma_{buc}|_{FEM}$ while $\sigma^p_{buc}|_{theory}$ obtained from the J_2 incremental
theory, namely Eq. (1.28), is larger than $\sigma_{buc}|_{FEM}$.

(4) When E_t/E is small, the J_2 deformation theory gives results closer to
the FEM numerical analysis than the J_2 incremental theory; when E_t/E
is large, the J_2 incremental theory gives results much closer to the FEM
simulation.

The research reports on the buckling stress in the plastic buckling appearing in the axially compressed circular tube (for example, the theoretical
work by Ore and Durban [159], who compared their results with experimental results obtained by Sobel and Newman [190], and the theoretical work

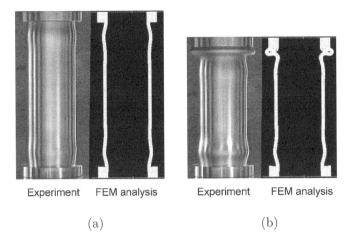

Experiment   FEM analysis   Experiment   FEM analysis

(a)   (b)

FIGURE 1.7
Comparison of deformation behavior obtained through experiment and FEM analysis: (a) at peak 'A' in Fig. 1.6; (b) at peak 'B' in Fig. 1.6.

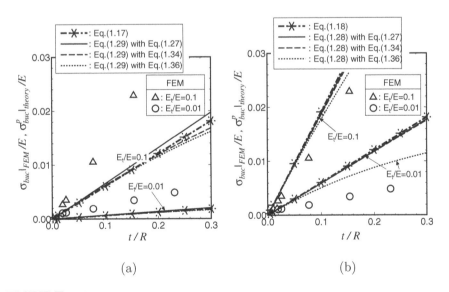

(a)   (b)

FIGURE 1.8
Comparison between $\sigma_{buc}|_{FEM}$ and the predicted values $\sigma^p_{buc}|_{theory}$ from the various theoretical formulas discussed in Section 1.1.2 for bilinear strain hardening[49]: (a) J_2 deformation theory; (b) J_2 incremental theory.

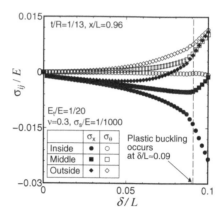

FIGURE 1.9

Stresses σ_x and σ_θ at the section corresponding to the apex of the first folding wrinkle under axial compression [49].

by Mao and Lu [142], who compared their results with the experimental results obtained by Lee [121]) show that the J_2 deformation theory gives better prediction results than the J_2 incremental theory. Note that the plastic strain hardening coefficient E_t is small in the cylindrical materials used in the experiments. In the experiment of Sobel and Newman [190], E_t was almost constant, $E_t/E \cong 0.004$, for the materials; in the experiment of Lee [121], although not constant in the materials tested, E_t was close to $E_t/E \cong 0.01$ for a stress value near the buckling stress of 90 MPa.

It is difficult to accept the conclusion that, in regard to the buckling stress in plastic axial compression of a circular tube, a better prediction is given by the J_2 deformation theory than the J_2 incremental theory; this is called the **plastic buckling paradox**, which is similarly found in the case of the plastic buckling of a plate. There are many reports (e.g., Hutchinson [96]) on this subject. Further, focusing on the biaxial state of stress at the buckling stress in a circular tube, Chen and Ushijima [49] reported another problem, namely, that the buckling stress in a compressed circular tube could not be handled as the plastic buckling stress under uniaxial compressive stress, as shown in Fig. 1.9. Fig. 1.9 shows the change in stresses σ_x and σ_θ at the section of $x/L = 0.96$ corresponding to the apex of the first folding wrinkle under an axial compression process of a circular tube with $\sigma_s/E = 0.001$, $E_t/E = 0.05$ and $t/R = 1/13$. It is seen from the figure that in the early stage before buckling (in this example plastic buckling occurs when $\delta/L = 0.09$) the stress state has stopped being in the uniaxial stress, and the stresses differ at the inner side, outside and center of the tube, respectively.

As discussed above, due to geometric imperfections, it is difficult to accurately predict the experimental buckling stress of elastic buckling in an

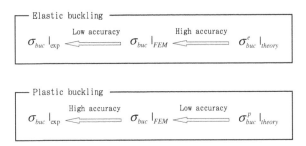

FIGURE 1.10
Method to evaluate buckling stress in an axially compressed circular tube.

axially compressed circular tube, as summarized in Fig. 1.10, even if one uses $\sigma_{buc}|_{FEM}$ obtained by FEM simulation or $\sigma_{buc}^e|_{theory}$ obtained from theoretical formulas.

On the other hand, in the case of plastic buckling, even though the experimental buckling stress $\sigma_{buc}|_{exp}$ is not accurately predicted by $\sigma_{buc}^p|_{theory}$ obtained from the theoretical formulas, it agrees well with $\sigma_{buc}|_{FEM}$ obtained by FEM simulation. Therefore, in evaluating the buckling stress in plastic buckling of an axially compressed circular tube, FEM simulation is widely used as a practical and effective tool. To evaluate the buckling stress under axial compression, Sobel and Newman [190] chose two different approaches—the evaluation method of using the explicit formula for buckling stress obtained by theoretical analyses and the evaluation method of numerical simulation on a computer—and called them the "simplified method of analysis" and "detailed method of analysis." To predict the buckling stress quickly and to qualitatively study the effect of buckling stress due to geometry and material properties, it is advised to use the simplified method of analysis, namely a simple formula supported by theoretical analyses. On the other hand, to obtain the buckling stress more accurately, the detailed method of analysis, namely a computer-based numerical simulation, is required even though it may need a large computational time.

Further, if there are approximation formulas based on the buckling stress $\sigma_{buc}|_{FEM}$ obtained by FEM simulation where the geometry and material properties of the circular tube are systematically changed as parameters, they can be used as a database in the early stage of designing the geometry and material properties. For example, the buckling stress in an axially compressed circular tube made of a material that obeys a practical stress-strain relation can be predicted by the two following steps:

(1) First, in Step 1, based on the accumulated buckling stress $\sigma_{buc}|_{FEM}$ data in the database for a material that obeys the bilinear hardening rule de-

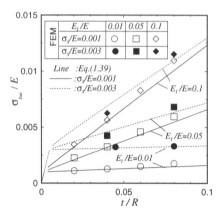

FIGURE 1.11
Prediction of buckling stress in an axially compressed circular tube using
Eq. (1.39).

scribed by the following stress-strain relation:

$$\sigma = \begin{cases} E\varepsilon & (\varepsilon < \sigma_s/E) \\ \sigma_s + E_t \left(\varepsilon - \dfrac{\sigma_s}{E} \right) & (\varepsilon \geq \sigma_s/E) \end{cases} \tag{1.38}$$

where σ_s is the yield strength, E is Young's modulus, and E_t is the strain
hardening coefficient, an approximation formula can be used, for example
[52]

$$\frac{\sigma_{buc}}{E} = \begin{cases} \dfrac{1}{\sqrt{3(1-\nu^2)}} \dfrac{t}{R} & \left(\dfrac{t}{R} \leq x_0 \right) \\ \dfrac{\sigma_s}{E} + \dfrac{t}{2R} \left(1 - e^{k_0} \right) \left(\dfrac{E_t}{E} \right)^{0.7(1-E_t/E)} & \left(\dfrac{t}{R} \geq x_0 \right) \end{cases} \tag{1.39}$$

where

$$x_0 = \frac{\sigma_s \sqrt{3(1-\nu^2)}}{E}, \quad k_0 = -22 \frac{E_t/E}{\sqrt{\sigma_s/E}} \left(\frac{t}{R} - x_0 \right)$$

Fig. 1.11 compares the approximate values of buckling stress from
Eq. (1.39) and from the FEM simulation, $\sigma_{buc}|_{FEM}$, for materials obeying
the bilinear strain hardening rule.

(2) The buckling stress is then predicted for a material that obeys an arbi-
trary stress-strain relation based on the buckling stress $\sigma_{buc}|_{FEM}$ data in
the material that obeys bilinear strain hardening.

For example, consider Ludwik's nth power hardening rule

$$\sigma = \sigma_s + E_a \left(\varepsilon - \frac{\sigma}{E} \right)^n \tag{1.40}$$

as a constitutive equation for a different elasto-plastic material.

In the bilinear hardening rule, the stress gradient $\partial\sigma/\partial\varepsilon$ is given by

$$\frac{\partial\sigma}{\partial\varepsilon} = E_t$$

and is equal to the tangent coefficient E_t, which is constant. On the other hand, in the nth power hardening rule, from Eq. (1.40), the stress gradient $\partial\sigma/\partial\varepsilon$ is derived as a function of stress σ:

$$\frac{\partial\sigma}{\partial\varepsilon} = \frac{E E_a n \left(\dfrac{\sigma - \sigma_s}{E_a}\right)^{\frac{n-1}{n}}}{E + E_a n \left(\dfrac{\sigma - \sigma_s}{E_a}\right)^{\frac{n-1}{n}}} \tag{1.41}$$

Here, in order to test whether the buckling stress $\sigma_{buc}|_{FEM(n)}$ appearing in the elasto-plastic circular tube that obeys the nth hardening rule can be approximated by the buckling stress $\sigma_{buc}|_{FEM(B)}$ under the bilinear hardening rule with the tangent coefficient E_t being identical to the stress gradient $\partial\sigma/\partial\varepsilon$ observed when $\sigma_x = \sigma_{buc}|_{FEM(n)}$ for the material of nth hardening rule, the relationship between the axial compressive stress $P/(2\pi Rt)$ and the relative crushing distance δ/L was evaluated by FEM simulation for a circular tube of wall thickness $t = 1$ mm and radius $R = 13$ mm with two different material parameters being $(n, E_a) = (0.5, 7.84 \text{ GPa})$ and $(0.5, 2.48 \text{ GPa})$, respectively, and is shown by the symbols ● and ■ in Fig. 1.12(a). Then, by substituting the obtained buckling stress $\sigma_{buc}|_{FEM(n)}$ into Eq. (1.41), the stress gradient $\partial\sigma/\partial\varepsilon$ ($\partial\sigma/\partial\varepsilon$=17.59 GPa for $(n, E_a) = (0.5, 7.84\text{GPa})$ and 5.713 GPa for $(n, E_a) = (0.5, 2.484\text{GPa})$) was evaluated at the buckling stress; FEM simulation of the axially compressed circular tube was performed also for the materials that obey the bilinear hardening rule and have the tangent coefficients E_t equal to these stress gradients, and the results are shown by the symbols ○ and □ in Fig. 1.12(a). The figure shows that the buckling stresses σ_{buc} are almost in agreement, respectively, though the compressive displacement δ/L in the axial direction may vary depending on the hardening rule.

Based on the above discussions, it becomes possible to predict the buckling stress of a material that obeys an arbitrary stress-strain relation from the buckling stress σ_{buc} in a material that obeys the bilinear hardening rule [208]. As an example, the solid line in Fig. 1.12(b) is an approximation curve to show the relationship between dimensionless stress σ_{buc}/E and hardening coefficient ratio E_t/E for a material obeying the bilinear hardening rule given by Eq. (1.39). Here, the relationship between stress σ and stress gradient $\partial\sigma/\partial\varepsilon$ is obtained from the stress-strain relationship assuming the nth hardening rule and is shown by the dotted line (n=0.5, E_a=2.48 GPa) and the broken line (n=0.5, E_a=7.84 GPa) for the two types of materials. At the intersection points A and B in the figure, the stress σ_{buc}/E can be obtained for the material that obeys the

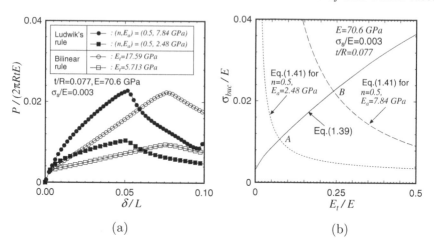

(a)   (b)

FIGURE 1.12
Prediction of the buckling stress based on Eq. (1.39): (a) buckling stress is governed by the stress gradient at the buckling stress irrespective of the hardening rule; (b) prediction of the buckling stress in materials obeying the nth power hardening rule [208].

nth hardening rule. These values are in good agreement with the result of elasto-plastic analysis assuming the nth hardening rule (see ■ and ● in Fig. 1.12(a)).

1.1.4   Effect of impact velocity on buckling stress

The dynamic compressive behavior of thin-walled tubes, which is sensitive to inertia and strain rate effects as shown by Tam and Calladine [196], Karagiozova and Jones [106, 107], Harrigan et al. [89] and Langseth et al. [120], is more complicated than the quasi-static response.

Abramowicz and Jones [6] explored the critical parameters which govern the transition from global bending to progressive collapse, for thin-walled mild-steel tubes subjected to quasi-static and dynamic axial loading. Karagiozova et al. [105] investigated the influence of inertia effects on the dynamic axisymmetric buckling of elastic-plastic cylindrical shells subjected to high-velocity axial impact, and clarified the particular conditions associated with the phenomena of dynamic plastic buckling and progressive buckling. Readers are advised to consult references [99, 101, 103] to acquire the fundamental results on the effects of dynamic loading. Only the investigation carried out by the author on the effect of impact velocity on buckling stress [52] is summarized herein.

In the following investigation, dynamic numerical simulations of the impact crush test were carried out using the non-linear FEM commercial code,

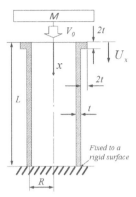

FIGURE 1.13
Tube geometry and loading condition used in the following investigation [52].

MSC.Dytran. The geometry of the FEM model and its boundary condition are shown in Fig. 1.13. The model is struck from the upper edge by a rigid mass with an initial velocity V_0 having sufficient kinetic energy for crushing the tested tube. The tube is made of an elastic plastic material that obeys the bilinear hardening rule.

Fig. 1.14 shows the reaction force obtained from striker for $V_0=40$m/s. As seen from Fig. 1.14, the peak load fluctuates so violently that an appropriate filter technique should be used to obtain an essential value of the peak load. However, the obtained value of the peak load depends on the type of filter used, as shown by Karagiozova and Jones [108]. On the contrary, the reaction force obtained from rigid surface is also shown in Fig. 1.14. It is seen from the figure that the fluctuation is very small and there is no need to use any filter to measure the peak load. Therefore, taking into consideration that the impulse force transmitted through a crushed tube is the reaction force from the tube bottom, here the peak load is output from the rigid surface.

Fig. 1.15(a) shows comparisons of axial compressive force and displacement diagram for a tube under different impact velocities of $V_0=5$, 180 and 360 km/h. The initial peak stress is associated with the initiation of local buckling deformation which occurs near the tube end. Fig. 1.15(b) shows the initial peak stress, namely the **dynamic buckling stress** σ^d_{buc}, for various impact velocities. Also in Fig. 1.15(b), quasi-static buckling stress for the tube is shown by a dashed line. It is found from this figure that the buckling stress σ^d_{buc} for a lower impact velocity is almost equal to the value of the quasi-static result, and the stress value becomes higher as the impact velocity V_0 increases.

Fig. 1.16 illustrates propagations of the stress wave in the axial direction for $V_0=5$ km/h (Fig. 1.16(a)) and 360 km/h (Fig. 1.16(b)) until the buckling stress occurs. Also in Fig. 1.16, values of the yield stress σ_s and the quasi-static

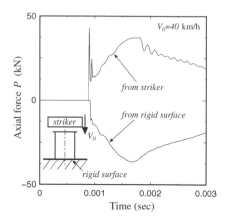

FIGURE 1.14
Comparison of the reaction force output from the upper striker and from the lower rigid surface [52].

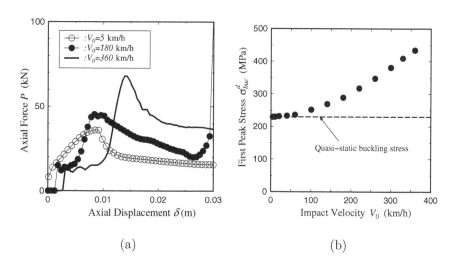

(a)   (b)

FIGURE 1.15
Axial compressive force during dynamic crushing [52]: (a) $P - \delta$ curves for different impact velocities; (b) variation of buckling stress σ_{buc}^d with impact velocity.

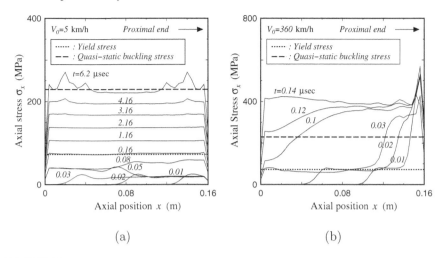

FIGURE 1.16
Axial stress distribution until an initiation of buckling [52]: (a) $V_0 = 5$ km/h;
(b) $V_0 = 360$ km/h.

buckling stress are shown by dotted and dashed lines, respectively. It is evident from Fig. 1.16(a) that for the case of a lower impact velocity ($V_0 = 5$ km/h), the amplitude of the stress wave in the axial direction at the initiation of impact is relatively small, and the stress wave travels and reflects many times along the tube until the local buckling deformation arises. In the end, the axial stress distribution developing over the tube becomes almost uniform at $t = 0.16$ μsec, and its amplitude increases stably to the quasi-static buckling stress. On the other hand, for the case of a higher impact velocity (Fig. 1.16(b)), the development of axial stress distribution apparently differs from that for a lower impact velocity (Fig. 1.16(a)). In Fig. 1.16(b), the stress concentration can be found at $x = 0.156$ m. Such a stress concentration occurs by the existence of the flange at the impacted end, and does not affect the overall buckling behavior of a circular tube. It is evident from Fig. 1.16(b) that the amplitude of the stress wave bigger than the value of quasi-static buckling stress develops near the impacted end at the beginning. However, such a high stress field is relatively narrow, and no buckling behavior seems to be observed even if the amplitude of the stress wave is bigger than that of the quasi-static result. Moreover, even though the axial stress developing over the tube becomes larger than the quasi-static result at $t = 0.12$ μsec, no buckling can be observed. Finally, the local buckling can be observed at $t = 0.14$ μsec when the amplitude of the axial stress is almost twice larger than the quasi-static buckling stress.

As shown in Fig. 1.7 in Section 1.1.3, which illustrates the quasi-static axial compressive behavior for a circular tube, the buckling stress is associated with a sufficient local bending deformation which is observed near the

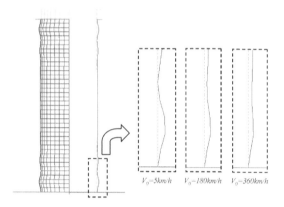

FIGURE 1.17
Comparisons of deformed shape near the fixed end at the initiation of buckling
for different impact velocities [52].

tube end. That is, during the axial compression, the tube seems to expand
in the radial direction, but near the tube end, such a movement is restricted
by the existence of the fixed boundary condition. As a result, the local bending deformation can be observed near the tube end, and the local bending
deformation is necessary for initiating the buckling behavior. Fig. 1.17 shows
comparisons of deformed shape near the fixed end when the buckling is observed for some cases of impact velocity V_0=5 km/h, 180 km/h and 360 km/h.
It is evident that for the case of a higher impact velocity, the amount of the
local bending deformation is smaller than that under a slower impact velocity.
In order to initiate buckling behavior where the smaller amount of the local
bending deformation occurs under higher impact velocity, that is, in order
to increase the radial displacement, a large amount of axial stress should be
needed. Therefore, it could be understood that the reason the buckling stress
increases as the impact velocity V_0 increases is because the impact for a higher
impact velocity causes a smaller radial displacement than that for a lower impact velocity. Moreover, the reason the bending deformation decreases as V_0
increases can be explained by the radial inertia effect. That is, the faster the
impact velocity becomes, the more rapidly the axial stress increases, but the
expansion in radial direction would be delayed by the radial inertia effect.

The relationship between the axial compressive force P and the radial
displacement U_r at the apex of wrinkle for four cases of V_0= 5, 180, 300 and
360 km/h are chased and the initiation of buckling for every V_0 is denoted
in Fig. 1.18 by the marks ●. From the figure, the reason the higher velocity
causes the larger buckling stress can be explained as follows. While the axial
compressive force increases by propagating the stress waves over the tube,
the radial displacement U_r develops by the axial compression, but the rate

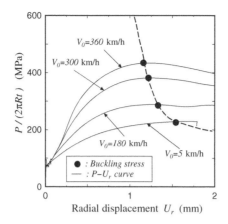

FIGURE 1.18
Variation of axial compressive force P with radial displacement U_r at the apex
of wrinkle for V_0=5, 180, 300 and 360 km/h [52].

of U_r relatively decreases as V_0 increases. As a result, the axial compressive
force increases by the stress wave reflecting many times until sufficient radial
displacement can be reached for initiating the local buckling.

In order to discuss the influence of V_0 on the buckling stress quantitatively,
an effective parameter considering the effect of mechanical properties on the
inertia effect is needed. Fig. 1.19 shows the relationship between the normalized axial compressive stress $P/(2\pi RtE)$ and displacement U_x/L for three
cases of a tube problem having different elastic modulus E, tube density ρ
and the impact velocity V_0. In the figure, parameter c represents the elastic
wave speed for a one-dimensional rod, and can be shown as $\sqrt{E/\rho}$. Here, all
models have different values of E, ρ and V_0, but keep the same ratio V_0/c.
It is evident that all models show the same response of the axial compressive
stress $P/(2\pi RtE)$ and U_x/L diagram. Strictly speaking, the elastic and the
plastic wave speed for a circular tube are distinct from the elastic wave speed
c for a one-dimensional rod. However, based on the fact that the relationship
between the normalized axial compressive stress $P/(2\pi RtE)$ and displacement
U_x/L is the same under the same V_0/c, the non-dimensional parameter V_0/c
can be used for evaluating the effect of impact velocity on the buckling stress
σ_{buc}^d.

Chen et al. [32] proposed an approximate equation for predicting the **dynamic buckling stress** of a circular tube made of a material that obeys the
bilinear hardening rule and is subjected to axial impact load as follows:

$$\frac{\sigma_{buc}^d}{E} = 60 \left(\frac{E_t}{E} \right)^{0.54} \left(\frac{V_0}{c} \right)^2 + \frac{\sigma_{buc}^s}{E} \qquad (1.42)$$

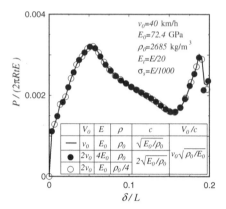

FIGURE 1.19
Normalized axial compressive stress and displacement behavior for tubes having the same ratio of V_0/c [52].

Here, the quasi-static term σ_{buc}^s/E is given by Eq. (1.39) shown in Section 1.1.3.

Fig. 1.20 shows the relationship between the dynamic buckling stress and the impact velocity for some cases of E_t/E for the ratio of $t/R=0.04$ (Fig. 1.20(a)) and 0.08 (Fig. 1.20(b)). In these figures, solid lines correspond to the approximate results obtained by Eq. (1.42), and the symbols show the numerical results obtained by FEM. It is clear from these figures that the predicted σ_{buc}^d agrees well with the numerical results for a wide range of impact velocity V_0.

Karagiozova and Jones [109] conducted a detailed analysis of biaxial stress wave in an elastic-plastic tube subjected to an impact and derived a theoretical solution of the peak load on the stuck end of the tube

$$P_{Mises} = 2\pi Rt \left(\frac{2\sigma_s}{\sqrt{3}} + V_0\sqrt{\frac{4E\rho}{3(\eta_T - 1) + 4(1 - \nu^2)}} \right) \qquad (1.43)$$

for tubes made of a material obeying the von Mises yield condition, and

$$P_{Tresca} = 2\pi Rt \left(\sigma_s + V_0\sqrt{\frac{2E\rho}{\sqrt{3}(\eta_T - 1) + 2(1 - \nu^2)}} \right) \qquad (1.44)$$

for tubes made of a material obeying the Tresca yield condition. Here, $\eta_T = E/E_t$. These formulae are well in agreement with numerical simulation using the FE code ABAQUS and are valid for impact velocities, which cause an instantaneous plastic deformation at the proximal end of the tube [109].

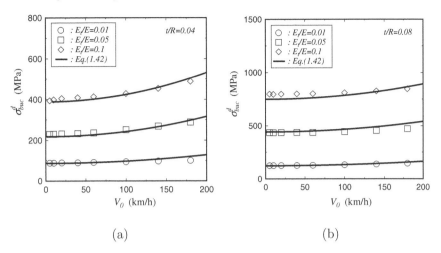

FIGURE 1.20

Estimation of the buckling stress σ_{buc}^d under some cases of impact velocity V_0:
(a) $t/R = 0.04$; (b) $t/R = 0.08$.

The above investigation is applied to tubes made of a strain rate insensitive material. However, many engineering materials are sensitive to strain rate. Here, the following empirical Cowper-Symonds uniaxial constitutive equation is used to assess material strain rate effects:

$$\frac{\sigma_s^d}{\sigma_s^s} = 1 + \left(\frac{\dot{\varepsilon}}{D}\right)^{1/q} \tag{1.45}$$

where σ_s^s and σ_s^d represent the static and the dynamic plastic stress respectively, $\dot{\varepsilon}$ is the **equivalent strain rate** and the parameters D and q are material constants. Five kinds of materials listed in Table 1.1 are used in the following investigation. Here, Comb. 0 is a strain rate insensitive material.

It is seen from the relations between σ_s^d and $\dot{\varepsilon}$ shown in Fig. 1.21 for the materials listed in Table 1.1 that the strain rate dependence of the dynamic yielding stress σ_s^d becomes remarkable in the order from Comb. 1 to Comb. 4.

In general, the dynamic buckling stress σ_{buc}^d for a tube made of a strain rate sensitive material is higher than that of a stain rate insensitive material. Fig. 1.22 shows the comparison of the axial stress and displacement relationship for two tubes made of Comb. 0 and Comb. 1 respectively, and subjected to the same impact velocity of V_0=100km/h. From Fig. 1.22, it can be seen that not only the buckling stress σ_{buc}^d but also the following stress variation for a strain rate sensitive material (solid line) becomes higher than that for a strain rate insensitive material (dotted line).

Table 1.2 shows the variation of the buckling stress σ_{buc}^d and the equivalent strain rate $\dot{\varepsilon}\big|_{buc}$ for some cases of impact velocity V_0. Here, the equivalent

TABLE 1.1

Five kinds of materials used in the following investigation.

	D (sec^{-1})	q	ρ (kg/m^3)	E (GPa)
Comb. 0	∞	∞	2685	72.4
Comb. 1	1288000	4.0	2685	72.4
Comb. 2	6844	3.91	7860	206
Comb. 3	802	3.585	7860	206
Comb. 4	40	5	7860	206

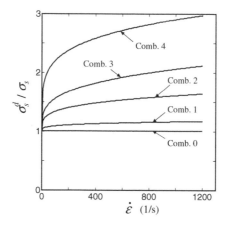

FIGURE 1.21

Effect of the strain rate $\dot{\varepsilon}$ on the dynamic yielding stress σ_s^d for the materials listed in Table 1.1.

TABLE 1.2

Comparison of the buckling stress σ_{buc}^d between strain rate sensitive and insensitive materials for various impact velocities V_0 [32].

	Comb. 1				Comb. 0		
V_0 (km)	σ_{buc}^d (MPa)	$\dot{\varepsilon}\vert_{buc}$ (1/s)	σ_s' (MPa)	E_t' (MPa)	$\sigma_s = \sigma_s'$ (MPa)	$E_t = E_t'$ (MPa)	σ_{buc}^d (MPa)
40	145.0	197.6	80.5	804.6	80.5	804.6	142.9
60	149.0	317.4	81.5	814.7	81.5	814.7	148.3
100	160.0	532.8	82.7	827.3	82.7	827.3	159.2
140	168.0	798.3	83.8	838.2	83.8	838.2	170.5

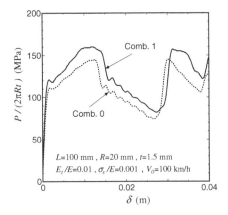

FIGURE 1.22
Comparison of $P - \delta$ curve for tubes made of Comb. 0 and Comb. 1 respectively, and subjected to the same impact velocity of $V_0 = 100$km/h [32].

strain rate $\dot{\varepsilon}\,|_{buc}$ is measured at the vertex of a wrinkle when the buckling stress σ_{buc}^d is observed.

As shown in Eq. (1.45), the dynamic plastic stress σ_s^d depends on the equivalent strain rate $\dot{\varepsilon}$. Therefore, the stress σ_s^d which corresponds to the strain rate $\dot{\varepsilon}\,|_{buc}$ at the vertex of a wrinkle can be calculated by substituting $\dot{\varepsilon}\,|_{buc}$ into Eq. (1.45) as follows:

$$
\begin{aligned}
\sigma_s^d &= \sigma_s^s \left\{ 1 + \left(\frac{\dot{\varepsilon}\,|_{buc}}{D} \right)^{1/q} \right\} \\
&= \left\{ \sigma_s + E_t \left(\varepsilon - \frac{\sigma_s}{E} \right) \right\} \left\{ 1 + \left(\frac{\dot{\varepsilon}\,|_{buc}}{D} \right)^{1/q} \right\} \\
&= \sigma_s' + E_t' \left(\varepsilon - \frac{\sigma_y}{E} \right)
\end{aligned}
\tag{1.46}
$$

Here σ_s' and E_t' represent the modifying initial yield stress and hardening coefficient for a strain rate insensitive material respectively. The values of σ_s' and E_t' for the cases of impact velocity, which are shown in the left part of Table 1.2, are obtained by Eq. (1.46).

The buckling stress σ_{buc}^d can be considered to be dominated by the stress and strain relationship near the vertex of a wrinkle when the stress σ_{buc}^d is observed [32]. In order to certify the effectiveness of this assumption, values of σ_s' and E_t' in Table 1.2 are used as the initial yield stress and hardening coefficient for material1 Comb. 0, and the buckling stress σ_{buc}^d for each velocity V_0 is calculated from FEM numerical simulation. These results of σ_{buc}^d are

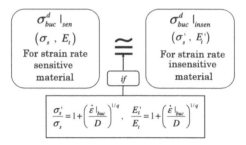

FIGURE 1.23
Evaluation of the buckling stress σ_{buc}^{d} for a tube made of strain rate sensitive material based on one made of strain rate insensitive material [32].

shown in Table 1.2. From Table 1.2, it can be seen that approximate values of the buckling stress σ_{buc}^{d} for a strain rate insensitive material with σ_s and E_t replaced by σ_s' and E_t' respectively, agree well with the value of σ_{buc}^{d} for a strain rate sensitive material, as shown in Fig. 1.23.

An estimate for the equivalent strain rate $\dot{\varepsilon}\big|_{buc}$ at the vertex of a wrinkle when the buckling stress σ_{buc}^{d} is observed in an impacted circular tube is given by Chen et al. [32] as follows.

$$\dot{\varepsilon}\big|_{buc} = 0.215\frac{V_0}{R}\left(1 + 2e^{-0.055L/\sqrt{tR}}\right) \tag{1.47}$$

Fig. 1.24 shows that Eq. (1.47) fits the numerical results of $\dot{\varepsilon}\big|_{buc}$ for various strain rate sensitive materials.

Thus, the dynamic buckling stress σ_{buc}^{d} for a stain rate sensitive material is found from Eq. (1.42) with σ_s and E_t replaced by σ_s' and E_t' from Eq. (1.46). The predicted values of the buckling stress σ_{buc}^{d} calculated in this way for an axially crushed circular tube made of various strain rate sensitive materials using the approximate values of $\dot{\varepsilon}\big|_{buc}$ from Eq. (1.47) are presented in Fig. 1.25, in which numerical results of σ_{buc}^{d} obtained from FEM analysis are also shown.

1.2   Axisymmetric crushing of circular tubes after buckling

Fig. 1.26 shows the deformation behavior of a tube wall corresponding to the local maximum and minimum points of compressive force during the compression of a circular tube analyzed by FEM. Although the circular tube expands

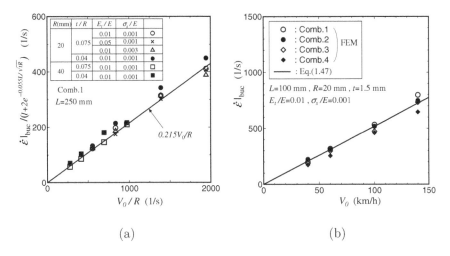

FIGURE 1.24

Comparisons of $\dot{\varepsilon}\left.\right|_{buc}$ obtained from Eq. (1.47) and FEM analysis: (a) $L = 250$ mm for Comb. 1 with various parameters of t/R, E_t/E and σ_s/E; (b) for Comb. 1, 2, 3 and 4 with $t = 1.5$ mm, $R = 20$ mm, $E_t/E = 0.01$ and $\sigma_s/E = 0.001$ [32].

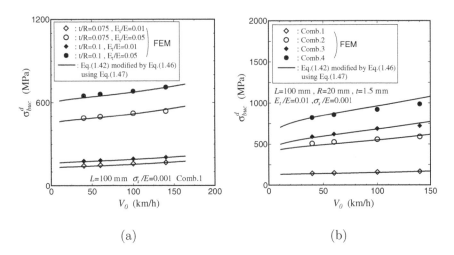

FIGURE 1.25

Comparisons of σ_{buc}^d obtained from FEM analysis and Eq. (1.42) modified by Eq. (1.46) with $\dot{\varepsilon}\left.\right|_{buc}$ from Eq. (1.47) [32]: (a) for Comb. 1 with various parameters of t/R, E_t/E and σ_s/E; (b) for Comb. 1, 2, 3 and 4 with $t = 1.5$ mm, $R = 20$ mm, $E_t/E = 0.01$ and $\sigma_s/E = 0.001$.

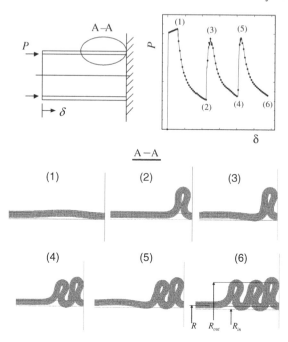

FIGURE 1.26
In the case of $R_{in} \cong R$, the formation of a wrinkle corresponds to one force peak on the force-displacement curve.

in the radial direction due to compressive force before the first wrinkle appears, the expansion in the radial direction is interrupted by the constrained end part, causing local bending (cross section A-A at process point (1) in the figure). Therefore, the first wrinkle appears near the end part of the circular tube in almost all cases. The size of the area where local bending occurs is of the order of $\sqrt{Rt}$, which is governed by the geometric size of the circular tube. After the initial peak P_{col} of compressive force, the compressive force P decreases until the wrinkles are folded to touch each other (cross section A-A at the process point (2) in the figure). As the compressive deformation further proceeds, the formation and folding of wrinkles are repeated, and the compressive force oscillates cyclically. In Fig. 1.26, the inner radius R_{in} of the compressed wrinkle is nearly equal to the radius R of the circular tube, and $R_{in} \cong R$ holds. Thereby, the formation of a wrinkle corresponds to a force peak on the force-displacement curve. However, $R_{in} < R$ usually holds. In that case, as shown in Fig. 1.27, the formation of a wrinkle corresponds to two force peaks on the force-displacement curve. These two force peaks correspond to the force appearing when the inner and outer wrinkles begin to form.

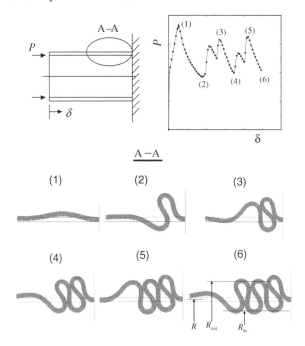

FIGURE 1.27
In the case of $R_{in} < R$, the formation of a wrinkle corresponds to two force peaks on the force-displacement curve.

1.2.1 Analytic model

As for the compressive force in the axial crushing of a circular tube, the most important parameters for collision energy absorbers are not only the peak force discussed in the previous section but also the **average compressive force** over the whole crushing process. This is a load parameter associated with the absorbed energy in the crushing of a circular tube.

To accurately derive an average force, it is necessary to evaluate the phenomenon either experimentally or by numerical simulation. However, in designing crash elements, a quick and simple solution is often needed, even though it may not be accurate. To meet such requirements, theoretical analysis of average force is needed.

The crushing of a crash element is a problem of very complex and large deformation involving large strain concentration and sometimes fracture of the element. A method called the "**kinematic approach**" is often used for analyzing such a complex behavior during crushing. In this method, to derive the average force during crushing of an element, (1) a plausible and analytically reasonable deformation model is assumed based on the actual crushing behavior, and (2) the average compressive force is calculated based on the balance between the rate of the strain energy in the adopted deformation model and the rate of the work of the compressive force. Many analytic models have been proposed based on these considerations. In this section, the method to theoretically analyze average compressive force is discussed.

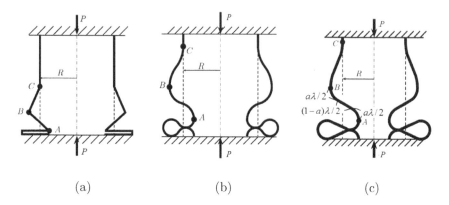

(a)   (b)   (c)

FIGURE 1.28
Model of wrinkle formation during axisymmetric compression of circular tubes: (a) Alexander's model; (b) Abramowicz and Jones' model; (c) Grzebieta's model.

Alexander [10] was the first to propose a deformation model of the axisymmetric crushing of a circular tube and to derive a theoretical solution of average crushing force. His approach was also the first **kinematic approach**. He proposed a deformation model for the crushing of a circular tube as shown in Fig. 1.28(a) based on an axial crushing experiment with metal circular tubes of $2R/t = 29$–89. In that analysis, assuming that the material is rigid perfectly plastic, and expressing the wrinkles as folding of the tube wall, the strain energy needed to form wrinkles was calculated from the bending energy associated with the folding of the tube wall and the membrane strain energy, and the compressive force of a circular tube was calculated. Alexander's analytic model is relatively simple yet properly reflects the features of axisymmetric crushing deformation of a circular tube; so this model is the most basic one for analyzing deformation behavior after the buckling of a circular tube.

After the proposal of Alexander's model, many improved models were reported. Johnson [99] proposed to improve on Alexander's model by considering, for example, the variation of circumferential strain ε_θ along the wrinkle and the effective crushing distance. Andronicou and Walker [12] considered the influence of biaxial stress (in the circumferential and axial directions) on yielding under the von Mises yield conditions. Further, Abramowicz and Jones [3, 5] modified the rotation angle associated with the folding of the tube wall by replacing the straight parts AB and BC of a wrinkle with oppositely placed two circular arcs of the same radius, as shown in Fig. 1.28(b). Furthermore, Grzebieta [85] proposed to put a straight part between the two circular arcs in Abramowicz and Jones' model [5](1986), as shown in Fig. 1.28(c). In the above models, it is assumed that the inner and outer wrinkle tips are formed simultaneously. On the other hand, Wierzbicki et al. [217] and Huang and Lu [95] proposed models assuming that the outer and inner wrinkles are alternately formed.

Among these proposed analytic models, representative examples are discussed in the following.

1.2.1.1 Alexander's model of simultaneous formation of inner and outer wrinkles

Fig. 1.29 shows the deformation behavior of the tube wall under Alexander's analytic model. Assuming that the material is rigid perfectly plastic with yield stress σ_s, as shown in the figure, a single wrinkle is made of two straight parts of length λ (assuming the half-wavelength of the wrinkle is λ) and three plastic hinges (A, B, and C in the figure). The bending energy E_b needed for forming a wrinkle is given by

$$
\begin{aligned}
E_b &= 2 \times \int_0^{\pi/2} M_P 2\pi R d\theta + \int_0^{\pi/2} M_P 2\pi (R + \lambda \sin\theta)(2d\theta) \\
&= 4\pi M_P (\pi R + \lambda)
\end{aligned}
\tag{1.48}
$$

where M_P is the fully plastic bending moment per unit circumferential length while considering the effect of plane strain, and $M_P = \dfrac{2}{\sqrt{3}} \dfrac{\sigma_s t^2}{4}$.

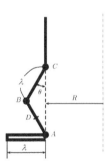

FIGURE 1.29
Alexander's analytic model.

On the other hand, the strain ε_θ in the circumferential direction caused by deformation of the wrinkle in the radial direction changes along the wrinkle; in Alexander's model, the strain $\varepsilon_\theta|_D$ in the middle part of the wrinkle (point

D in the figure) is used as the average strain. Then, the strain $\varepsilon_\theta|_D$ can be obtained by using

$$\varepsilon_\theta|_D = \log\left(\frac{R + (\lambda/2)\sin\theta}{R}\right)$$

from the displacement in the radial direction, and thus, the membrane energy E_m necessary for forming a wrinkle is approximately given by

$$E_m = \int_0^{\pi/2} \pi(2R + \lambda\sin\theta)2\lambda t\sigma_s \frac{d\varepsilon_\theta|_D}{d\theta}d\theta = 2\pi\sigma_s\lambda^2 t \qquad (1.49)$$

Therefore, assuming the crushing distance δ to be

$$\delta = 2\lambda,$$

the average force P_{ave} is given by

$$\frac{P_{ave}}{M_P} = \frac{(E_b + E_m)/2\lambda}{M_P} = 2\pi\left(\frac{\pi R}{\lambda} + 1\right) + 2\sqrt{3}\pi\frac{\lambda}{t} \qquad (1.50)$$

The half-wavelength λ of wrinkles, which is an unknown parameter in Eq. (1.50), can be determined based on the assumption that the average force should be minimized, that is, from the equation $\dfrac{\partial(P_{ave}/M_P)}{\partial\lambda} = 0$. Therefore,

$$\frac{\lambda}{R} = \sqrt{\frac{\pi}{\sqrt{3}}}\sqrt{\frac{t}{R}} \cong 1.347\sqrt{\frac{t}{R}} \qquad (1.51)$$

is obtained. Substituting Eq. (1.51) into Eq. (1.50) yields

$$\frac{P_{ave}}{M_P} = 29.31\sqrt{\frac{R}{t}} + 2\pi \qquad (1.52)$$

This equation was obtained under the assumption that the wrinkles are folded outside the circular tube. In contrast, if all wrinkles are folded inside the tube, the average force is given by

$$\frac{P_{ave}}{M_P} = 29.31\sqrt{\frac{R}{t}} - 2\pi \qquad (1.53)$$

Alexander proposed the average force

$$\frac{P_{ave}}{M_P} = 29.31\sqrt{\frac{R}{t}} \qquad (1.54)$$

by averaging the two equations given above.

Alexander's formula generally gives an underestimated value. Even though Alexander's formula is very simple, it covers the major features of the crushing of a circular tube, and is a basic analytic model that expresses the post-buckling behavior of a circular tube.

Modifications have been made to improve Alexander's analytic formula. Representative refined models are as follows:

(a) In deriving membrane energy E_m, Johnson [99] and Abramowicz and Jones [3] did not use the approximation that sets the strain at point D to the averaged strain of wrinkle, and integrated the distributed strain rate

$$\dot{\varepsilon} = \frac{d}{dt}\left[\frac{\left\{2\pi(R + s\sin\theta) - 2\pi R\right\}}{2\pi R}\right] \tag{1.55}$$

along the wrinkle. Here, s is the distance from the hinge point A or C. This changes Eq. (1.49) into

$$E_m = 2\pi\sigma_s\lambda^2 t\left(1 + \frac{\lambda}{3R}\right) \tag{1.56}$$

and then Eq. (1.50) for the average force is rewritten as

$$\frac{P_{ave}}{M_P} = 2\pi\left(\frac{\pi R}{\lambda} + 1\right) + 2\sqrt{3}\pi\frac{\lambda}{t}\left(1 + \frac{\lambda}{3R}\right) \tag{1.57}$$

Here, as for the half-wavelength λ, Abramowicz and Jones, assuming $\lambda/R = x_0$, derived an equation that x_0 should satisfy as follows:

$$x_0 = \frac{\lambda}{R} = \sqrt{\frac{\pi t}{\sqrt{3}R(1 + 2x_0/3)}} \tag{1.58}$$

from $\dfrac{\partial(P_{ave}/M_p)}{\partial\lambda} = 0$. Abramowicz and Jones [3] derived the solution of Eq. (1.58) for λ/R as follows:

$$\frac{\lambda}{R} = 1.760(t/2R)^{1/2} \tag{1.59}$$

by substituting $x_0 = 0.256$ into the right side of Eq. (1.58) because the root of Eq. (1.58) for $R/t = 23.36$ is given by $x_0 = 0.256$. As shown by Abramowicz and Jones [3], Eq. (1.59) gives good agreement with the exact predictions from Eq. (1.58) for a wide range of R/t values. Further, by substituting Eq. (1.59) into Eq. (1.57),

$$\frac{P_{ave}}{M_P} = 20.79(2R/t)^{1/2} + 11.90 \tag{1.60}$$

is obtained.

(b) Abramowicz and Jones [3] further included, in their model analysis, the decrease of **effective crushing distance** that was dependent on the wrinkle thickness. Based on Fig. 1.30, Abramowicz and Jones [3] proposed that the effective crushing distance δ_e is given by

$$\delta_e = 2\lambda - 2x_m - t = 1.72\lambda - t \tag{1.61}$$

where $x_m \cong 0.14\lambda$ from [2]. Eq. (1.61) gives

$$\frac{\delta_e}{2\lambda} = 0.86 - 0.568\sqrt{\frac{t}{2R}} \tag{1.62}$$

By considering this, the average force, Eq. (1.60), is given by

$$\frac{P_{ave}}{M_P} = \frac{20.79(2R/t)^{1/2} + 11.90}{0.86 - 0.568(t/2R)^{1/2}} \tag{1.63}$$

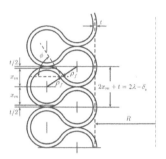

FIGURE 1.30
Analytic model of compression distance in the crushing of a circular tube.

(c) Two years later, Abramowicz and Jones [5], arguing that the rotation angle of a plastic hinge is not π but $\pi + 2\phi$, as shown in Fig. 1.30, defined the bending energy E_b by multiplying the bending energy given by Eq. (1.48) by a factor of $\dfrac{\pi + 2\phi}{\pi}$:

$$E_b = 4\pi M_0(\pi R + \lambda)\frac{\pi + 2\phi}{\pi}$$

Note that, compared with Eq. (1.48), here the stress σ_0, called the **energy equivalent flow stress**, is used instead of the plastic yield stress σ_s, for plastic deformation and so $M_0 = \sigma_0 t^2/4$ instead of M_P as the fully plastic bending moment per unit length of a plate. Further, using the flow stress σ_0, Abramowicz and Jones [5] rewrote the membrane energy E_m, Eq. (1.56), as

$$E_m = 2\pi\sigma_0\lambda^2 t\left(1 + \frac{\lambda}{3R}\right)$$

and the average force, Eq. (1.57), as

$$\frac{P_{ave}}{M_0} = 2\pi\left(\frac{\pi R}{\lambda} + 1\right)\frac{\pi + 2\phi}{\pi} + 4\pi\frac{\lambda}{t}\left(1 + \frac{\lambda}{3R}\right) \tag{1.64}$$

From Fig. 1.30,

$$\phi = \lambda/2\rho_f - \pi/2 \tag{1.65}$$

and from the experimental approximation formula $\delta_e/(2\lambda) \cong 0.75$ for the **effective crushing distance**, $\lambda/\rho_f \cong 4$ holds. By substituting this into Eq. (1.65), $\phi \cong 0.43$. Therefore,

$$\frac{\lambda}{R} = 1.84(t/2R)^{1/2} \tag{1.66}$$

$$\frac{P_{ave}}{M_0} = \frac{25.23(2R/t)^{1/2} + 15.09}{0.86 - 0.568(t/2R_0)^{1/2}} \tag{1.67}$$

(d) Grzebieta [85] modified the model of Alexander [10] by replacing the hinge with circular arcs, and proposed to use Abramowicz and Jones' model as shown in Fig. 1.28(b) where the half-wavelengths AB and BC are each composed of two circular arcs for cases in which the tube wall is thick, and also proposed a model as shown in Fig. 1.28(c) where the half-wavelengths AB and BC are each composed of two circular arcs and a straight part for cases in which the tube wall is thin. Further, Grzebieta [85] used the von Mises yield condition to consider the effect of stress in the axial direction on the membrane energy E_m. As a result, the compressive force cannot be given by a simple formula, and is obtained from the root of the following nonlinear equation:

$$
\begin{aligned}
&\frac{t}{\sqrt{3}}\left(1 - \frac{3}{4}\bar{P}^2\right)\left(2R + \frac{2\lambda(1 - \cos\alpha)}{3\alpha} + \frac{\lambda\sin\alpha}{3}\right) \\
&+ \left(-\bar{P}\cos\alpha + \sqrt{4 - 3\bar{P}^2\cos^2\alpha}\right)\left(\frac{2\lambda\sin\alpha}{3\alpha} + \frac{\lambda\cos\alpha}{3}\right)L\frac{d\beta}{d\alpha} \\
&- \frac{2\bar{P}\lambda R}{3\alpha}\left(\frac{2\sin\alpha}{\alpha} - 2\cos\alpha + \alpha\sin\alpha\right) = 0
\end{aligned}
\tag{1.68}
$$

where

$$\bar{P} = \frac{P}{2\pi Rt\sigma_s}$$

$$L = \sqrt{\left(\frac{2\lambda\sin\alpha}{3\alpha} + \frac{\lambda\cos\alpha}{3}\right)^2 + \left(\frac{2\lambda(1 - \cos\alpha)}{3\alpha} + \frac{\lambda\sin\alpha}{3}\right)^2}$$

$$\beta = \cos^{-1}\left\{\frac{\dfrac{2\sin\alpha}{\alpha} + \cos\alpha}{\sqrt{\dfrac{3}{4}\left(\dfrac{2(1 - \cos\alpha)}{\alpha} + \sin\alpha\right) + 1}}\right\}$$

Further, α is a parameter of the crushing process, and the equation

$$\frac{\delta}{2} = \lambda - \frac{2\lambda\sin\alpha}{3\alpha} - \frac{\lambda\cos\alpha}{3}$$

holds for the compressive displacement δ in the crushing of a circular tube.

1.2.1.2 Wierzbicki et al.'s model of alternately formed inner and outer wrinkles

It is assumed that the outer and inner wrinkles are simultaneously formed in Alexander's model. This assumption does not agree with the experimental data on the crushing of a circular tube. To solve this problem, Wierzbicki et al. [217] proposed a model in which the outer and inner wrinkles are alternately formed.

(a) Wierzbicki et al.'s linear model

Wierzbicki et al. [217] assumed that a wrinkle consists of two straight lines as shown in Fig. 1.31, and additionally that the formation of a wrinkle consists of two steps: the total crushing of the inner part as shown in Fig. 1.31(a) (the angle α becomes zero), and the total crushing of the outer part as shown in Fig. 1.31(b) (the angle β becomes zero), and the crushing process is described by the angles α and β. Due to the total crushing of the inner wrinkle, the angle α changes from $\cos^{-1}(1 - m)$ to 0. During this process, an outer wrinkle is formed, and the angle β changes from $\pi/2$ to $\cos^{-1} m$. The relationship

$$\cos\alpha - \cos\beta = \frac{R - R_{in}}{\lambda} \tag{1.69}$$

holds between α and β from Fig. 1.31(a). On the other hand, during the total crushing of the outer wrinkle, the angle β changes from $\cos^{-1} m$ to 0, and at the same time the inner wrinkle is formed, and the angle α changes from $\pi/2$ to $\cos^{-1}(1 - m)$. Here, m is the **eccentricity ratio** as defined by $m = (R_{out} - R)/(R_{out} - R_{in})$ where R_{out} and R_{in} are outer and inner radii of the wrinkle. The relation

$$\cos\alpha - \cos\beta = \frac{R - R_{out}}{\lambda} \tag{1.70}$$

holds between α and β from Fig. 1.31(b).

Considering that the angular velocity of rotation at the plastic hinges (points A, B, and C in Fig. 1.31(a)) is given by $\dot{\alpha}$, $\dot{\alpha}+\dot{\beta}$, $\dot{\beta}$, respectively, the bending energy rate needed for forming a wrinkle is given by

$$\dot{E}_b = 2\pi RM_0\Big[|\dot{\alpha}| + |\dot{\alpha} + \dot{\beta}| + |\dot{\beta}|\Big] = 4\pi RM_0(|\dot{\alpha}| + |\dot{\beta}|) \tag{1.71}$$

and the strain rate $\dot{\varepsilon}_\theta$ in the circumferential direction is, both at AB and BC of the wrinkle, given by

$$|\dot{\varepsilon}_\theta| = \frac{s\sin\alpha}{R}|\dot{\alpha}|$$

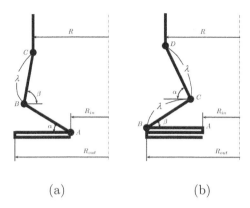

(a)   (b)

FIGURE 1.31
Linear model of the alternate formation of inner and outer wrinkles: (a) the total crushing of the inner part of a wrinkle; (b) the total crushing of the outer part of a wrinkle.

where s is the distance from A or C. Therefore, the rate of membrane energy to form a wrinkle is given by

$$\dot{E}_m = 2 \int_0^\lambda \sigma_s |\dot{\varepsilon}_\theta| 2\pi R t ds = 2\pi \sigma_s t \lambda^2 \sin \alpha |\dot{\alpha}| \qquad (1.72)$$

On the other hand, the rate of the tube shortening $\dot{\delta}$ is given by

$$|\dot{\delta}| = \lambda \left[|\dot{\alpha}| \cos \alpha + |\dot{\beta}| \cos \beta \right]$$

and the force P is given by

$$P = \frac{\dot{E}_b + \dot{E}_m}{|\dot{\delta}|} \qquad (1.73)$$

In particular, by averaging over a cycle, the average force P_{ave} is given by

$$P_{ave} = 2M_0 \left(\frac{2\pi}{t} \lambda + \frac{\pi^2 R}{\lambda} \right) \qquad (1.74)$$

From the condition that the average force becomes minimal, as Alexander assumed, the unknown parameter λ is given by

$$\frac{\lambda}{R} = \sqrt{\pi} \sqrt{\frac{t}{2R}} \cong 1.772 \sqrt{\frac{t}{2R}} \qquad (1.75)$$

and substituting Eq. (1.75) into Eq. (1.74) yields

$$\frac{P_{ave}}{M_0} = 4\pi^{3/2} \sqrt{\frac{2R}{t}} \cong 22.27 \sqrt{\frac{2R}{t}} \qquad (1.76)$$

Note here that the average force does not depend on the eccentricity ratio m. Namely, Wierzbicki et al. introduced an eccentricity ratio m into the formula for wrinkle deformation energy, but failed to obtain the eccentricity ratio m because the parameter disappeared while deriving his final energy equation. Whereas, Singace et al. [187] modified Wierzbicki et al.'s model and considered the difference between the inner and outer radii of the wrinkle so that the eccentricity ratio m remains in the final energy equation. Thus, the **eccentricity ratio** m is determined from the local minimum of the wrinkle deformation energy, and is always 0.65 irrespective of the geometry and material properties of the circular tube. Therefore, the distance between the hinges is given by

$$\frac{\lambda}{R} = \sqrt{\pi}\sqrt{\frac{t}{2R}} \cong 1.772\sqrt{\frac{t}{2R}} \tag{1.77}$$

exactly as Wierzbicki et al. indicated, and the average force

$$\frac{P_{ave}}{M_0} \cong 22.27\sqrt{\frac{2R}{t}} + 5.632 \tag{1.78}$$

is the same as the one given by Wierzbicki et al. except for the added constant term.

(b) Wierzbicki et al.'s curved model

As shown in Fig. 1.31, it was assumed that the wrinkle is composed of two straight parts; instead, to better fit the wrinkle formed in the crushing of a circular tube, Wierzbicki et al.[217] approximated the AB and BC parts of the wrinkle, as shown in Fig. 1.32, by the two circular arcs of radii r_1 and r_2, respectively. From the geometry shown in the figure,

$$r_1 = \frac{\lambda}{2\alpha}, \quad r_2 = \frac{\lambda}{2\beta} \tag{1.79}$$

Further, the process of wrinkle formation is divided into two steps: the total crushing of inner wrinkle (where the angle α reaches the limit value α_f) and the total crushing of outer wrinkle (where the angle α reaches the limit value β_f); and the crushing process is described by the angles α and β. In the total crushing process of the inner wrinkle, the angle α changes from α_1 to its limit value α_f, and the angle β changes from 0 to β_1. Thereby, α and β satisfy

$$\frac{\lambda}{\alpha}(1 - \cos\alpha) - \frac{\lambda}{\beta}(1 - \cos\beta) = R_{out} - R \tag{1.80}$$

On the other hand, in the total crushing of the outer wrinkle, the angle β changes from β_1 to its limit value β_f, and the angle α changes from 0 to α_1. Hence, α and β satisfy

$$\frac{\lambda}{\alpha}(1 - \cos\alpha) - \frac{\lambda}{\beta}(1 - \cos\beta) = R - R_{in} \tag{1.81}$$

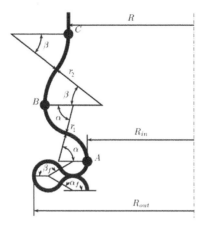

FIGURE 1.32
Curved model of the alternate formation of inner and outer wrinkles.

Here, α_f and β_f are the central angles of the circular arcs of the crushed wrinkle, and $\alpha_f = \beta_f = 5\pi/6$ as shown in Fig. 1.32. Further, β_1 is the maximum angle β during the total crushing of the inner wrinkle, and is given by the value of β obtained by substituting $\alpha = \alpha_f$ into Eq. (1.80). α_1 is the maximum angle α during the total crushing of the outer wrinkle, and is given by the value of α obtained by substituting $\beta = \beta_f$ into Eq. (1.81).

Note that the compressive force P is obtained from

$$P = \frac{\dot{E}_b + \dot{E}_m}{\dot{\delta}} \tag{1.82}$$

in the same way as from the linear model. Here,

$$\delta = \lambda \left(2 - \frac{\sin\alpha}{\alpha} - \frac{\sin\beta}{\beta} \right)$$

The formulae for the rates of bending energy $\dot{E}_b$ and membrane energy $\dot{E}_m$ to form a wrinkle are omitted here because they are complicated and their solution requires the use of numerical methods. The numerical calculation gives

$$\frac{\lambda}{R} \cong 2.62 \sqrt{\frac{t}{2R}} \tag{1.83}$$

$$\frac{P_{ave}}{M_0} \cong 31.74 \sqrt{\frac{2R}{t}} \tag{1.84}$$

(c) Huang and Lu's model.

   In modifying Wierzbicki et al.'s model [217], Huang and Lu [95] assumed that the wrinkle is composed of a linear part of length $(1 - a)\lambda$ and two circular arcs of length $a\lambda/2$, as shown in Fig. 1.28(c). Here, the coefficient a is given by

$$a = b\frac{t}{\lambda} \tag{1.85}$$

where $b \cong 2\text{–}4$ is assumed.

   The analysis can be conducted similarly to Wierzbicki et al.'s model, and the average compressive force is given by

$$\frac{P_{ave}}{M_0} = 16\pi \frac{\sqrt{\alpha_f \int_0^{\alpha_f} f(\alpha)d\alpha}}{\Delta(a)} \sqrt{\frac{2R}{t}} \tag{1.86}$$

where

$$\lambda = \frac{1}{2}\sqrt{\frac{\alpha_f}{\int_0^{\alpha_f} f(\alpha)d\alpha}}\sqrt{2Rt} \tag{1.87}$$

$$\Delta(a) = 2 - \frac{a}{\alpha_f} \tag{1.88}$$

$$\cos\alpha_f = \frac{a}{2(1-a)}\frac{1 - 2\sin\alpha_f}{\alpha_f} \tag{1.89}$$

$$
\begin{aligned}
f(\alpha) &= \int_0^\alpha \left[\frac{a^2}{4\alpha^3}(\phi\sin\phi - 1 + \cos\phi)\right]d\phi \\
&+ \int_0^1 \left[\frac{a}{2\alpha^2}(1 - \cos\alpha) - \frac{a}{2\alpha}\sin\alpha - (1 - a)x\cos\alpha\right] \\
&\quad \times(1 - a)dx \\
&+ \int_0^\alpha \left[\frac{a^2}{4\alpha^3}(1 - 2\cos\alpha + \cos\phi + \phi\sin\phi - 2\alpha\sin\alpha) \right. \\
&\quad \left. - \frac{a(1 - a)}{2\alpha}\cos\alpha\right]d\phi
\end{aligned}
\tag{1.90}
$$

   Fig. 1.33 compares the three models assuming alternately formed inner and outer wrinkles—the linear model and curved line model of Wierzbicki et al. [217], and the linear-curved line model of Huang and Lu [95]—on the relationship between compressive force P and compressive displacement δ; FEM simulation is also shown by a thin curve for comparison. The relationship between P and δ evaluated by using the three models, as shown by the thick curves in the figure, corresponds to the formation of a single wrinkle; note that the curves are adjusted to the positions A, B and C, respectively, for easier comparison. As the figure shows, although the peak force obtained from the

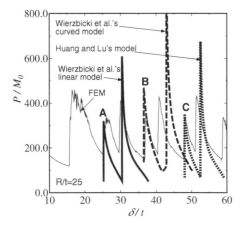

FIGURE 1.33
Comparison of $P - \delta$ curves obtained from FEM and three theoretical models.

alternate formation models of wrinkles does not quantitatively agree with the FEM simulation, the conclusion that there are two force peaks per wrinkle qualitatively agrees with the experimental data and the FEM simulation. Note however that analytic accuracy is not clearly improved even by Huang and Lu's linear-curved line model [95], which is the most complex.

1.2.2 Average crushing force

Major analytic formulas proposed for the average crushing force are summarized in Table 1.3.

In the table, M_0 and M_P indicate the fully plastic bending moment per unit length, where $M_0 = \dfrac{\sigma_0 t^2}{4}$ and $M_P = \dfrac{2}{\sqrt{3}} \times \dfrac{\sigma_s t^2}{4}$. Here, the coefficient $\dfrac{2}{\sqrt{3}}$ considers the effect of plane strain. In the empirical formula based on the experiment, η is called the **structural effectiveness** and ϕ is called the **solidity ratio**, where $\eta = P_{ave}/A\sigma_u$ and $\phi = A/A_s$. Here, σ_u is the tensile strength of the material, and A and A_s are the area of the circular tube and the area surrounded by the circular tube, respectively, where $A = 2\pi Rt$ and $A_s = \pi R^2$.

Fig. 1.34 shows the average crushing force obtained by using these formulas. Here, the numbers (1)–(10) correspond to the numbers of the equations in Table 1.3. The figure also shows the average force obtained through FEM simulation and experiment for comparison [10, 130, 136, 198, 3, 5, 88, 95]. The results of numerical simulation were obtained here by assuming the ratio between Young's modulus and plastic yield stress to be $E/\sigma_s = 1000$ and

TABLE 1.3

Formulas for the average crushing force and half-wavelength in the axisymmetric crushing of a circular tube.

Eq. Num.	Author(s) (year)	Average force P_{ave}	Half-wave length λ
(1)	Alexander (1960) [10]	$\dfrac{P_{ave}}{M_P} = 20.73\sqrt{\dfrac{2R}{t}} + 6.283$	$\lambda = 0.952\sqrt{2Rt}$
(2)	Abramowicz and Jones (1984) [3]	$\dfrac{P_{ave}}{M_P} = \dfrac{20.79\sqrt{2R/t} + 11.9}{0.86 - 0.568\sqrt{t/2R}}$	$\lambda = 0.880\sqrt{2Rt}$
(3)	Abramowicz and Jones (1986) [5]	$\dfrac{P_{ave}}{M_0} = \dfrac{25.23\sqrt{2R/t} + 15.09}{0.86 - 0.568\sqrt{t/2R}}$	$\lambda = 0.920\sqrt{2Rt}$
(4)	Wierzbicki et al. (1992) [217]	$\dfrac{P_{ave}}{M_0} = 22.27\sqrt{\dfrac{2R}{t}}$	$\lambda = 0.886\sqrt{2Rt}$
(5)	Wierzbicki et al. (1992) [217]	$\dfrac{P_{ave}}{M_0} = 31.74\sqrt{\dfrac{2R}{t}}$	$\lambda = 1.31\sqrt{2Rt}$
(6)	Singace et al. (1995) [187]	$\dfrac{P_{ave}}{M_0} = 22.27\sqrt{\dfrac{2R}{t}} + 5.632$	$\lambda = 0.886\sqrt{2Rt}$
(7)	Guillow et al. (2001) [88]	$\dfrac{P_{ave}}{M_0} = 72.3\left(\dfrac{2R}{t}\right)^{0.32}$	
(8)	Thornton et al. (1983) [198]	$\eta = 2\phi^{0.7}$	
(9)	Mamalis and Johnson (1983) [136]	$\eta = \dfrac{7\phi}{4 + \phi} + 0.07$	
(10)	Huang and Lu (2003) [95]	$\dfrac{P_{ave}}{M_0} = 16\pi\dfrac{\sqrt{\alpha_f A(a)}}{\Delta(a)}\sqrt{\dfrac{2R}{t}}$ $A(a) = \displaystyle\int_0^{\alpha_f} f(\alpha)d\alpha$	$\lambda = \sqrt{\dfrac{Rt\alpha_f}{2\,A(a)}}$

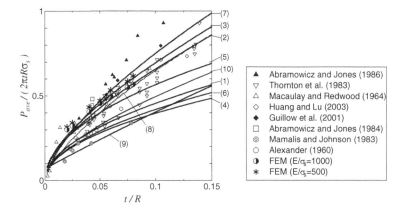

FIGURE 1.34
Comparison of average force obtained from the models.

$E/\sigma_s = 500$, respectively, in a material without strain hardening ($E_t/E = 0$). Although dispersion is great depending on experimental conditions, the experimental results of various researchers are generally larger than the theoretical prediction from the formulas, largely because of the strain hardening of materials. Further, among the average force obtained from the formulas, except for the formula of Guillow et al. [88] (the equation number (7) in Table 1.3) based on the experimental result, the values given by the formula of Abramowicz and Jones [5] (the equation number (3) in Table 1.3) are the closest to the average force obtained by FEM simulation and agree roughly with the average of all experimental results.

1.2.2.1 Effects of strain hardening

The above-mentioned analytic models assume a perfect elasto-plastic material and do not consider the effect of strain hardening of the material. However, as widely known, the average force of a circular tube under axial compression increases with the strain hardening of the material. Fig. 1.35 shows the relationship between average compressive stress $P_{ave}/(2\pi tR)$ and thickness ratio t/R of circular tubes made of materials that follow the bilinear hardening rule as evaluated by FEM simulation. Here, the numerical results with the three strain hardening coefficients $E_t/E = 0$, $1/100$ and $1/20$ are shown by the symbols □, ○ and ●. As Fig. 1.35 shows, the average compressive stress $P_{ave}/(2\pi tR)$ increases with the strain hardening coefficient E_t, and the departure from the average stress when $E_t = 0$ is assumed becomes more evident as the thickness ratio t/R increases.

As the process of wrinkle formation in the axial crushing of a circular tube suggests, the strain-related strain hardening of a material is difficult to han-

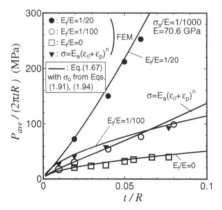

FIGURE 1.35
Average compressive force of circular tubes made of materials with different strain hardening coefficients for a material obeying the bilinear hardening rule.

dle by theory because the strain occurs variously along the wrinkle. Therefore, to consider the effect of strain hardening of a material on the average stress $P_{ave}/(2\pi tR)$, it is recommended to use, as the material property, the **energy equivalent flow stress** σ_0, which reflects the energy absorbed in the crushing process, instead of the yield stress σ_s. Two approaches were reported for determining the energy equivalent flow stress σ_0. One is to use the tensile strength σ_u of the material. For example, in evaluating the compressive force of a mild steel circular tube using a theoretical model, Wierzbicki [217] noted that the energy equivalent flow stress σ_0 is equivalent to 92% of the tensile strength σ_u; Huang and Lu [95] reported that the flow stress σ_0 is equivalent to 95% of the tensile strength σ_u of the aluminum alloy A6060-T5; and Galib et al. [77] defined the flow stress σ_0 by the average of 0.2% yield strength, $\sigma_{0.2}$, and tensile strength, σ_u, for studying the experimental results on the crushing of thin-walled circular tubes made of aluminum alloy A6060-T5. As mentioned above, although the flow stress σ_0 is defined according to the tensile strength σ_u, there is no consensus on the coefficient that should be applied to the tensile strength σ_u. Tensile strength, which is the maximum nominal stress in the uniaxial tensile test of a round bar test piece and is determined from the competition between the decrease of the cross section of the test piece and the strain hardening of the material under tension, generally lacks a physical basis for being applied to the analysis of strain hardening of a material.

Another definition of the energy equivalent flow stress σ_0 was proposed by Abramowicz and Jones [5]:

$$\sigma_0 = \frac{\int_0^{\varepsilon_f} \sigma d\varepsilon}{\varepsilon_f} \tag{1.91}$$

Here, ε_f is the average final strain after the wrinkle is folded. Considering that the average final strain in the folded wrinkle is about $\lambda/2R$ in Alexander's model [10], Abramowicz and Jones [3] defined ε_f as

$$\varepsilon_f \cong 0.88\sqrt{\frac{t}{2R}} \tag{1.92}$$

from Eq. (1.59). Further, assuming the eccentricity ratio to be $m = 0.5$, and the average final strain in the folded wrinkle to be about $\lambda/4R$, Wierzbicki [217] derived ε_f as

$$\varepsilon_f \cong 0.443\sqrt{\frac{t}{2R}} \tag{1.93}$$

from Eq. (1.77). On the other hand, focusing on the folded wrinkle, Ushijima et al. [207] proposed that the average $\bar{\varepsilon}^p_{eq}$ of the distributed equivalent plastic strain ε^p_{eq} along the path 1-2-3 of the wrinkle shown in Fig. 1.36(a) could be used as the average final strain ε_f to derive the plastic flow stress σ_0. Fig. 1.36(b) shows the FEM simulation of the relationship between average equivalent plastic strain $\bar{\varepsilon}^p_{eq}$ and thickness ratio t/R of a circular tube. The figure shows that irrespective of the level of strain hardening, a one-to-one correspondence exists approximately between the average equivalent plastic strain $\bar{\varepsilon}^p_{eq}$ and the thickness ratio t/R. Fig. 1.36(b) also shows the strain given by Eqs. (1.92) and (1.93). Furthermore, based on the numerical results shown in Fig. 1.36(b), $\bar{\varepsilon}^p_{eq}$ is approximated by

$$\bar{\varepsilon}^p_{eq} = \frac{5}{2}\left(\frac{t}{R}\right)^{4/5} \tag{1.94}$$

The energy equivalent flow stress σ_0 was determined by substituting $\bar{\varepsilon}^p_{eq}$, which was obtained by using Eq. (1.94), into Eq. (1.91) as ε_f; the average compressive stress $P_{ave}/(2\pi t R)$ considering the effect of strain hardening was obtained by substituting it into Eq. (1.67), the theoretical solution of Abramowicz and Jones [5]. These values are shown by solid lines in Fig. 1.35. The figure shows that irrespective of the magnitude of the strain hardening coefficient E_t, the result agrees very well with the analytic result of average stress $P_{ave}/(2\pi t R)$ in circular tubes with various values of thickness ratio t/R.

Further, in Fig. 1.35, the relationship between average stress $P_{ave}/(2\pi t R)$ and thickness ratio t/R obtained from FEM analysis is shown by ▼ for the material obeying Swift's nth power hardening rule ($E_a = 214$MPa, $\varepsilon_0 = 1.0 \times 10^{-4}$, $n = 0.098$), which is given by

$$\sigma = E_a(\varepsilon_0 + \varepsilon^p)^n \tag{1.95}$$

For this type of material, the average compressive stress $P_{ave}/(2\pi t R)$ was obtained by substituting, not yield stress σ_s, but the flow stress σ_0 obtained from Eqs. (1.91) and (1.94) into Eq. (1.67). As shown in the figure, the values

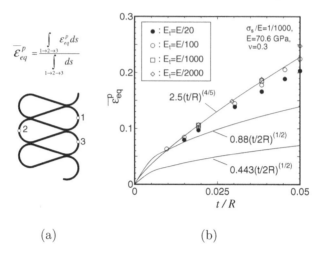

(a)   (b)

FIGURE 1.36
Average equivalent plastic strain $\bar{\varepsilon}^p_{eq}$: (a) definition of $\bar{\varepsilon}^p_{eq}$; (b) $\bar{\varepsilon}^p_{eq}$ obtained
from FEM.

agree very well with the numerical simulation even in the case of the nth
power hardening rule.

However, in considering the increase of average force due to strain hardening, it is necessary to consider the possibility that the deformation mode
becomes non-axisymmetric due to the increase of strain hardening of the material, as shown by Ushijima et al. [207].

1.2.2.2   Material strain rate effects

In order to use the Cowper-Symonds equation, Eq. (1.45), to assess material
strain rate effects in dynamic crush of cylindrical tubes, an estimate for average strain rate is required. It is shown by Abramowicz and Jones [3] that
$\dot{\varepsilon}$ in Eq. (1.45) can be obtained from the circumferential strain generated in
formation of wrinkles, which has an approximate average final value ε_f given
by Eq. (1.92), and is expressed as

$$\dot{\varepsilon} = 2\varepsilon_f V_m / \delta_e \tag{1.96}$$

where V_m is the mean velocity of the striker during crushing and is given by

$$V_m = V_0 / 2 \tag{1.97}$$

Eq. (1.96) together with Eqs. (1.62), (1.92) and (1.97) gives

$$\dot{\varepsilon} = \frac{0.25}{0.86 - 0.568\sqrt{t/2R}} \frac{V_0}{R} \tag{1.98}$$

It is shown in [4] that the equation

$$\frac{\sigma_u^d}{\sigma_u} = 1 + \left(\frac{\dot{\varepsilon}}{6844}\right)^{1/3.91} \tag{1.99}$$

fits the experimental data for the ultimate tensile strength of steel specimens under dynamic impact. In [3], Eq. (1.63) with the ultimate tensile strength σ_u instead of σ_0 is used to estimate the static average crushing force P_{ave} for the axisymmetric crushing of a cylindrical tube. Thus, the dynamic average crushing force P_{ave}^d satisfies

$$\frac{P_{ave}^d}{P_{ave}} = 1 + \left(\frac{\dot{\varepsilon}}{6844}\right)^{1/3.91} \tag{1.100}$$

Substituting Eqs. (1.63), (1.98) and (1.99) into Eq. (1.100), Abramowicz and Jones [3] derived

$$\frac{P_{ave}^d}{M_0} = \left\{\frac{20.79(2R/t)^{1/2} + 11.90}{0.86 - 0.568(t/2R)^{1/2}}\right\}\left[1 + \left(\frac{0.25V_0/R}{6844[0.86 - 0.568(t/2R)^{1/2}]}\right)^{1/3.91}\right] \tag{1.101}$$

for axisymmetric crushing mode of cylindrical tubes.

1.2.3 Half-wavelength, curvature radius of wrinkle tip and eccentricity ratio

Half-wavelength λ, curvature radius of the wrinkle tip ρ_f and eccentricity ratio m are the parameters of the geometry of the wrinkle formed by the crushing of a circular tube. This section inclusively discusses the research results for these parameters.

1.2.3.1 Half-wavelength and curvature radius of the wrinkle tip in the crushing of circular tubes

The shape of the wrinkle formed by the axisymmetric crushing of a circular tube can be expressed by the half-wavelength λ and curvature radius of the wrinkle tip ρ_f. Half-wavelength λ is the curve length between the inner and outer vertexes of the wrinkle. Fig. 1.37 shows the shape of the wrinkle obtained by FEM simulation; as the figure shows, the half-wavelength of the wrinkle can be roughly divided into a linear part and curved sections. The curved section is also called the effective part of a plastic hinge, and can be approximated by a circular arc of radius of curvature ρ_f.

Fig. 1.38 shows FEM simulation results for the half-wavelength λ of a circular tube made of a material obeying the bilinear hardening rule. The figure shows that although the **half-wavelength** λ depends on the strain hardening properties of the material, it is mainly a function of the geometry of the circular tube, and λ/R increases with t/R. The figure also shows the experimental

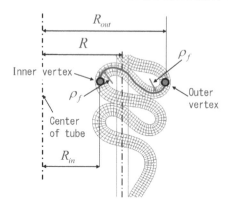

FIGURE 1.37
Shape of the wrinkle formed by the crushing of circular tube.

results of Singace et al. [187] and several values from the formulae for the half-wavelength λ listed in Table 1.3 (numbers in the figure such as (1) and (3) correspond to the equation numbers in Table 1.3). The figure shows that Eq. (1.83), which is based on Wierzbicki et al.'s curved model [217] under the assumption that inner and outer wrinkles are alternately formed, agrees well with the FEM simulation result.

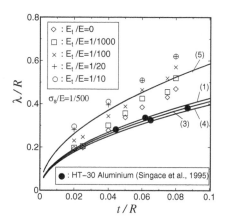

FIGURE 1.38
Half-wavelength λ of the wrinkle in the crushing of circular tubes.

There are few reports on the **curvature radius** ρ_f of the wrinkles formed by the axisymmetric crushing of a circular tube. When Abramowicz and Jones

[5] proposed the model shown in Fig. 1.28(b) where the wrinkle was approximated by two circular arcs, the curvature radius ρ_f of the arcs was assumed to be

$$\rho_f/\lambda \cong 0.25 \tag{1.102}$$

based on the approximate expression of the effective compression distance: $\delta_e/2\lambda \cong 0.75$, or

$$\rho_f/\lambda \cong 0.14 + 0.568\sqrt{\frac{t}{2R}} - \frac{t/2}{\lambda} \tag{1.103}$$

based on the approximate expression of the effective compression distance: $\delta_e/2\lambda \cong 0.86 - 0.568\sqrt{t/2R}$.

On the other hand, Grzebieta [85] proposed the model shown in Fig. 1.28(c) where the wrinkle was approximated by a linear part and two circular arcs, and pointed out that, as the thickness ratio t/R of the circular tube decreases, the plastic hinge became more concentrated in a shorter area and the ratio of curvature radius ρ_f and half-wavelength λ decreased. Furthermore, Stronge and Yu [195] reported that the length of the plastic hinge was 2–5 times greater than the thickness. Further, Huang and Lu [95] proposed a model similar to that of Grzebieta [85], assuming that the wrinkle was composed of a linear part and two circular arcs. The curvature radius ρ_f was not explicitly discussed, and the ratio between the total length $a\lambda$ of the two circular arcs and the wall thickness t of a circular tube was assumed to be constant b of 2–4, based on the experimental result. Fig. 1.39 shows the ratio between the curvature radius ρ_f obtained from the model of Huang and Lu [95] and the half-wavelength λ assuming $b = 2$. The figure also shows the values of ρ_f/λ obtained by FEM simulation for the three types of circular tubes of $t = 1, 2$ and 3 mm. The figure shows that, as the thickness ratio t/R of the circular tube increases, the ratio ρ_f/λ of curvature radius and half-wavelength increases. Note that, based on FEM simulations, the following relation approximately holds for the ratio between the curvature radius ρ_f of the wrinkle and the wall thickness t of the circular tube:

$$\rho_f/t \cong 0.8 \tag{1.104}$$

Using this equation and the one formulated by Wierzbicki et al. [217], Eq. (1.83) that gives the half-wavelength λ,

$$\rho_f/\lambda = 0.432\sqrt{\frac{t}{R}} \tag{1.105}$$

is obtained.

Fig. 1.39 also shows the calculated curvature radius ρ_f obtained by using Eq. (1.102), Eq. (1.103) and Eq. (1.105).

1.2.3.2 Eccentricity in crushing of circular tube

The **eccentricity ratio** is a parameter to show the outer expanse of a wrinkle formed in the continuous folding process of the tube wall; assuming the outer

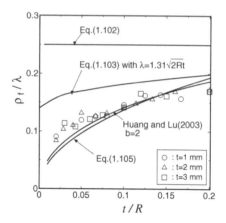

FIGURE 1.39
Curvature radius ρ_f of the wrinkle formed by the crushing of a circular tube.

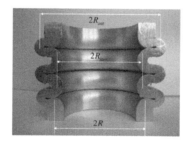

FIGURE 1.40
Cross section of an axially crushed circular tube.

and inner radiuses R_{out} and R_{in} as shown in Fig. 1.40, the eccentricity ratio is defined as

$$m = \frac{R_{out} - R}{R_{out} - R_{in}} \tag{1.106}$$

The eccentricity ratio m is a parameter needed to devise a mechanical model for calculating important values such as the average crushing force and energy absorption in the crushing of circular tubes. Alexander [10] used the two values $m = 1$ and $m = 0$ as the eccentricity ratio m to derive average forces for the two geometrically possible extreme cases, and stated that the actual average force should be an intermediate value, but without mentioning specific values of the eccentricity ratio m. In deriving a theoretical formula of the average force, Wierzbicki et al. [217] introduced the eccentricity ratio m as an unknown parameter in the energy equation for wrinkle deformation, and attempted to theoretically determine the eccentricity ratio m from the local minimum of energy; however, those efforts were unsuccessful because the parameter disappeared when the energy equation was derived. Singace et al. [187] revised Wierzbicki's model so that the eccentricity ratio m could remain in the energy equation, considering the difference between the inner and outer radii of wrinkles. From the local energy minimum for the deformation of a wrinkle, a constant eccentricity ratio m of 0.65 was determined, regardless of the geometry or material properties of the circular tube. However, that finding contradicts the experimental results of Grzebieta and Murray [86], who reported that the eccentricity ratio m is greater in a thick circular tube than in a thin circular tube. The FEM simulation confirms this problem, too.

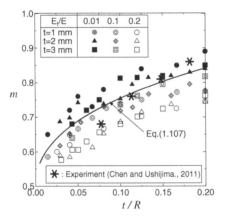

FIGURE 1.41
Eccentricity ratio of a wrinkle formed by axial crushing of various types of circular tubes.

Fig. 1.41 shows the results of FEM analysis on the relationship of eccentricity ratio m with circular tube thickness t and radius R. Here, the vertical axis shows the eccentricity ratio m, and the horizontal axis shows the thickness/radius ratio t/R; the calculation was performed for three strain hardening coefficients ($E_t/E = 0.01$, 0.1, and 0.2) for materials obeying the bilinear strain hardening rule and three tube thicknesses ($t = 1$ mm, 2 mm and 3 mm), and the tube radius R was changed as a parameter. In the figure are also shown experimental results on the axial crushing of circular tubes made of aluminum alloy $A5052$-$H112$ [51] by asterisk ($*$).

As the figure shows, the eccentricity ratio m is always in the range $0.5 < m < 1$. The eccentricity ratio m can be summarized as a function of thickness/radius ratio for any materials with various strain hardening coefficients and increases with t/R. In addition, the eccentricity ratio m depends also on the strain hardening coefficient E_t/E, and decreases as E_t/E increases.

By considering that the relations of the circumferential stress σ_θ and the circumferential strain ε_θ differ greatly between the inner and outer folds, and by minimizing the deformation energy with respect to expanding and contracting in the radial direction, Chen and Ushijima (2011b) proposed the following approximation equation for the eccentricity m:

$$m \cong \frac{1 + 4\sqrt{\dfrac{t}{R}} + 2\dfrac{t}{R}}{2 + 4\sqrt{\dfrac{t}{R}}} \tag{1.107}$$

Fig. 1.41 also shows the values given by the approximation equation, Eq. (1.107). As the figure shows, although the effect of the strain hardening coefficient of a material cannot be evaluated, Eq. (1.107) can at least show the change in the eccentricity ratio m with thickness, which qualitatively agrees with the numerical analysis by FEM.

1.2.4   Strain concentration in crushing process

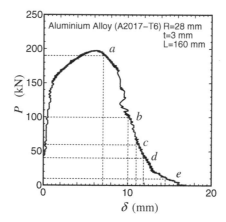

FIGURE 1.42
Relation between compressive force and displacement in the crushing of a tube accompanied by cracks (experimental results for A2017-T6) [210].

In the axial compression of a circular tube, the energy absorption depends on the formation process of the wrinkle. However, the desired energy absorption properties are, at times, not obtained because a crack appears in the tube wall when wrinkles are formed [210]. Fig. 1.42 shows the relationship between compressive force P and displacement δ in the axial direction as obtained in

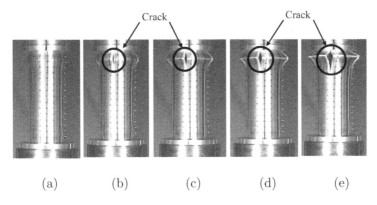

FIGURE 1.43
Deformation behavior of a circular tube at the points a, b, c, d and e in Fig. 1.42 [210].

a quasi-static crushing experiment using thin-walled circular tubes made of aluminum alloy ($A2017$-$T6$). Fig. 1.43(a)–(e) shows the deformation behavior of a circular tube at several points (a–e) in Fig. 1.42. During the formation of the wrinkle in the axial crushing deformation, the strain ε_θ in the circumferential direction is in most cases concentrated at the tip of the wrinkle, and a crack appears in the axial direction starting from the wrinkle surface (see Fig. 1.43(b)); in this way, the compressive force P greatly decreases as the crack develops. To improve energy absorption in the axial crushing of a circular tube, it is therefore important to control the strain concentration in the circumferential direction at the tip of the wrinkle.

Fig. 1.44(b) shows the variation of strain ε_θ in the circumferential direction at the tip of outer wrinkles, A, B and C, as shown in Fig. 1.44(a). Here, the strain ε_θ at the tip of the first-formed wrinkle, A, increases until the wrinkle is completely folded in the crushing process and then remains almost unchanged. On the other hand, the strain ε_θ in the circumferential direction in the second- and third-formed wrinkles at the tips B and C increases as the crushing proceeds, but does not exceed the maximum value appearing at the point A.

Let $\varepsilon_\theta|_{max}$ be **the maximum strain in the circumferential direction** at the outer tip of the wrinkle that folds first; based on the FEM simulation, it is found that $\varepsilon_\theta|_{max}$ does not strongly depend on the material's stress-strain characteristics (e.g., yield stress σ_s and strain hardening coefficient E_t for the case of the bilinear hardening rule), and depends mainly on the geometry of the circular tube. Note that the relationship between $\varepsilon_\theta|_{max}$ and the geometry of the circular tube is such that, as shown in Fig. 1.45, $\varepsilon_\theta|_{max}$ is almost an increasing function of the dimensionless thickness ratio t/R.

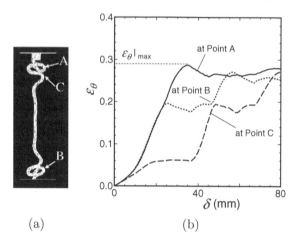

(a)   (b)

FIGURE 1.44

Variation of strain ε_θ at the wrinkle tips A, B and C during axial compression
[210]: (a) vertical section of the compressed circular tube; (b) ε_θ at A, B and
C.

The maximum strain $\varepsilon_\theta|_{max}$ at the outer tip of a wrinkle can be estimated
as

$$\varepsilon_\theta|_{max} = \frac{R_{out} - R}{R} \tag{1.108}$$

Therefore, from the definition of eccentricity ratio m,

$$R_{out} - R = m(R_{out} - R_{in}) \tag{1.109}$$

is obtained, and by considering the geometry of the wrinkle,

$$(R_{out} - R_{in}) = \lambda - (\pi - 2)\rho_f \tag{1.110}$$

is obtained. Substituting Eqs. (1.109) and (1.110) into Eq. (1.108), the formula
for maximum strain $\varepsilon_\theta|_{max}$ is derived:

$$\varepsilon_\theta|_{max} = m\left[\left(\frac{\lambda}{R}\right) - (\pi - 2)\left(\frac{\rho_f}{R}\right)\right] \tag{1.111}$$

Therefore, by using the formulas for half-wavelength λ, curvature radius ρ_f,
and eccentricity ratio m as discussed in the previous section (Eqs. (1.83),
(1.104) and (1.107)), the maximum strain $\varepsilon_\theta|_{max}$ is obtained from Eq. (1.111),
which agrees well with the FEM simulation result as shown by the solid line
in Fig. 1.45.

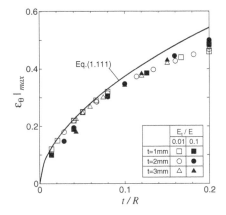

FIGURE 1.45

Maximum strain $\varepsilon_\theta|_{max}$ in the circumferential direction is a function of t/R.

1.3 Non-axisymmetric crushing of circular tubes

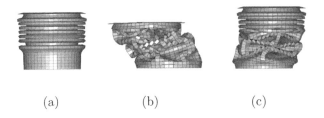

(a)   (b)   (c)

FIGURE 1.46

Axisymmetric and non-axisymmetric deformation modes in the axial crushing of circular tubes [38]: (a) axisymmetric; (b) non-axisymmetric; (c) transition from axisymmetric to non-axisymmetric.

From the deformation behavior in the circumferential direction (Fig. 1.46), the **progressive folding collapse** of the circular tube can be classified into the following groups: **axisymmetric mode** (Fig. 1.46(a)); **non-axisymmetric mode** (Fig. 1.46(b)), where waves are also observed in the circumferential direction; and **mixed mode** where a transition is observed from the axisymmetric to non-axisymmetric mode (Fig. 1.46(c)). In addition, in the cross sections of circular tubes of the non-axisymmetric deformation mode, as shown in Fig. 1.47, there are various n-gonal cross sections, where $n = 2, 3, 4, 5, 6$ indicates the number of polygon sides.

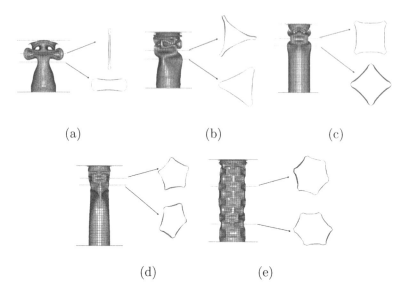

(a)   (b)   (c)

(d)   (e)

FIGURE 1.47
Non-axisymmetric crushing deformation modes of circular tubes ($R =$ 25 mm, $\sigma_s/E = 0.001$, $E_t/E = 0.05$): (a) $n = 2$ ($t = 2$ mm); (b) $n = 3$ ($t = 1$ mm); (c) $n = 4$ ($t = 0.6$ mm); (d) $n = 5$ ($t = 0.4$ mm); (e) $n = 6$ ($t = 0.3$ mm).

Fig. 1.48 shows the behavior of the compressive force in the non-axisymmetric crushing of a circular tube for the case that $n = 4$ non-axisymmetric wrinkles are formed. A force peak appears in the first crushing, and then the first wrinkle appears near the end part; and the compressive force P decreases until the wrinkle is completely folded and makes contact with itself. As the compressive deformation further proceeds, the formation and folding of wrinkles are repeated and therefore periodic oscillation of the compressive force is observed. Thereby, a single peak force corresponds to the formation of a single wrinkle in the force-displacement diagram.

In the following, non-axisymmetric compressive deformation is discussed with the subtitles on non-axisymmetric buckling and non-axisymmetric crushing after buckling.

1.3.1   Buckling stress

On the non-axisymmetric buckling of circular tubes, it is first necessary to recall the following two points.

(1) As shown by Timoshenko and Gere [203], the **non-axisymmetric elastic buckling stress** is equal to the axisymmetric elastic buckling stress

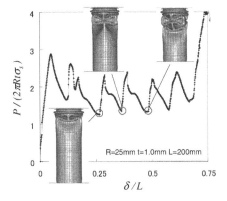

FIGURE 1.48
Variation of compressive force in the non-axisymmetric crushing of circular tubes.

(see Eq. (1.1)) and is given by

$$\sigma_{buc}^{e} = \frac{Et}{R\sqrt{3(1-\nu^2)}} \tag{1.112}$$

Therefore, Batterman and Lee [21] attributed the non-axisymmetric elastic buckling always found in the axial crushing of thin-walled circular tubes to initial geometric imperfections. In other words, in the elastic buckling of thin-walled circular tubes, the non-axisymmetric buckling mode appears because the non-axisymmetric buckling stress decreases much faster than the axisymmetric buckling stress due to the initial imperfections that inevitably occur in an experiment.

(2) Non-axisymmetric buckling stress is greater than axisymmetric buckling stress in the case of plastic buckling.

Bijlaard first pointed this out [22]. Then, Batterman and Lee [21] showed this based on the result of a theoretical analysis by Lee [121]. The non-axisymmetric plastic buckling stress under the J_2 **incremental theory** is given by

$$\sigma_{buc}^{p}|_{CM} = \frac{2E}{3}\frac{t}{R}\left\{\frac{3}{(5-4\nu)\eta_t-(1-2\nu)^2}\times\right.$$

$$\left.\left[\frac{[\eta_t(\eta_t+3)]^{1/2}+\dfrac{(7-2\nu)\eta_t-3(1-2\nu)}{2(1+\nu)}}{[\eta_t(\eta_t+3)]^{1/2}+3-\eta_t}\right]^{1/2}\right\} \tag{1.113}$$

and under the J_2 **deformation theory** (assuming $\nu = 0.5$ for simplicity) is given by

$$\sigma_{buc}^{p}|_{CM} = \frac{2}{3}\sqrt{E_t E_s}\frac{t}{R}\left\{\frac{2 + [1 + (3\eta_s/\eta_t)]^{1/2}}{(3\eta_s/\eta_t) - 1 + [1 + (3\eta_s/\eta_t)]^{1/2}}\right\}^{1/2} \quad (1.114)$$

where

$$\eta_t = \frac{E}{E_t}, \quad \eta_s = \frac{E}{E_s}$$

The buckling stress given by Eq. (1.113) is greater than the axisymmetric plastic buckling stress under the J_2 incremental theory (see Eq. (1.24)):

$$\sigma_{buc}^{p}|_{AX} = \frac{2E}{\sqrt{3}} \cdot \frac{1}{\sqrt{(5 - 4\nu)\eta_T - (1 - 2\nu)^2}}\frac{t}{R} \quad (1.115)$$

Moreover, the buckling stress given by Eq. (1.114) is greater than the axisymmetric plastic buckling stress using the J_2 deformation theory (see Eq. (1.21)):

$$\sigma_{buc}^{p}|_{AX} = \frac{2}{\sqrt{3}}\sqrt{E_t E_s}\frac{t}{R} \quad (1.116)$$

Therefore, Batterman and Lee [21] pointed out that the plastic buckling in the axial crushing of a circular tube usually becomes axisymmetric.

If so, however, a problem arises regarding how to explain the non-axisymmetric deformation appearing in the plastic buckling as shown in Fig. 1.48. Here, it is necessary to examine when the non-axisymmetric deformation starts in the plastic buckling. When the peak stress appears or slightly thereafter, the deformation of the tube wall is still very small and it is very difficult to detect non-axisymmetric deformation behavior; therefore, to decide whether the deformation is non-axisymmetric, here it is reasonable to examine the compressive stress distribution in the circumferential direction as in Fig. 1.49(b). Here, when peak stress appears at point A or slightly thereafter at point B, the distribution of compressive stress is uniform in the cross section of the circular tube. The deformation proceeds after buckling, and the distribution of compressive stress ceases to be uniform at point C, and four waves appear in the distribution of compressive stress in the circumferential direction corresponding to the $n = 4$ non-axisymmetric deformation, suggesting that the deformation mode has become non-axisymmetric. Therefore, it is considered that, in the non-axisymmetric deformation mode in Fig. 1.48, strictly speaking, axisymmetric plastic buckling occurs first, and a transition to the non-axisymmetric deformation mode occurs after buckling.

Fig. 1.50 shows the buckling stress in the axial compression obtained by FEM simulation for circular tubes of various thicknesses t and radii R. In the simulation, a two-dimensional FEM model, which is only designed for axisymmetric deformation, and a three-dimensional FEM model, which

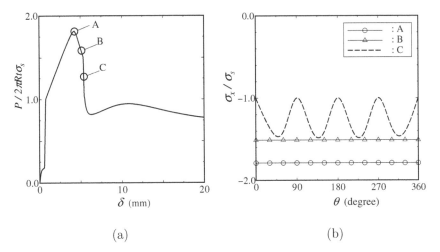

(a)   (b)

FIGURE 1.49
Sample of transition to non-axisymmetric mode from axisymmetric mode in plastic buckling of a tube ($R = 25$ mm, $t = 0.4$ mm, $\sigma_s/E = 0.001$, $E_t/E = 0.05$): (a) curve of $P - \delta$; (b) distribution of compressive stress in the cross section of the tube at time points A, B, C in Fig. 1.49(a).

is capable of handling non-axisymmetric deformation, were used. In the three-dimensional model, the buckling behavior of the circular tube depends on its thickness/radius ratio t/R. The figure also shows the number n of deformation modes in the circumferential direction in the case that buckling appears non-axisymmetric ($n = 0$ is axisymmetric). The figure shows that, independently of the buckling mode, the buckling stress roughly fits a simple straight line, roughly agreeing with the approximation formula Eq. (1.39). This also suggests that the non-axisymmetric deformation is due to a mode transition after axisymmetric buckling.

For the theoretical analysis of non-axisymmetric buckling stress of circular tubes, the method is essentially the same as that for axisymmetric buckling as discussed in Sections 1.1.1 and 1.1.2. Here, as an example, the analysis of elastic buckling stress is briefly discussed based on the **Donnell's equations** shown in Section 1.1.1.

Since the deformation of circular tubes is non-axisymmetric, it is necessary to consider the variation of stress and displacement in the circumferential direction (θ direction). Assuming that during buckling the generators of the shell subdivide into m half-waves and the circumference into $2n$ half-waves, a separated form is used for the solutions of $\hat{w}$ and $\hat{f}$. Here, $\hat{w}$, $\hat{f}$ are small perturbations at buckling for the radial displacement w and the Airy stress

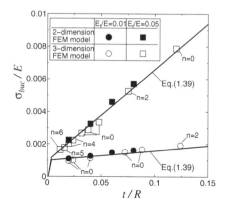

FIGURE 1.50
Buckling stress appearing in axial compression of a circular tube, including the non-axisymmetric deformation mode.

function f; see Eq. (1.5).

$$\hat{w} = A_w \sin \frac{m\pi x}{L} \cos \frac{ny}{R}, \quad \hat{f} = A_f \sin \frac{m\pi x}{L} \cos \frac{ny}{R} \tag{1.117}$$

Substituting Eq. (1.117) into Eq. (1.8) yields

$$-N_{x0} = \frac{1}{2} \left\{ \frac{(\alpha_m^2 + \beta_n^2)^2}{\alpha_m^2} + \frac{\alpha_m^2}{(\alpha_m^2 + \beta_n^2)^2} \right\} \times \frac{Et^2}{R\sqrt{3(1 - \nu^2)}} \tag{1.118}$$

where

$$\alpha_m^2 = \frac{Rt}{2\sqrt{3(1 - \nu^2)}} \left(\frac{m\pi}{L}\right)^2, \quad \beta_n^2 = \frac{Rt}{2\sqrt{3(1 - \nu^2)}} \left(\frac{n}{R}\right)^2 \tag{1.119}$$

$-N_{x0}$ is a minimum for

$$\frac{(\alpha_m^2 + \beta_n^2)^2}{\alpha_m^2} = 1 \tag{1.120}$$

Thus for a cylinder of intermediate length all mode shapes that satisfy Eq. (1.120) have the same buckling stress

$$\sigma_{buc}^e = \frac{-N_{x0}}{t} = \frac{Et}{R\sqrt{3(1 - \nu^2)}} \tag{1.121}$$

or

$$\sigma_{buc}^e = \sigma_{buc}^e \big|_{AX} \tag{1.122}$$

However, Donnell's approach is viewed as approximated and therefore less accurate, as shown in Fig. 1.51. Fig. 1.51 shows the relationship between half-wavelength L/mR of wrinkles in the axial direction and the elastic buckling stress σ^e_{buc} for the various deformation modes n obtained from Donnell's theory and Love's theory [203].

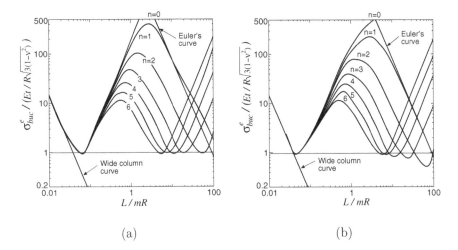

(a)   (b)

FIGURE 1.51

Relationship between the elastic buckling stress and the half-wavelength H/R in the axial direction for axial compression of tube $(t/R = 0.001)$: (a) Donnell's theory; (b) Love's theory.

1.3.2 Average crushing force

Pugsley and Macaulay [165] examined the **average crushing force in the non-axisymmetric crushing**. On the basis of their experiment, they proposed a formula,

$$\frac{P_{ave}}{2\sigma_s \pi Rt} = 10t/R + 0.03 \quad (t/R \leq 0.02) \tag{1.123}$$

to calculate the average force in the non-axisymmetric crushing of a thin-walled circular tube. Eq. (1.123) cannot be applied to the circular tube of thickness of $t/R > 0.05$ as shown in Fig. 1.54 presented later in this section.

After the research report of Pugsley and Macaulay [165], many theoretical reports have been published on the average force of the non-axisymmetric crushing in circular tubes. However, the study of an analytic model for the non-axisymmetric crushing of circular tubes seems to be much less successful than that for axisymmetric crushing.

The *origami* **model** proposed by Johnson et al. [100] is a representative analytic model for predicting the average force during the non-axisymmetric crushing of circular tubes. Based on experiments on the crushing of PVC circular tubes, Johnson et al. [100] proposed a model of non-axisymmetric wrinkle formation described by the folded origami shown in Fig. 1.52, assuming that the neutral plane of the circular tube does not extend or contract in the non-axisymmetric crushing of circular tubes, that is, the tube wall is inextensional, and assuming that the formation of a non-axisymmetric wrinkle is due to rotation about plastic hinges.

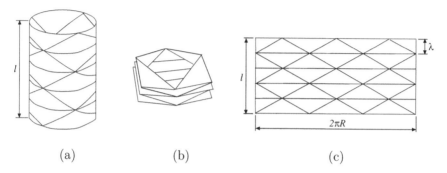

(a)   (b)   (c)

FIGURE 1.52
Johnson et al.'s model of non-axisymmetric crushing of circular tubes [100]: (a) hinge lines on the tube surface; (b) final shape upon crushing; (c) developed form of hinge lines.

Fig. 1.52 shows a model of non-axisymmetric wrinkle formation (in the figure, $n = 3$) in which a thin-walled circular tube of length l, radius R, and thickness t is under axial compression. Fig. 1.52(a) shows the hinge lines on the tube wall where non-axisymmetric wrinkles are formed. Fig. 1.52(b) shows the final shape upon crushing by non-axisymmetric compressive deformation, and Fig. 1.52(c) shows the bend lines spread on a flat plane. Non-axisymmetric wrinkles are formed when the n upward triangles ($\triangle$) and n downward triangles ($\triangledown$), aligned in series along the circumferential direction on the circular tube with length of half-wavelength λ, are folded alternately along the base and oblique lines. Thereby, it is assumed that the material is perfectly elasto-plastic, and that plastic deformation is concentrated on the hinge lines. Therefore, the possible sources of strain energy to form a wrinkle of half-wavelength λ are as follows:

(1) the strain energy U_1 needed for spreading the curved surface of a circular tube on the flat plane (with rotation angle 2π),

$$U_1 = 2\pi M_p \lambda \tag{1.124}$$

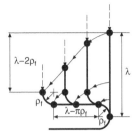

FIGURE 1.53
Movement of a plastic hinge in the non-axisymmetric crushing of a circular tube.

(2) the strain energy U_2 needed for folding the triangle along the oblique line of the triangle, that is, along the inclined hinge in the figure (with rotation angle π),

$$U_2 = 2n \left(\pi M_p \frac{\lambda}{\sin\left(\dfrac{\pi}{2n}\right)} \right) \qquad (1.125)$$

(3) the strain energy U_3 needed for folding the triangle along the base of the triangle, that is, along the horizontal hinge in the figure (with rotation angle π),

$$U_3 = n \left(\pi M_p \frac{2\lambda}{\tan\left(\dfrac{\pi}{2n}\right)} \right) \qquad (1.126)$$

From the equilibrium between strain energy to form a wrinkle and the work of the compressive force, the average force P_{ave} is given by [100]

$$P_{ave} = \frac{U_1 + U_2 + U_3}{\lambda} = 2\pi M_p \times \left[1 + \frac{n}{\sin\dfrac{\pi}{2n}} + \frac{n}{\tan\dfrac{\pi}{2n}} \right] \qquad (1.127)$$

where

$$M_p = \frac{2}{\sqrt{3}} \frac{\sigma_s t^2}{4}$$

and

$$\lambda = \frac{2\pi R}{2n} \tan\left(\frac{\pi}{2n}\right)$$

Considering the movement of a hinge in the formation of wrinkles as shown in Fig. 1.53, Johnson et al. [100] modified Eq. (1.127). In the formation of wrinkles, as the figure shows, a horizontal hinge travels the distance $\lambda - \pi\rho_f$. Therefore, assuming that the radius of curvature ρ_f of the hinge is constant, since the work by the movement of a hinge can be calculated by $2M_p \times$ (area swept by the hinge)$/\rho_f$, the strain energy U_4 needed for the movement of the hinge is given by

$$U_4 = \frac{2M_p \times 2\pi R(\lambda - \pi\rho_f)}{\rho_f} \tag{1.128}$$

Therefore, further considering the decrease of crushing distance due to the radius of curvature ρ_f (see $\lambda - 2\rho_f$ in the figure), the average force P_{ave} is given by

$$P_{ave} = \frac{U_1 + U_2 + U_3 + U_4}{\lambda - 2\rho_f} = \frac{2\pi M_p}{1 - \frac{2\rho_f}{\lambda}} \left\{ 1 + \frac{n}{\sin\left(\frac{\pi}{2n}\right)} - \frac{n}{\tan\left(\frac{\pi}{2n}\right)} + \frac{2R}{\rho_f} \right\} \tag{1.129}$$

From the requirement that P_{ave} in Eq. (1.129) be minimized, the radius of curvature ρ_f of the hinge is determined using the following equation:

$$\rho_f = \frac{2R}{A} \left(\sqrt{1 + \frac{A}{2}\frac{\pi}{2n}\tan(\pi/2n)} - 1 \right) \tag{1.130}$$

where

$$A = 1 + \frac{n}{\sin\dfrac{\pi}{2n}} - \frac{n}{\tan\dfrac{\pi}{2n}}$$

Besides the above-mentioned formula proposed by Johnson et al. [100], many formulas have been reported for calculating the **average force in the non-axisymmetric crushing** of a circular tube as follows.

(**1**) Using an effective yield stress σ_a that takes into account the strain hardening of the material instead of yield stress σ_s, Pugsley [164] proposed the formula

$$\frac{P_{ave}}{2\sigma_s \pi Rt} = 9.097 \left(\frac{t}{R}\right)\left(\frac{\sigma_a}{\sigma_s}\right) \tag{1.131}$$

However, this formula is basically a simplified version of Eq. (1.127) proposed by Johnson [100]. $P_{ave}/(2\sigma_s \pi Rt) = 9.097(t/R)$ is obtained by assuming (1) $n = 5$, (2) $U_1 \cong 0$, and (3) by replacing the trigonometric function $(1 + \cos\theta)/\sin\theta$ with the approximation $1.98/\theta$ in Eq. (1.127).

(**2**) Wierzbicki [215] proposed the equation

$$\frac{P_{ave}}{2\sigma_s \pi Rt} = 3.64 \left(\frac{t}{R}\right)^{2/3} \tag{1.132}$$

Further, Abramowicz and Jones [3] assumed the effective crushing distance $\delta_e/2\lambda = 0.73$ in the axial crushing of a circular tube, and modified Eq. (1.132) as follows:

$$\frac{P_{ave}}{M_p} = 86.14 \left(\frac{2R}{t}\right)^{0.33} \tag{1.133}$$

(3) Later, Abramowicz and Jones [5] proposed the formula

$$\frac{P_{ave}}{M_0} = A_{1n}\sqrt{\frac{2R}{t}} + A_{2n} \tag{1.134}$$

to give the average force in non-axisymmetric crushing as a function of the number n of the sides of the formed polygon. Here, A_{1n} and A_{2n} are $A_{1n} = 22.64, 21.07, 20.61, 20.40, 20.30$, and $A_{2n} = 4\pi, 32.66, 60.70, 96.72, 140.74$ for $n = 2, 3, 4, 5, 6$, respectively.

Further, there was another proposal where Eq. (1.134) is replaced by

$$\frac{P_{ave}}{M_0} = \frac{A_{1n}}{0.73}\sqrt{\frac{2R}{t}} + \frac{A_{2n}}{0.73} \tag{1.135}$$

assuming an effective crushing distance of $\delta_e/2\lambda = 0.73$.

(4) Further, based on the alternate formation model of inner and outer wrinkles, Singace [184] proposed the following formula for the average force:

$$\frac{P_{ave}}{M_0} = \frac{4\pi^2}{n}\tan\left(\frac{\pi}{2n}\right)\frac{R}{t} - \frac{\pi}{3}n \tag{1.136}$$

(5) An experiment by Guillow et al. [88] on the axial crushing of a circular tube made of aluminum alloy 6060-T5 showed that the average force P_{ave} is approximately linearly proportional to $(2R/t)^{0.32}$, regardless of the axisymmetry. Accordingly, Guillow et al. [88] proposed a formula for the non-axisymmetric average force of circular tubes as given by

$$\frac{P_{ave}}{M_0} = 72.3\left(\frac{2R}{t}\right)^{0.32} \tag{1.137}$$

Table 1.4 summarizes the formulas for the average force in the non-axisymmetric crushing of circular tubes proposed by the researchers discussed above.

In Fig. 1.54, the curves show the **average crushing force** obtained from the formulas in Table 1.4, and the symbols indicate the experimental results reported by Abramowicz and Jones [5], Guillow et al. [88], and Johnson et al. [100]. The numbers in parentheses indicate the equation numbers in Table 1.4. The experimental results regarding the axial crushing deformation mode include not only the non-axisymmetric mode but also the axisymmetric mode and the mixed mode. As the figure shows, the dimensionless average crushing stress normalized to $P_{ave}/(2\pi Rt\sigma_s)$ increases with the relative thickness t/R of circular tubes. In contrast, although not shown in this figure, the equation of number (8) in Table 1.4 [184] shows a different behavior. By rewriting this equation to give the average crushing stress,

$$\frac{P_{ave}}{2\pi Rt\sigma_s} = \frac{1}{4\sqrt{3}}\left[\frac{4\pi}{n}\tan\left(\frac{\pi}{2n}\right) - \frac{n}{3}\left(\frac{t}{R}\right)\right] \tag{1.138}$$

TABLE 1.4
Formulas for the average crushing force in the non-axisymmetric crushing of a circular tube.

Eq. Num.	Author(s) (year)	Average crushing force P_{ave}
(1)	Pugsley and Macaulay (1960) [165]	$\dfrac{P_{ave}}{2\sigma_s \pi R t} = 10t/R + 0.03 \quad (t/R \le 0.02)$
(2)	Johnson et al. (1977) [100]	$\dfrac{P_{ave}}{M_p} = 2\pi \left[1 + \dfrac{n}{\sin \dfrac{\pi}{2n}} + \dfrac{n}{\tan \dfrac{\pi}{2n}} \right]$
(3)	Johnson et al. (1977) [100]	$\dfrac{P_{ave}}{M_p} = \dfrac{2\pi}{1 - (2\rho_f/\lambda)} \left(A + \dfrac{2R}{\rho_f} \right)$ $A = 1 + \dfrac{n}{\sin(\pi/2n)} - \dfrac{n}{\tan(\pi/2n)},$ $\lambda = \dfrac{\pi R}{n} \tan(\pi/2n), \; \rho_f = \text{Eq. (1.130)}$
(4)	Wierzbicki (1983) [215]	$\dfrac{P_{ave}}{2\sigma_s \pi R t} = 3.64 \left(\dfrac{t}{R} \right)^{2/3}$
(5)	Abramowicz and Jones (1984) [3]	$\dfrac{P_{ave}}{M_p} = 86.14 \left(\dfrac{2R}{t} \right)^{1/3}$
(6)	Abramowicz and Jones (1986) [5]	$\dfrac{P_{ave}}{M_0} = A_{1n} \sqrt{\dfrac{2R}{t}} + A_{2n}$
(7)	Abramowicz and Jones (1986) [5]	$\dfrac{P_{ave}}{M_0} = \dfrac{A_{1n}}{0.73} \sqrt{\dfrac{2R}{t}} + \dfrac{A_{2n}}{0.73}$
(8)	Singace (1999) [184]	$\dfrac{P_{ave}}{M_0} = \dfrac{4\pi^2}{n} \tan \left(\dfrac{\pi}{2n} \right) \dfrac{R}{t} - \dfrac{\pi}{3}n$
(9)	Guillow et al. (2001) [88]	$\dfrac{P_{ave}}{M_0} = 72.3 \left(\dfrac{2R}{t} \right)^{0.32}$

is obtained. This shows that the average crushing stress decreases against the relative thickness t/R.

Further, as the equations of number (2), (3), (6), (7) and (8) in Table 1.4 and the PVC experiment of Johnson et al. [100] show, the average crushing force depends on the value n of the crushing mode. However, as the equations of number (1), (4), (5) and (9) in Table 1.4 and the experiment of Abramowicz and Jones [5] for mild steel and of Guillow et al. [88] for aluminum alloy 6060-T5 show, the average crushing force is apparently a function of the thickness/radius ratio t/R of circular tubes irrespective of the resulting (axisymmetric or non-axisymmetric) deformation mode.

These suggest that it is difficult to accurately predict the average force in the non-axisymmetric crushing of a circular tube. There are mainly two reasons for this:

(1) In both the experiment and the FEM simulation, it is rare to obtain the non-axisymmetric deformation mode with a single value of n; the mixed mode is often observed where a transition occurs from axisymmetric to non-axisymmetric deformation.

(2) As discussed in the following section, the compressive deformation mode depends also on the strain hardening properties of the material. For example, in the FEM simulation on the perfect elasto-plastic material without strain hardening ($E_t = 0$), the non-axisymmetric mode with a single value of n is not observed. For this reason, the comparison with the FEM analysis is not shown in Fig. 1.54.

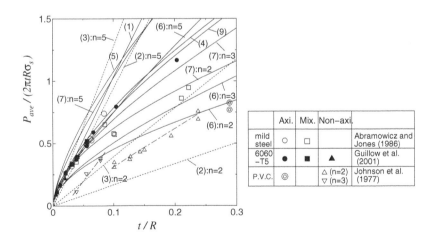

FIGURE 1.54
Average force in the non-axisymmetric crushing of circular tubes.

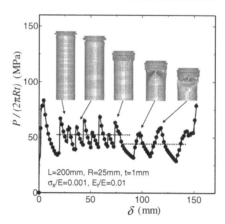

FIGURE 1.55
Variation of average crushing stress in the mixed compressive mode of circular tubes [38].

However, in the axial crushing of a circular tube, although a mixed mode where transition occurs from axisymmetric to non-axisymmetric deformation is observed, a mixed mode transition from non-axisymmetric to axisymmetric deformation does not exist; this suggests that the average crushing stress in the axial symmetric deformation is larger than the non-axisymmetric average crushing stress. Fig. 1.55 shows the relationship between the crushing stress of the mixed deformation mode and the deformation behavior in the crushing process. The average crushing stress (dotted line) in each of the axisymmetric and non-axisymmetric deformation areas in the figure indicates that the average crushing stress decreases with the transition from the axisymmetric to the non-axisymmetric modes.

Further, as shown by Johnson et al.'s model [100] in Fig. 1.52, as for the **half-wavelength** λ of wrinkles in the non-axisymmetric compression of a circular tube, the half-wavelength λ depends on geometry shown in Fig. 1.52(c) and is given by

$$\frac{\lambda}{R} = \frac{\pi}{n} \tan \left(\frac{\pi}{2n} \right) \tag{1.139}$$

Fig. 1.56 compares the half-wavelength of the non-axisymmetric wrinkle of the circular tubes as obtained by FEM simulation and Eq. (1.139). The figure also shows the result of calculation using the formula

$$\frac{\lambda}{R} = 1.632 \left(\frac{t}{2R} \right)^{1/3} \tag{1.140}$$

which was proposed by Wierzbicki [215]. The FEM analysis shows that λ/R increases as the number n of the sides of the polygon in the non-axisymmetrically

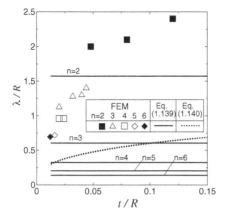

FIGURE 1.56
Half-wavelength of wrinkles for the non-axisymmetric compression of circular
tubes.

deformed cross section decreases and as the thickness t/R of the circular
tube increases. On the other hand, Eq. (1.139) shows that, although λ/R
increases as n decreases, it does not depend on the thickness t/R, and the
half-wavelength is considerably smaller than the value in the FEM analysis.
It was assumed in the origami model that the bending at the plastic hinge
is concentrated at the hinge line; however, in the actual non-axisymmetric
compressive deformation, the bent part occupies a considerably large area
depending on the thickness. Therefore, accurate analysis is difficult by the
origami model for the non-axisymmetric compressive deformation of a circular tube.

1.4 Deformation modes in the axial crushing of circular tubes

In this section, various deformation modes found in the axial crushing of
circular tubes shall be discussed.

The deformation modes in the axial crushing of circular tubes are classified by the deformation behavior in the axial and circumferential directions.
Fig. 1.57 shows the representative deformation modes summarized by focusing
on the deformation behavior in the axial direction. As the figure shows, the
deformation of circular tubes under compressive force in the axial direction is
classified first by the criterion of whether the folding deformation (formation

of wrinkles) continues while the entire circular tube stands upright until it is finally crushed. Here, the mode in which the circular tube stands upright until it is finally crushed is called the **stable type** ((a) in Fig. 1.57), and the mode in which the circular tube bends from the beginning or midway without achieving uniform crushing is called the **unstable type** ((b) and (c) in Fig. 1.57).

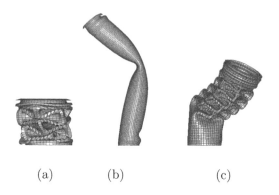

(a)   (b)   (c)

FIGURE 1.57
Stable and unstable deformation modes ($R = 25$ mm) [38]: (a) stable type ($t = 1$ mm, $L = 20$ mm); (b) unstable type (bending mode) ($t = 3$ mm, $L = 400$ mm); (c) unstable type (transition mode) ($t = 1$ mm, $L = 400$ mm).

The **progressive folding collapse** is a typical stable type. As discussed in Section 1.3, based on the deformation behavior in the circumferential direction, the progressive folding collapse deformations are further classified into an **axisymmetric mode** maintaining a circular cross section, a **non-axisymmetric mode** with a polygonal cross section, and a **mixed mode** (Fig. 1.46(c)) where transition from axisymmetric to non-axisymmetric mode occurs.

Fig. 1.58(a), (b) shows the cross-section forms of the wrinkles formed by the axial compression of circular tubes of different thicknesses t assuming that the materials obey the bilinear hardening rule with $E_t/E = 1/20$ and $E_t/E = 1/100$ [38]. Here, to test the effect of a flange at the end part of a circular tube, the setup of circular tubes with flanges at both ends shown in Fig. 1.59 is considered. In Fig. 1.58, the mark O shows axisymmetric wrinkles, and the polygonal symbols ($\triangle$, $\square$, etc.) show non-axisymmetric wrinkles with the number of the sides of the polygon $n = 2$, 3, 4, 5 and 6 as a parameter. In the figure, the horizontal axis shows the logarithmic relative thickness t/R of the circular tube, and the vertical axis shows the ratio s/R between the distance s and the radius R of the circular tube. Here, s is the distance from the position of each wrinkle to the tube end of the side where the wrinkle initially appears, measured in the axial direction before the tube has deformed. As the figure shows, the thickness/radius ratio t/R is a key factor to determine whether

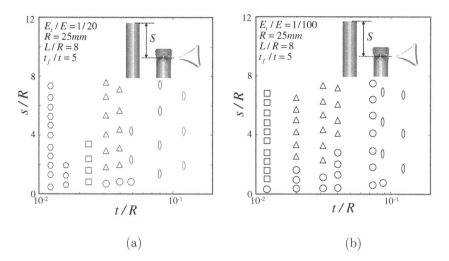

(a)   (b)

FIGURE 1.58
Formation of wrinkles with various cross sections in the axial compression of circular tubes for materials obeying the bilinear strain hardening rule (O axisymmetric wrinkles; other symbols including △, □, etc. show non-axisymmetric wrinkles with the number indicating the number of sides of the polygon) [38]: (a) $E_t/E = 1/20$; (b) $E_t/E = 1/100$.

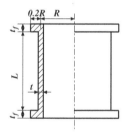

FIGURE 1.59
Geometry of the circular tubes with flanges at both ends [38].

the wrinkle becomes axisymmetric or non-axisymmetric, and the number n of polygon sides in the cross section of the non-axisymmetric wrinkle decreases as t/R increases. However, not axisymmetric but rather non-axisymmetric wrinkles for $n = 2$ are found, even in the thick-walled (large t/R) circular tube ($t/R > 0.048$ at $E_h/E = 1/20$, $t/R > 0.084$ at $E_h/E = 1/100$). Experimental studies (e.g., study by Guillow et al. [88]) show that axisymmetric wrinkles appear if t/R is large ($t/R > 1/25$), apparently contradicting the data in Fig. 1.58. This contradiction can be explained by the following three reasons.

(1) As understood from the comparison between Fig. 1.58(a) and (b) and Fig. 1.62 shown below, the strain hardening properties of the material greatly affect the axisymmetric deformation mode of circular tubes; a smaller strain hardening coefficient E_t results more in axisymmetric deformation. Therefore, if the strain hardening coefficient E_t of the material is sufficiently small, it is possible that non-axisymmetric wrinkles of $n = 2$ do not appear even if t/R is large.

(2) As discussed later in this section, axisymmetric constraint at the end part of a circular tube, for example, brought about by flanges at both ends, suppresses the non-axisymmetric mode and tends to induce the axisymmetric compression mode. Therefore, in the experiment using thick-walled circular tubes (for thin-walled circular tubes, non-axisymmetric wrinkles tend to appear because the influence of initial imperfections is large), axisymmetric modes tend to appear due to the constraint at the end part.

(3) As Fig. 1.65 shows below, the formation of axisymmetric deformation depends on the circular tube length, too. The circular tubes discussed in Fig. 1.58 have length $L/R = 8$. On the other hand, the circular tubes used in experimental studies are often shorter than this (e.g., $L/R \leq 5$ for the circular tube in the study of Guillow et al. [88] in which it was found that axisymmetric wrinkles tended to appear).

The flanges at both ends of a circular tube greatly affect the deformation mode. Fig. 1.58 shows that even if the relative thickness t/R is the same, the deformation more non-axisymmetric than axisymmetric is observed as the distance s/R from the end part— from the flange—increases. This is attributed to the influence of constraint by the flange at the end part on the displacement in the circumferential direction. Furthermore, in Fig. 1.58(b), before the crushing, the wrinkles are always axisymmetric in circular tubes of $t/R = 0.072$ ($R = 25$ mm, $t = 1.8$ mm); this depends on the flange at the end part, too. In the calculation in the figure, the flange thickness is $t_f = 9$ mm, and the deformation is shown in Fig. 1.60(a). If the flange is as thin as $t_f = 3.6$ mm (Fig. 1.60(b)), four axisymmetric wrinkles (three from the upper end, one from the lower end) appear first and then non-axisymmetric wrinkles of $n = 2$ appear. This suggests that, since the average compressive stress in the axisymmetric deformation is greater than that in the non-axisymmetric deformation, the formation of axisymmetric wrinkles is difficult without the constraint by the flange at the end part.

The effect of the flange on the cross-sectional geometry of the wrinkled tube can be understood from Fig. 1.61 as well. Fig. 1.61 shows the variation of the deformation mode of the circular tubes of $L = 150$ mm, $R = 25$ mm, $t = 1$ mm, $E_t/E = 1/20$ with the flange thicknesses of $t_f = 5$ and 2 mm. In the axial crushing of the circular tubes, when the shortening of the tube is $\delta = 35$ mm, the deformation modes of $n = 3$ and $n = 2$ appear at $t_f = 5$ mm and $t_f = 2$ mm, respectively. This shows that, in relation to what is shown

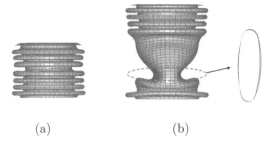

(a)   (b)

FIGURE 1.60
Effects of flange thickness t_f at the end part on the deformation mode ($L = 200$ mm, $R = 25$ mm, $t = 1.8$ mm, $E_t/E = 1/100$) [38]: (a) $t_f = 9$ mm; (b) $t_f = 3.6$ mm.

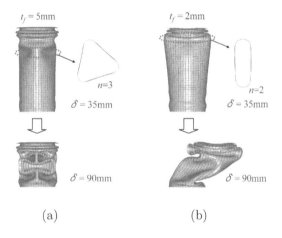

(a)   (b)

FIGURE 1.61
Effects of flange thickness t_f at the end part on the deformation mode ($L = 150$ mm, $R = 25$ mm, $t = 1$ mm, $E_t/E = 1/20$) [38]: (a) $t_f = 5$ mm; (b) $t_f = 2$ mm.

in Fig. 1.60, the wrinkle of $n = 2$ tends to appear when the constraint by the flange at the end part is reduced by decreasing the flange thickness ($t_f = 2$ mm). Further, while the circular tube of $t_f = 5$ mm remains upright for the shortening of $\delta = 90$ mm, the circular tube of $t_f = 2$ mm is bent by unstable deformation type, suggesting that the constraint at the end part affects the deformation type.

Further, non-axisymmetric wrinkles appear more easily as strain hardening of material increases. This is understood from a comparison between Fig. 1.62(a) and (b). These figures compare the behavior of wrinkles viewed from the axial direction (x direction) in the circular tubes of the size $L =$

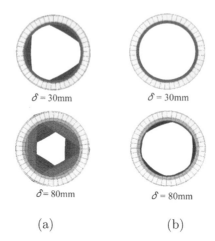

$\delta = 30$mm $\delta = 30$mm

$\delta = 80$mm $\delta = 80$mm

(a)   (b)

FIGURE 1.62
Effects of strain hardening coefficient E_t on the deformation mode ($R = 25$ mm, $t = 1$ mm, $L = 200$ mm)[38]: (a) $E_h/E = 1/20$; (b) $E_h/E = 1/100$.

200 mm, $R = 25$ mm, $t = 1$ mm. First, when the strain hardening coefficient is as large as $E_t/E = 1/20$, a non-axisymmetric (triangular) wrinkle of $n = 3$ appears at the shortening $\delta = 30$ mm. On the other hand, when the strain hardening coefficient is as small as $E_t/E = 1/100$, the wrinkle is axisymmetric at $\delta = 30$ mm, and then a non-axisymmetric wrinkle appears at the shortening of about $\delta = 80$ mm.

These discussions show that the non-axisymmetric deformation mode tends to appear in a long circular tube because the constraint of the flange is less strong in the central part of such a tube. That is, although the axisymmetric wrinkle appears first due to the constraint at the end part, and then the constraint of the flange becomes less effective as the wrinkle propagates, the deformation mode changes into the one in which the circular tube deforms more freely and as a result the non-axisymmetric wrinkle appears. In this case, the deformation mode can be controlled by placing rings or circular disks, which act like flanges, in the middle part of the circular tube. Fig. 1.63 compares the deformation behavior (b) with and (a) without a circular disk in the middle part of the circular tube of $L = 200$ mm, $R = 25$ mm, $t = 1$ mm and $E_t/E = 1/100$. As the figure shows, the non-axisymmetric mode appears distant from the flange in Fig. 1.63(a) while the non-axisymmetric mode is not observed in Fig. 1.63(b).

On the other hand, **unstable type** can be classified into two modes as shown in Fig. 1.57(b) and (c). One is deformation where a global bending occurs in the lateral direction without local wrinkles (hereinafter, the **bending mode**), and the other is deformation where global bending occurs in the lateral direction after several wrinkles are formed by local buckling (the **tran-**

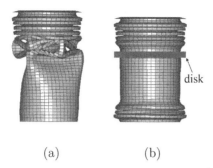

(a)   (b)

FIGURE 1.63
Effects of a circular disk in the middle part of a circular tube on the deformation mode [38]: (a) without disk; (b) with disk.

sition mode). In the transition mode, why did global bending failure occur
after the tube was shortened by the formation of several wrinkles, rather than
at the start of compression, when the tube had a longer initial length and
Euler buckling was more likely to occur? Abramowicz and Jones [6] pointed
out that the transition mode could be observed in a circular tube under a
dynamic axial crushing force because the effect of inertia decreases as the
dynamic axial crushing proceeds. However, there should be no inertial effect
on the quasi-static axial crushing as shown in Fig. 1.57(c). The transition
mode in the quasi-static axial crushing is attributable to the variation of the
boundary conditions due to the formation of wrinkles at the end part. That
is, if the deformation is carefully observed in a circular tube that shows the
transition mode, it is found that several non-axisymmetric wrinkles always
appear before the circular tube bends, and the end part of the circular tube
moves horizontally, though only slightly, while non-axisymmetric wrinkles are
formed [38]. There has been no report of the bending of a circular tube after
only the formation of axisymmetric wrinkles. Fig. 1.64 shows the horizontal
movement U_y of the center of the upper end of the circular tubes of $t = 1$ mm,
$R = 25$ mm, $E_t/E = 1/20$ and lengths $L = 200$ and 400 mm. In the axial
crushing of these circular tubes, 2–3 axisymmetric wrinkles appear first, and
then $n = 3$ non-axisymmetric wrinkles appear, and accordingly the movement
U_y gradually increases. A stable type deformation was found in the circular
tube of length $L = 200$ mm because it was completely crushed before the
movement U_y grew large enough; a transition deformation mode was found in
the circular tube of length $L = 400$ mm in which the movement U_y grew large
enough and the circular tube bent before the crushing was completed. On the
other hand, if the horizontal movement of the upper end face is prevented,
for example, if a large friction coefficient of $\mu_f = 0.3$ is assumed between the
upper end face and the rigid body for compression, the simulation analysis
of the circular tube of $L = 400$ mm shows that although the formed wrinkle
is non-axisymmetric as in the case of $\mu_f = 0.03$, a stable type deformation

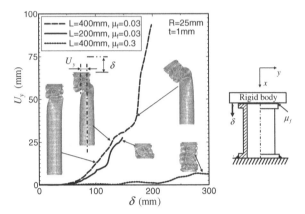

FIGURE 1.64
Sliding displacement in the horizontal direction of a tube end causes a global
bending of the tube [38].

is observed as shown in Fig. 1.64 because the horizontal movement of the
upper end face is prevented. Therefore, in the transition deformation mode
Euler buckling did not occur in the early stage of the compression since the
length of the circular tube is shorter than the limit length of Euler buckling; however, non-axisymmetric wrinkles are formed, shifting the center of
the crushing force, and resulting in a global bending of the shortened tube. A
similar phenomenon is also observed in the crushing process of a square tube
showing the transition deformation mode, as shown by Jensen et al. [98]. This
demonstrates that the lobes started to develop eccentrically before generating
of transition from progressive to global buckling in experimental investigations
on the behavior of square aluminium tubes subjected to axial loading.

As the above discussions suggest, the deformation mode in the axial crushing of a circular tube is strongly affected by not only the thickness and length
of the circular tube but also the flange thickness and the frictional force at
the end part. For circular tubes using materials that obey the bilinear hardening rule with the strain hardening coefficient of $E_t/E = 1/100$, assuming
the flange thicknesses $t_f/t = 5$ and $t_f/t = 2$, the deformation mode maps
for various geometry combinations of R/t and L/R are shown in Fig. 1.65.
From this figure, unstable deformation occurs if the non-dimensional length
is large, and stable deformation occurs if L/R is small; transition deformation
occurs if the non-dimensional radius R/t is large, and bending deformation
occurs if R/t is small. When the flange thickness is changed from $t_f/t = 5$
to $t_f/t = 2$, the boundary between deformation modes moves from the solid
lines to the dotted lines. As the figure shows, the flange at the end part of a
circular tube does not affect the tube's bending mode, but, through the effect
on the cross-sectional geometry of the wrinkle, affects regions of the stable

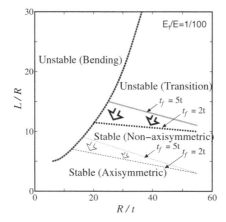

FIGURE 1.65
Relationship between deformation mode and geometry in axially crushed circular tubes [38].

and axial symmetrical modes, and these regions become small if the flange is made thin.

1.5 Effects of stiffeners

The buckling strength of a thin-walled circular tube is greatly improved by attaching stiffeners in the axial or circumferential direction; stiffeners are widely used in many types of structures, and have been investigated by many researchers (e.g., Milligan et al. [149], Singer [188]). Further, stiffeners are useful for improving the energy absorbing performance of a circular tube. Jones and Papageorgiou [102] studied the effect of stiffeners attached inside the circular tube on the plastic buckling in the axial direction, and pointed out that the best stiffener size is about $b_2/(2\pi R/n) \cong 5/16$ (where b_2 is the width of the rectangular cross section of the stiffener, and $2\pi R/n$ is the distance between stiffeners in the circumferential direction). Further, Birch and Jones [24] conducted experiments on the quasi-static and dynamic crushing when stiffeners were attached inside or outside a steel circular tube in the axial direction; they found an optimized ratio between stiffener thickness t_2 and circular tube radius R to maximize energy absorption, thereby showing that the further increase of stiffener size does not improve energy absorption properties.

In this section, based on FEM analysis [39] of the axial crushing deformation of circular tubes to which stiffeners are attached (a) in the circumferential direction, (b) in the axial direction or (c) in both the axial and circumferential directions, as shown in Fig. 1.66, the effect of stiffeners attached in the axial or circumferential direction (to be called ribs or rings, respectively, in the following) inside or outside the circular tube on the deformation mode and the crushing force of the circular tube under axial compression will be methodically discussed. Note that the geometry of ribs and rings is shown in Fig. 1.67, and the radius of the circular tube is always $R = 25$ mm in the following.

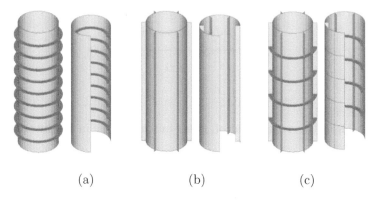

(a)   (b)   (c)

FIGURE 1.66
Stiffened cylindrical tubes [39]: (a) ring stiffeners; (b) rib stiffeners; (c) combination of ring and rib stiffeners.

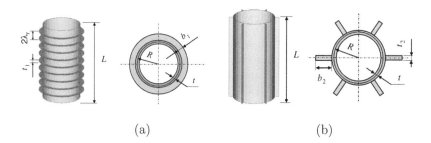

(a)   (b)

FIGURE 1.67
Geometry of stiffened cylindrical tube [39]: (a) circumferentially stiffened ringed tube; (b) axially stiffened ribbed tube.

1.5.1   Effects of rings

For the circular tubes with different lengths L and the thickness $t = 1$ mm to which rings are attached to the inner wall with intervals of $2\lambda_r = 20$ mm,

TABLE 1.5

Deformation modes of circumferentially stiffened ringed tubes ($2\lambda_r = 20$ mm) [39].

L (mm)	b_1 or t_1 (mm) constant	$b_1 \times t_1$ (mm^2) 0	1–3	4–10
L = 400	$t_1 = 1$	▲	▲	▲
	$b_1 = 5$	▲	▲	▲
L = 300	$t_1 = 1$	☐	☐	●
	$b_1 = 5$	☐	☐	●
L = 200	$t_1 = 1$	■	●	●
	$b_1 = 5$	■	●	●
L = 100	$t_1 = 1$	●	●	●
	$b_1 = 5$	●	●	●

Table 1.5 shows the variation of deformation modes with rings for the one case assuming the ring width of $b_1 = 5$ mm and changing the thickness t_1, and the other case assuming the ring width of $t_1 = 1$ mm and changing the width b_1, by four symbols (▲: bending mode; ☐: transition mode; ●: axisymmetric mode; ■: non-axisymmetric mode). A column in the table shows the product of width and thickness $b_1 \times t_1$; the value 0 shows a simple circular tube without rings; the deformation behavior is classified into bending mode, transition mode, axisymmetric mode and non-axisymmetric mode as shown in Fig. 1.68. As the table shows, when rings are attached to a circular tube, variations may or may not be found in the deformation mode. That is, the rings are not advantageous in the bending mode for $L = 400$ mm and the axisymmetric mode for $L = 100$ mm, but the rings are advantageous in the non-axisymmetric mode for $L = 200$ mm and the transition mode for $L = 300$ mm. In the circular tube of $L = 200$ mm, by adding rings, the deformation mode changes into the axisymmetric mode, while in the circular tube of $L = 300$ mm, the deformation mode could change into the axisymmetric mode depending on the values of b_1 and t_1.

The effect of rings depends not only on their size (b_1 and t_1) but also on the interval between the rings. Fig. 1.69 shows a deformation mode map assuming various types of rings attached to a circular tube ($L = 200$ mm and $t = 1$ mm) in which only non-axisymmetric modes appear without rings. Here, in the mode map, the classification parameters are the interval between rings $2\lambda_r$ and the size of rings $b_1 \times t_1$. The figure shows that axisymmetric wrinkles tend to appear if the interval between the rings is roughly $2\lambda_r = 20$ mm for attachment inside or outside the circular tube. Considering that in the circular tube without stiffeners of $R = 25$ mm and $t = 1$ mm, the wrinkle wavelength is about $2\lambda_0 = 20$ mm in the case of axisymmetric deformation, it is assumed that the effect of rings would become the greatest when the interval between rings $2\lambda_r$ is equal to the wavelength of the axisymmetric wrinkle $2\lambda_0$. Further,

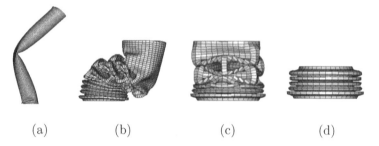

(a)   (b)   (c)   (d)

FIGURE 1.68
Deformation modes for tubes without stiffeners [39]: (a) unstable type: bending mode ($L = 400$ mm); (b) unstable type: transition mode ($L = 300$ mm); (c) stable type: non-axisymmetric mode ($L = 200$ mm); (d) stable type: axisymmetric mode: ($L = 100$ mm).

the symbol ○ in the figure, similar to the mark ●, shows the axisymmetric mode, but there the wrinkle wavelength 2λ is not equal to the interval between rings $2\lambda_r$. Comparing (a) and (b) in Fig. 1.69, it is clear that the axisymmetric deformation with $\lambda = \lambda_r$ is more easily induced if the rings are attached inside the circular tube.

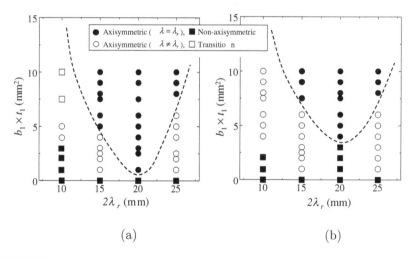

(a)   (b)

FIGURE 1.69
Mode map for circumferentially stiffened ringed circular tubes ($L = 200$ mm) [39]: (a) internal stiffener; (b) external stiffener.

Fig. 1.70(a) shows the crushing force in the axial crushing of a circular tube, when the rings are attached inside the circular tube. The tube size is $R = 25$ mm, $t = 1$ mm and $L = 100$ mm, and its deformation mode is

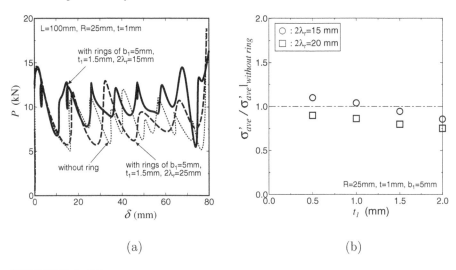

(a)   (b)

FIGURE 1.70
Effect of rings on compressive force [39]: (a) relation of P and δ; (b) average crushing stress σ'_{ave}.

axisymmetric. The axisymmetric deformation mode is also observed in the case that rings of $b_1 = 5$ mm and $t_1 = 1.5$ mm are attached as shown in the figure at the intervals $2\lambda_r = 15$ mm and 25 mm. Here, the wrinkle wavelength can be controlled by the rings, and as a result, the variation in the magnitude of the crushing force can be reduced and the average crushing force can be made greater than that of the circular tube without rings. However, if the equivalent area of cross section $A' = V/L$ (defined as the ratio of volume V and length L of the model) is used to evaluate the average crushing stress $\sigma'_{ave} = P/A'$, as Fig. 1.70(b) shows, even though in some cases σ'_{ave} may become only slightly greater than that of a circular tube without rings, σ'_{ave} usually decreases because the equivalent cross section A' is increased by the use of rings. Therefore, the basic role of rings will be to induce axisymmetric deformation of wrinkles.

1.5.2   Effects of ribs

Fig. 1.71 shows the deformation mode map where the parameters are the thickness t and the length L for a circular tube of $R = 25$ mm without and with ribs ($b_2 = 6$ mm, $t_2 = 1$ mm and $n = 6$) attached outside. In the figure, the results on the circular tubes with ribs are shown by symbols (▲: bending mode; □: transition mode; ●: stable type), and the modes for the circular tubes without ribs are shown by the hatched areas. The figure shows that the area of the bending mode increases by adding ribs as compared with the circular tube without ribs. Therefore, ribs are applicable only to short

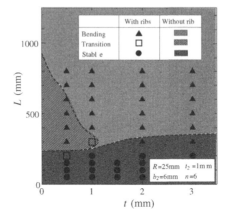

FIGURE 1.71
Deformation mode map of tubes with and without external ribs [39].

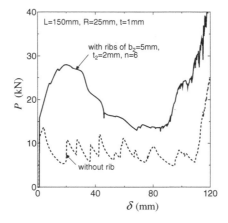

FIGURE 1.72
Comparison of crushing forces between tubes with and without longitudinal ribs [39].

tubes. Fig. 1.72 shows the crushing force of a circular tube ($L = 150$ mm, $t = 1$ mm) to which ribs ($b_2 = 5$ mm, $t_2 = 2$ mm, $n = 6$) are attached outside. The figure shows that by attaching ribs, the peak force is higher than that of a circular tube without ribs, and the compressive force begins to decrease rapidly after the crushing proceeds to a certain extent. The increase of the peak force and subsequent decrease of the crushing force after peak force is caused by the increase of wrinkle wavelength due to the enhanced bending rigidity by attaching ribs. Further, the corner in the joint area between tube and rib maintains high compressive stress during the crushing, contributing to the increase of the crushing force [54]. Note that the phenomenon as above was also observed in the case that the ribs were attached inside the circular tube, and there was no great difference from the case that the ribs were attached outside.

To test the effect of ribs on the first peak force and average crushing force in the axial crushing of circular tubes, the rib width b_2 was changed as a parameter assuming a constant rib thickness ($t_2 = 1$ mm) and a circular tube of $L = 150$ mm and $t = 1$ mm; Fig. 1.73 shows the collapse force and average crushing force for various rib widths b_2. In the figure, "Extension" and "Rotation" indicate the deformation modes of ribs [24] during the crushing as shown in Fig. 1.74. Fig. 1.73 shows that crushing force increases with the rib width b_2 when ribs are attached either inside or outside but that, if the rib width b_2 becomes too great, the deformation mode changes from "Extension" to "Rotation," and the crushing force ceases to increase further.

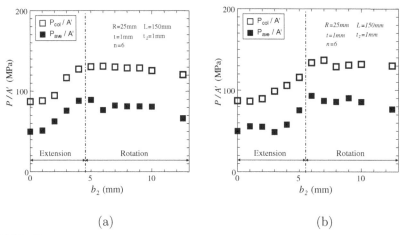

FIGURE 1.73
Effect of b_2 on peak force and average force for stiffened tubes [39]: (a) internal stiffener; (b) external stiffener.

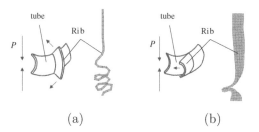

FIGURE 1.74
Buckling modes of ribs [39]: (a) extension; (b) rotation.

1.5.3   Combined effects of rings and ribs

Rings change the deformation mode of circular tubes from non-axisymmetric to axisymmetric, and it becomes possible to control the wrinkle wavelength; on the other hand, ribs increase both the crushing force and the wavelength of wrinkles. Therefore, by combining them, positive coupling effects would be expected such that the increase of wrinkle wavelength by the use of ribs is partly cancelled out by the use of rings, and the synergistic improvement of energy absorption properties becomes possible.

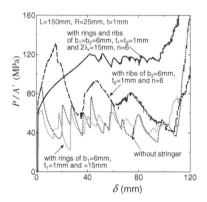

FIGURE 1.75
Coupling effect of axially and circumferentially stiffened tube on crushing force [39].

Fig. 1.75 compares the crushing force between the case that both ribs and rings are attached inside a circular tube of $L = 150$ mm and $t = 1$ mm ($b_1 = b_2 = 6$ mm, $t_1 = t_2 = 1$ mm, $2\lambda_r = 15$ mm, $n = 6$) and the case that either rings or ribs of the same size are attached inside the circular tubes. Through the combined use of ribs and rings, the corner effect becomes available, and the increase of wrinkle wavelength due to ribs is suppressed by the interval between the rings; as a result, an excellent level of energy absorption can be realized under the stable axisymmetric deformation process.

Fig. 1.76 shows the crushing force when the rings and ribs of $b_1 = b_2 = 5$ mm, $2\lambda_r = 25$ mm, and $n = 6$ are attached inside a circular tube of $R = 25$ mm, $t = 1$ mm, and $L = 100$ mm. In Fig. 1.76(a), $t_2 = 1$ mm is assumed and the ring thickness t_1 is varied while in Fig. 1.76(b), $t_1 = 1$ mm is assumed and the rib thickness t_2 is varied. These figures show that, insofar as the rings have a reasonable size $b_1 \times t_1$ that is sufficient to form wrinkles across the interval between them, they do not appreciably affect the crushing force and thus the crushing force depends solely on the ribs.

The above-mentioned result suggests that ideal impact energy absorbers can be developed by using the coupled effects of ribs and rings.

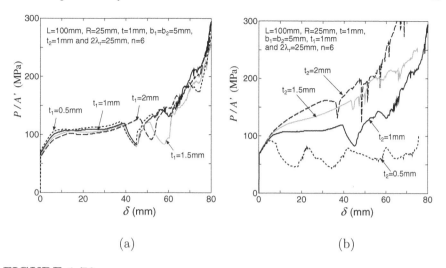

FIGURE 1.76

Effect of thickness of rings or ribs on crushing force [39]: (a) effect of t_1 of rings; (b) effect of t_2 of ribs.

2

Axial compression of square tubes

This chapter discusses the deformation of polygonal tubes under axial compressive force. Here, we consider mainly rectangular tubes such as the one shown in Fig. 2.1 and particularly, unless otherwise noted, a square tube ($C_1 = C_2 = C$ in the figure).

FIGURE 2.1
Geometry of a rectangular tube.

2.1 Characteristics of axial compressive deformation of square tubes

As in the case of thin-walled cylinders, there are multiple crushing deformation modes of square tubes under axial compressive force. Fig. 2.2 shows examples of typical deformation behavior of square tubes as analyzed by numerical simulation using a finite-element method (FEM). Fig. 2.2 shows that a short square tube, standing upright, may be crushed through the formation of many buckles (Fig. 2.2(a)); a long square tube may simply bend in the middle through Euler buckling (Fig. 2.2(c)); and a square tube may bend and fall over after having several buckles form (Fig. 2.2(b)). The type of crushing mode that occurs is dependent on factors such as geometry and the material

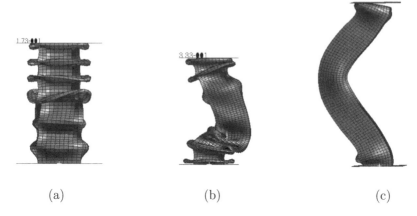

(a)   (b)   (c)

FIGURE 2.2
Deformation modes of square tubes under axial compression: (a) stable type;
(b) unstable type (transition mode); (c) unstable type (bending mode).

of the square tube. Here, as in the case of circular tubes, the deformation
modes shown in Figs. 2.2(a), (b) and (c) are called **stable type, transition
mode** and **bending mode**; the bending and transition modes are also called
unstable type because falling over occurs in both modes. The geometric conditions to produce the crushing modes are summarized in the map (Fig. 2.3)
based on the results of FEM analysis for square tubes consisting of materials
that obey the bilinear hardening rule with a constant coefficient of plastic
strain hardening E_t of $E_t/E = 0.01$. The dotted curve in Fig. 2.3 shows the
boundary between unstable and stable areas derived by using the empirical
formula

$$\left(\frac{L}{C}\right)_{cr} = 2.482 \exp\left(0.0409\frac{C}{t}\right) \tag{2.1}$$

proposed by Abramowicz and Jones [6] on the basis of the experimental results
for mild steel.

If square tubes are to be used as collision energy absorbers, clearly they
should be designed to ensure stable axial crushing. The stable deformation
mode of square tubes can be classified into various sub-modes according to
buckle shape. Among them, the most common deformation modes are the
extensional and **inextensional** modes [216]. In the extensional mode shown
in Fig. 2.4(a), either the crests or valleys of the buckle are created in the same
position of the axial direction on each side plane of a square tube; circumferential extension should occur, and thus the mode is called extensional. On
the other hand, in the inextensional mode shown in Fig. 2.4(b), the crests
and valleys of the buckles appear alternately in the adjacent side planes of
the square tube; circumferential extension hardly occurs, and so the mode
is called inextensional. The extensional mode corresponds to the axisymmet-

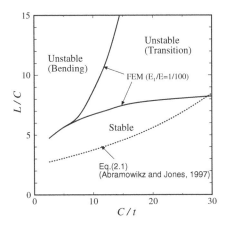

FIGURE 2.3
Relation between axial crushing modes and geometry of square tubes.

ric mode, and the inextensional mode corresponds to the non-axisymmetric mode of a circular tube. Furthermore, in the axial crushing of square tubes, although there are cases where all buckles are simply extensional or inextensional, there could be cases where some extensional buckling appears first and then becomes inextensional; this deformation mode is called the **mixed mode**.

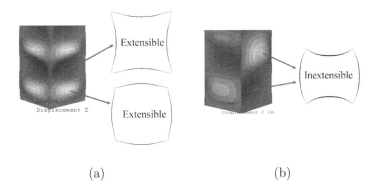

(a)   (b)

FIGURE 2.4
Deformation modes: (a) extensional mode; (b) inextensional mode.

In this chapter, square tubes subjected to axial compression are mainly discussed. For such tubes, the inextensional mode of deformation appears, in general, when the tube wall is thin, whereas the extensional mode appears when the tube wall is thick. For example, a study by Abramowicz and Jones

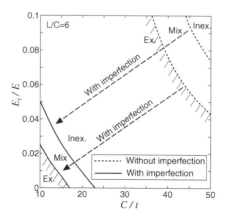

FIGURE 2.5
Deformation mode map of axial crushing of square tubes.

[5] showed that, in square tubes, the inextensional mode appears at $C/t > 20$ while the extensional mode appears at $C/t < 10$. Note, however, that initial geometric imperfections greatly influence the deformation mode of the tubes. Fig. 2.5 shows a map of the geometric conditions of square tubes and the strain hardening coefficient E_t of the material and summarizes the conditions for extensional, inextensional and mixed modes based on FEM numerical analysis of square tubes of length $L/C = 6$ with and without the initial irregularity for a material obeying the bilinear hardening rule. Under the geometric and material conditions assumed in the figure, the extensional mode almost always appears in a square tube without initial geometric imperfection, but when a very small initial irregularity is introduced, the deformation switches from the extensional to the inextensional mode. Here, as an initial irregularity, an inclination of about $(t/C)/100$ was attached near the edge of a side plate of the square tube. Further, as in a circular tube, the deformation mode of a square tube is strongly affected by the boundary conditions at the edges. For example, although extensional deformation is induced by the presence of a flange at the end part, the effect is reduced as the distance of the buckle from the edge is increased.

Note that, in the case of a hexagonal tube, which is nearly a circular tube, the deformation resembles that of a circular tube rather than a square tube. For example, in the axial crushing deformation of a regular hexagonal tube, although the deformation mode shown in Fig. 2.6(a) is called "extensional" as in the case of the square tube shown in Fig. 2.4(a), the deformation instead resembles the axisymmetric mode of a circular tube. Further, in the deformation shown in Fig. 2.6(b), the crests and valleys of buckles appear alternately in the adjacent side planes of the tube, similarly to in the square tube shown

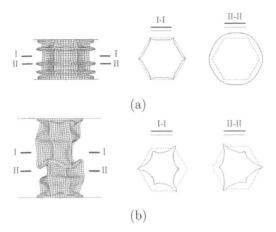

FIGURE 2.6
Deformation behavior in axial crushing of hexagonal tubes: (a) extensional mode; (b) inextensional mode.

in Fig. 2.4(b); however, the cross section of the buckle formed is closer to a triangle rather than the original hexagon and here the horizontal hinges and the inclined travelling hinges, which are normally observed in the inextensional mode of the square tube (see Section 2.4.1), are not observed clearly; all this suggests that the deformation is closer to the $n = 3$ non-axisymmetric mode of a circular tube.

In comparison with a circular tube, there are two main features in the relation between crushing force and compressive displacement in the axial crushing of square tubes.

(1) Generally, in the case of circular tubes, the crushing force increases and decreases cyclically together with sequential formation of buckles after the first peak force. On the other hand, the deformation of square tubes is characterized by two stages, as shown in Fig. 2.7: first, shallow buckles are created over the entire length of the square tube, and secondly, they are then crushed one by one. At the first stage, the force changes with the compressive deformation; it is not that subsequent buckles appear after the earlier buckles are completely crushed as in a circular tube; rather, new buckles appear continuously while the earlier buckles remain relatively shallow, and they prevail over the entire length of the tube. Then, at the second stage, the created buckles are crushed one by one (normally starting from the end part), and the force increases and decreases cyclically.

Such a feature is observed also by experiments. DiPaolo and Tom [64] conducted an experimental investigation on the axial crush response of steel, square tubes under quasi-static testing conditions, and pointed out that the axial crush response can be divided into an "initial" phase and a "secondary" phase: the initial phase includes the pre-collapse response prior to the occurrence of the peak or maximum load, the change from axial to bending load-resistance in the sidewalls, and the formation of the first few interior and exterior folds on sets of opposite sidewalls with corresponding increases and decreases in the load-displacement curve; the secondary folding phase consists of the "steady state" fold formation process.

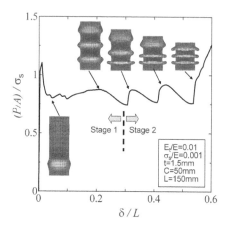

FIGURE 2.7
Two stages in the axial crushing of square tubes.

(2) In the deformation of a circular tube under axial compression, generally, the buckling force is the first peak of the compressive force. On the other hand, in a square tube, the buckling force is not necessarily the first peak of compressive force. Fig. 2.8 shows the variation of force and the deformation behavior in the axial crushing of square tubes made of an elasto-plastic material that obeys the bilinear hardening rule with $E_t/E = 0.01$ or $E_t/E = 0.1$. Fig. 2.8(a) shows that, in the case of $E_t/E = 0.01$, the plastic buckling force is the peak force in the square tube, as in the case of a circular tube. Here, the force decreases after buckling but the variation of force is not as large as in a circular tube, and the deformation enters the first stage where the buckles prevail over the entire axial length as mentioned in item (1). In the case of $E_t/E = 0.1$, as shown in Fig. 2.8(b), the force decreases after plastic buckling and soon begins to increase. While the force increases, the buckles prevail over the entire length, and compression deformation is almost uniform along the axial direction. After shallow buckles are formed over the entire length of the square tube, the compressive deformation begins to concentrate on a certain buckle, and the force takes its peak value. Therefore, in this case, the peak force corresponds to the collapse force of the buckle.

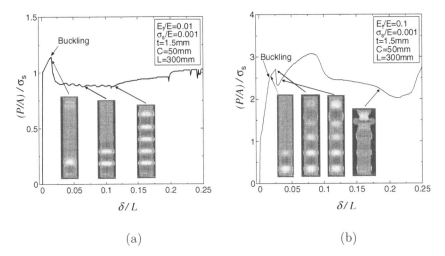

(a)   (b)

FIGURE 2.8
Peak force in the initial stage of axial compression in square tubes: (a) $E_t/E = 0.01$, (b) $E_t/E = 0.1$.

Since the above-mentioned two aspects of the relation between compressive force and displacement in a square tube are strongly related to the collapse behavior of a plate under axial compression, the buckling of a plate under axial compression and the behavior after buckling will be discussed first in Section 2.2, and then on that basis, the buckling and collapse force in the axial compression of square tubes will be discussed in Section 2.3. Further, as will be discussed in Section 2.4, the average crushing force in the axial compression of square tubes depends on primarily the deformation mode after buckling; the inextensional mode will be discussed in Section 2.4.1 and the extensional mode will be discussed in Section 2.4.2.

2.2   Buckling and collapse force of a plate

The peak force in the axial crushing of a plate corresponds to either the buckling force of the plate or to the collapse force after buckling. To discuss the matter in more detail, first, it is necessary to discuss the buckling of a plate under an axial compressive force.

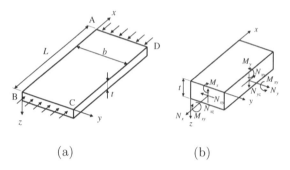

(a)   (b)

FIGURE 2.9
Plate to which axial compressive force is uniformly applied: (a) plate; (b) force and moment components acting on a small element of the plate.

2.2.1 Buckling force of a plate

Let us assume a plate with width b, thickness t and length L as shown in Fig. 2.9(a) to study buckling under uniform compressive force.

Expressing the force and moment components acting on a small element by the symbols shown in Fig. 2.9(b), the equations of equilibrium between in-plane (in the x, y plane) and out-of-plane (in the z direction) force and moment components are given by

$$\begin{cases} \dfrac{\partial N_x}{\partial x} + \dfrac{\partial N_{xy}}{\partial y} = 0 \\[2mm] \dfrac{\partial N_{xy}}{\partial x} + \dfrac{\partial N_y}{\partial y} = 0 \\[2mm] \dfrac{\partial N_{xz}}{\partial x} + \dfrac{\partial N_{yz}}{\partial y} + N_x \dfrac{\partial^2 w}{\partial x^2} = 0 \\[2mm] \dfrac{\partial M_x}{\partial x} - \dfrac{\partial M_{xy}}{\partial y} - N_{xz} = 0 \\[2mm] \dfrac{\partial M_y}{\partial y} - \dfrac{\partial M_{xy}}{\partial x} - N_{yz} = 0 \end{cases} \tag{2.2}$$

where w is the out-of-plane displacement in the z direction.

Deriving N_{xz}, N_{yz} from the fourth and fifth equations of Eq. (2.2), and substituting them into the third equation in Eq. (2.2),

$$\frac{\partial^2 M_x}{\partial x^2} - 2\frac{\partial^2 M_{xy}}{\partial x \partial y} + \frac{\partial^2 M_y}{\partial y^2} + N_x \frac{\partial^2 w}{\partial x^2} = 0 \tag{2.3}$$

is obtained.

By using the simplest **Kirchhoff-Love's theory** as the bending deformation theory,[1] the moment is expressed as a function of w in the z direction,

---

[1] The Mindlin theory instead of the Kirchhoff-Love theory can also be used to consider the shear strain of a plate of moderate thickness. For detailed analysis of buckling, refer to studies of Brunelle [27] and Brunelle and Robertson [28].

and then w is the only unknown parameter for analyzing plate buckling. For the case of **elastic buckling**, for example, the moments are given by

$$
\begin{aligned}
M_x &= \int_{-t/2}^{+t/2} \sigma_x z dz &= -D \left[\frac{\partial^2 w}{\partial x^2} + \nu \frac{\partial^2 w}{\partial y^2} \right] \\
M_y &= \int_{-t/2}^{+t/2} \sigma_y z dz &= -D \left[\nu \frac{\partial^2 w}{\partial x^2} + \frac{\partial^2 w}{\partial y^2} \right] \\
M_{xy} &= -\int_{-t/2}^{+t/2} \tau_{xy} z dz &= D(1 - \nu) \frac{\partial^2 w}{\partial x \partial y}
\end{aligned}
\tag{2.4}
$$

as a function of w. Here,

$$
D = \frac{Et^3}{12(1 - \nu^2)}
$$

Substituting Eq. (2.4) into Eq. (2.3) yields

$$
-D \left[\frac{\partial^4 w}{\partial x^4} + 2 \frac{\partial^4 w}{\partial x^2 \partial y^2} + \frac{\partial^4 w}{\partial y^4} \right] + N_x \frac{\partial^2 w}{\partial x^2} = 0
\tag{2.5}
$$

Here, the buckling problem with a boundary condition where all four edges are simply supported is discussed first. By considering the boundary condition of simple support at the plate edges, the buckling displacement can be assumed to be

$$
w = A_w \sin \left(\frac{m \pi x}{L} \right) \sin \left(\frac{n \pi y}{b} \right)
\tag{2.6}
$$

By substituting Eq. (2.6) into Eq. (2.5), the axial force N_x that induces buckling is given by

$$
D \left[\left(\frac{m \pi}{L} \right)^2 + \left(\frac{n \pi}{b} \right)^2 \right]^2 + N_x \left(\frac{m \pi}{L} \right)^2 = 0
\tag{2.7}
$$

Then, the **elastic buckling stress** σ_{buc}^e (here, compressive stress is taken to be positive) can be obtained as

$$
\sigma_{buc}^e = \frac{\pi^2 E}{12(1 - \nu^2)} \left(\frac{t}{b} \right)^2 \left[\left(\frac{b}{L/m} \right) + n^2 \left(\frac{L/m}{b} \right) \right]^2
\tag{2.8}
$$

The buckling stress σ_{buc}^e is found by minimizing Eq. (2.8). So that Eq. (2.8) has a minimum, n must be 1. This means that in the lateral direction there is only one buckle with the half-wavelength.

Further, if $L/b < 1$ for the length of the plate, $m = 1$ for the number m of half-wavelengths in the length direction. From this,

$$
\sigma_{buc}^e = \frac{\pi^2 E}{12(1 - \nu^2)} \left(\frac{t}{b} \right)^2 \left[\left(\frac{b}{L} \right) + \left(\frac{L}{b} \right) \right]^2
\tag{2.9}
$$

If the plate is sufficiently long, $m = L/b$ approximately holds. Then, the **buckling stress** is given by

$$\sigma_{buc}^e = \frac{\pi^2 E}{3(1 - \nu^2)} \left(\frac{t}{b}\right)^2 \tag{2.10}$$

From $L/m = b$, it can be concluded that the half-wavelength in the length direction of the plate is equal to the plate width.

When the boundary condition for edges AB and CD is not simple support, the buckling displacement w is given not by Eq. (2.6) but by

$$w = f(y) \sin\left(\frac{m\pi x}{L}\right) \tag{2.11}$$

Then a buckling stress from the solution of the differential equation Eq. (2.5) satisfying the boundary conditions at $y = 0$ and $y = b$ can be derived, as done in the case of simple support. The analysis is not discussed here because it is detailed in a book by Timoshenko and Gere [203].

The **elastic buckling stress** is generally expressed for all boundary conditions by using the **buckling coefficient** k as follows:

$$\sigma_{buc}^e = k \frac{\pi^2 E}{12(1 - \nu^2)} \left(\frac{t}{b}\right)^2 \tag{2.12}$$

The buckling coefficient k depends on the boundary conditions and the aspect ratio of the plate. Here, both forced edges BC and DA are assumed to be simply supported. If both unforced edges AB and CD are also simply supported, Eqs. (2.9) and (2.10) show that the buckling coefficient k is given by

$$k = \begin{cases} 4 & \text{for} \quad L/b \geq 1 \\ 2 + \alpha^2 + \dfrac{1}{\alpha^2} & \text{for} \quad \alpha = L/b < 1 \end{cases} \tag{2.13}$$

On the other hand, if both edges AB and CD are clamped, k is

$$k = 6.98 \tag{2.14}$$

and the half-wavelength of the buckle in the plate length direction is about 0.7 times the plate width.

Further, if one edge is fixed and the other is free,

$$k = 1.328 \tag{2.15}$$

and the half-wavelength of the buckle in the plate length direction is about 1.64 times the plate width.

Furthermore, in the edges AB and CD, one edge is simply supported and the other is free, k is

$$k = \frac{6(1 - \nu)}{\pi^2} + \left(\frac{b}{L}\right)^2 \tag{2.16}$$

and only a single buckle appears in the plate length direction.

The square tubes used in collision energy absorbers have thick walls, and buckling can occur only after plastic yield over the full cross section. The analysis of the **plastic buckling** of the plates is basically the same as an analysis of the elastic buckling discussed above. Here, the relational expression between incremental stress and strain after the material has yielded is necessary. As in Section 1.1.2, through analysis using the relational expression between incremental stress and strain, Eq. (1.26), an equation for the incremental buckling displacement $\dot{w}$ can be obtained as

$$-\frac{t^3}{12}\left[E_{xx}\frac{\partial^4 \dot{w}}{\partial x^4} + \left(4G_{xy} + 2E_{xy}\right)\frac{\partial^4 \dot{w}}{\partial x^2 \partial y^2} + E_{yy}\frac{\partial^4 \dot{w}}{\partial y^4}\right] + N_x\frac{\partial^2 \dot{w}}{\partial x^2} = 0 \quad (2.17)$$

By substituting Eq. (2.6) into this, the conditions needed to induce buckling are expressed as

$$\frac{t^3}{12}\left[E_{xx}\left(\frac{m\pi}{L}\right)^4 + \left(4G_{xy} + 2E_{xy}\right)\left(\frac{m\pi}{L}\right)^2\left(\frac{n\pi}{b}\right)^2 + E_{yy}\left(\frac{n\pi}{b}\right)^4\right]$$
$$+N_x\left(\frac{m\pi}{L}\right)^2 = 0 \quad (2.18)$$

and the buckling stress is given by

$$\sigma^p_{buc} = \frac{\pi^2}{12}\left(\frac{t}{b}\right)^2\left[E_{xx}\left(\frac{b}{L/m}\right)^2 + n^2\left(4G_{xy} + 2E_{xy}\right) + n^4 E_{yy}\left(\frac{L/m}{b}\right)^2\right] \quad (2.19)$$

For Eq. (2.19) to have a minimum, n should be 1. Then, if the plate is long, in order to obtain a minimum,

$$\frac{L/m}{b} = \sqrt[4]{\frac{E_{xx}}{E_{yy}}} \quad (2.20)$$

should hold. Therefore, the **plastic buckling stress** of the plate, σ^p_{buc}, is given by

$$\sigma^p_{buc} = \frac{\pi^2}{6}\left(\frac{t}{b}\right)^2\left(\sqrt{E_{xx}E_{yy}} + E_{xy} + 2G_{xy}\right) \quad (2.21)$$

The coefficients E_{xx}, E_{yy}, E_{xy} and G_{xy} in the equation depend on which plastic deformation theory is applied; if the J_2 incremental theory of plasticity is assumed, these coefficients are

$$E_{xx} = H_T\frac{\eta_T + 3}{4}, \qquad\qquad E_{yy} = H_T\eta_T$$
$$E_{xy} = H_T\frac{\eta_T + 2\nu - 1}{2}, \qquad G_{xy} = G \quad (2.22)$$
$$H_T = \frac{4E}{(5 - 4\nu)\eta_T - (1 - 2\nu)^2}, \qquad \eta_T = \frac{E}{E_t}$$

and if the J_2 deformation theory of plasticity is assumed, these coefficients are

$$
\begin{aligned}
E_{xx} &= H_s \frac{\eta_T + 3\eta_s}{4}, & E_{yy} &= H_s \eta_T \\
E_{xy} &= H_s \frac{\eta_T + 2\nu - 1}{2}, & G_{xy} &= G_s \\
H_s &= \frac{4E}{(3\eta_s + 2 - 4\nu)\eta_T - (1 - 2\nu)^2}, & \eta_s &= \frac{E}{E_s} \\
G_s &= \frac{E}{3\eta_s - (1 - 2\nu)}
\end{aligned}
\tag{2.23}
$$

In general, compared with experimental results, theoretical analysis of the **plastic buckling stress** of a plate has large errors similar to the theoretical analysis of the plastic buckling stress of a circular tube. In particular, errors of analysis are greater when using the J_2 incremental theory than the J_2 deformation theory. Analysis using J_2 incremental theory always produces an overestimate relative to experimental data [82]. Such overestimation is difficult to understand from the viewpoint of plasticity theory. Although efforts have been made to solve the problem by using a more appropriate yield surface [17, 57], the formulation of the yield surface is complex, and it is practically difficult to use the theory for predicting plastic buckling stress. As a result, the J_2 deformation theory is generally used to predict the plastic buckling stress of a plate. In particular, Stowell's theory [194], which is based on the J_2 deformation theory, is widely used. Employing the J_2 deformation theory and assuming $\nu = 0.5$, Stowell [194] derived the coefficients

$$
\begin{aligned}
E_{xx} &= \frac{4E_s}{3} e, & E_{xy} &= \frac{2E_s}{3} \\
E_{yy} &= \frac{4E_s}{3}, & G_{xy} &= \frac{E_s}{3}
\end{aligned}
\tag{2.24}
$$

from Eq. (2.23), where

$$
e = \frac{1}{4} + \frac{3E_t}{4E_s}
\tag{2.25}
$$

Substituting Eq. (2.24) into Eq. (2.21), the buckling stress is given by

$$
\sigma_{buc}^p = \frac{\pi^2 E_s}{9} \left(\frac{t}{b}\right)^2 \left[2 + \sqrt{1 + 3\frac{E_t}{E_s}}\right]
\tag{2.26}
$$

where, from Eq. (2.20),

$$
\frac{L/m}{b} = \sqrt[4]{e}
\tag{2.27}
$$

is obtained.

The range of variation of e is $\frac{1}{4} \le e \le 1$. $e = \frac{1}{4}$ holds for a perfect elasto-plastic material without strain hardening. e increases with the strain hardening coefficient, and the maximum value is $e = 1$ for an elastic material. Therefore, Eq. (2.27) means that the half-wavelength of the plastic buckle in the length direction increases with the strain hardening coefficient; $\dfrac{L/m}{b} = 1/\sqrt{2}$ for a perfect elasto-plastic material, and $\dfrac{L/m}{b} = 1$ for an elastic material.

Further, assuming the ratio η_p between plastic and elastic buckling stresses, the expression for plastic buckling stress

$$\sigma_{buc}^p = \eta_p \, \sigma_{buc}^e \tag{2.28}$$

is often found in the literature. Stowell [194] showed that η_p is given by

$$\eta_p = \frac{E_s}{E}\left(B_1 + B_2 \sqrt{\frac{1}{4} + \frac{3E_t}{4E_s}}\right) \tag{2.29}$$

where B_1 and B_2 are coefficients that depend on the boundary conditions as shown in Table 2.1.

TABLE 2.1
Coefficients B_1 and B_2 in Eq. (2.29) for long plate.

Boundary conditions		B_1	B_2
AB	CD		
Simply supported	Free	1	0
Simply supported	Simply supported	0.5	0.5
Clamped	Free	0.428	0.572
Clamped	Clamped	0.352	0.648

There are not many evaluation formulas based on the J_2 incremental theory; among them is a formula for plastic buckling proposed by Inoue [97]. Based on the study of slip distribution assuming Tresca's yield condition and plastic deformation due to slip, Inoue [97] proposed

$$E_{xx} = E_{xy} = E_{yy} = 0, \qquad G_{xy} = \frac{1}{2}G \tag{2.30}$$

for materials without strain hardening, and

$$E_{xx} = E_t, \qquad E_{xy} = E_{yy} = 0, \qquad G_{xy} = \frac{1}{2}G \tag{2.31}$$

for materials with strain hardening. Here, Eq. (2.20) cannot be used because $E_{yy} = 0$, and thus buckling stress cannot be derived directly from Eq. (2.21).

In this case, substituting Eq. (2.30) into Eq. (2.19) (assuming $n = 1$), plastic buckling stress σ_{buc}^p is derived as follows for a material without strain hardening:

$$\sigma_{buc}^p = \frac{\pi^2}{6}\left(\frac{t}{b}\right)^2 G \tag{2.32}$$

Note here that the half-wavelength of the plastic buckle in the length direction is not determined in this analysis.

Further, substituting Eq. (2.31) into Eq. (2.19) (assuming $n = 1$), the plastic buckling stress in consideration of strain hardening is given by

$$\sigma_{buc}^p = \frac{\pi^2}{12} \left(\frac{t}{b}\right)^2 \left[E_t \left(\frac{b}{L/m}\right)^2 + 2G\right] \tag{2.33}$$

and $m = 1$ should hold assuming this has a minimum. Therefore, the plastic buckling stress σ_{buc}^p is given by

$$\sigma_{buc}^p = \frac{\pi^2}{12} \left(\frac{t}{b}\right)^2 \left[E_t \left(\frac{b}{L}\right)^2 + 2G\right] \tag{2.34}$$

Note here that only a single buckle of a half-wavelength is created in the length direction of the plate.

On the other hand, the author proposed a formula for the plastic buckling stress of a plate:

$$\sigma_{buc}^p = k\frac{\pi^2 E_r}{12(1 - \nu_r^2)} \left(\frac{t}{b}\right)^2 + \sigma_c \tag{2.35}$$

by extending Eq. (2.12) for elastic buckling stress, where

$$\sigma_c = \sigma_s \left[1 - \left(\frac{E_r}{E}\right)\left(\frac{1 - \nu^2}{1 - \nu_r^2}\right)\right], \quad E_r = \frac{4EE_t}{(\sqrt{E} + \sqrt{E_t})^2}, \quad \nu_r = 0.5 \tag{2.36}$$

The first term of the right side of Eq. (2.35) is derived by replacing E and ν in Eq. (2.12) by E_r and ν_r, respectively, and the second term σ_c is introduced so that the buckling stress of Eq. (2.35) equals the yield stress σ_s for a particular plate thickness t where the elastic buckling stress given by Eq. (2.12) becomes equal to the yield stress of the material σ_s.

Fig. 2.10 compares the plastic buckling stress of the plate σ_{buc}^p derived theoretically using Eq. (2.26) and Eq. (2.35), and that found by FEM analysis. Here, the values of plastic buckling stress derived by using Eq. (2.29) and Eq. (2.21) with Eq. (2.23) are almost equal to those derived by using Eq. (2.26); so the values are not additionally shown in Fig. 2.10. Fig. 2.10 shows that Eq. (2.26) for plastic buckling stress gives underestimates while the values derived by using Eq. (2.35) approximately agree with those of FEM analysis.

2.2.2 Collapse force after elastic buckling (for perfect elasto-plastic materials)

After elastic buckling due to axial compression of a thin plate, the compressive strain and hence stress in the axial direction locally decrease at the center of the plate, owing to the out-of-plane buckling deformation; at both edges,

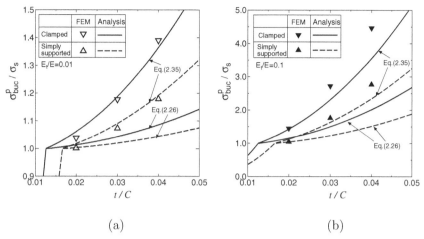

FIGURE 2.10
Plastic buckling stress of plates σ_{buc}^{p}: (a) $E_t/E = 0.01$; (b) $E_t/E = 0.1$.

however, out-of-plane deformation is limited, and consequently compressive strain and hence stress locally increase with compression. Therefore, on the whole, the compressive force of the plate steadily increases even after buckling. In other words, under the axial compression, the most important feature of the buckling of a thin plate is that the force can increase after buckling and reach the **collapse force**, which is greater than the buckling force. After Schuman and Back [179] first reported this in 1930, many researchers investigated this phenomenon. Here, in order to understand the features of the buckling of a plate, the results of theoretical analysis of the deformation behavior after plate buckling are described.

In Section 2.2.1, to derive buckling force it is assumed that the stress immediately before buckling was uniaxial ($N_y = N_{xy} = 0$, $N_x \neq 0$); because the equilibrium equation for in-plane stress and also the equilibrium equation for out-of-plane stress and moment are independent of each other, the analysis was based on only the moment, or the 3rd, 4th and 5th equations in Eq. (2.2); here, however, to derive the deformation behavior after buckling, it is necessary to consider that the stress after buckling contains not only N_x but also N_y and N_{xy} in a general case, and to solve the five equations in Eq. (2.2) simultaneously.

Deformation analysis after buckling was first done by Cox [60]. However, since Cox did not consider the influence of shear stress, the obtained result was lower than that of rigorous analysis, being at the "lower boundary." Later, for plate compression with four simply supported edges, Marguerre [143] assumed, after buckling, the displacement w in the out-of-plane direction to be given by

$$w = W_{11} \sin \frac{\pi x}{L} \sin \frac{\pi y}{b} \qquad (2.37)$$

for deriving the stress distribution after buckling. As a result, for a square plate of $L = b$ for example, if $N_y = 0$ is assumed as the boundary condition at the unforced edges (edges AB and CD in Fig. 2.9), the amplitude W_{11} of the displacement w in the out-of-plane direction is given by

$$\left(\frac{W_{11}}{b}\right) = \frac{2}{\pi}\sqrt{\varepsilon_{x0} - \varepsilon_{buc}^e} \tag{2.38}$$

as a function of compressive strain ε_{x0} (which is given by the displacement at the forced edges divided by the plate length). Thus, the distribution of compressive stress σ_x across the plate width (here, compressive stress is taken to be positive) is non-uniform in the width direction as given by

$$\sigma_x = \sigma_{buc}^e|_{N_y=0} + \frac{E}{2}\left(\varepsilon_{x0} - \varepsilon_{buc}^e|_{N_y=0}\right)\left(1 + \cos\frac{2\pi y}{b}\right) \tag{2.39}$$

where the absolute value of compressive stress becomes the largest at both edges ($y = 0$ and $y = b$) as given by

$$\sigma_x|_{y=0} = \sigma_x|_{y=b} = E\varepsilon_{x0} \tag{2.40}$$

while the absolute value of compressive stress becomes the smallest at the center ($y = b/2$) as given by

$$\sigma_x|_{y=b/2} = E\varepsilon_{buc}^e|_{N_y=0} \tag{2.41}$$

where $\varepsilon_{buc}^e|_{N_y=0}$ is the strain at buckling of the plate.

$$\varepsilon_{buc}^e|_{N_y=0} = \frac{\sigma_{buc}^e|_{N_y=0}}{E} = \frac{\pi^2}{3(1-\nu^2)}\left(\frac{t}{b}\right)^2 \tag{2.42}$$

Fig. 2.11(a) shows the distribution of compressive stress σ_x after plate buckling, as obtained from Eq. (2.39). Corresponding to the obtained stress distribution, the average compressive stress $\bar{\sigma}_x$ is defined as $\bar{\sigma}_x = P_x/(tb)$ where P_x is the compressive force, and is given by

$$\bar{\sigma}_x|_{N_y=0} = \frac{\int_0^b \sigma_x dy}{b} = \sigma_{buc}^e|_{N_y=0} + \frac{E}{2}\left(\varepsilon_{x0} - \varepsilon_{buc}^e|_{N_y=0}\right) \tag{2.43}$$

As this equation shows, the stiffness of the plate after buckling (increasing rate of average compressive stress against the increase of compressive strain ε_{x0}) is $E/2$, which is half that before buckling. Since only a part of the plate loses force capacity after buckling, the compressive force of the plate increases and the stiffness of the plate decreases on the whole.

Eq. (2.39) was derived assuming $N_y = 0$ at the edges AB and CD. On the other hand, if it is assumed that the displacement is zero in the y direction

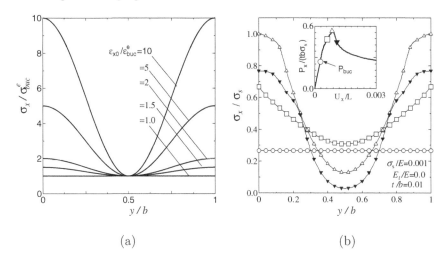

FIGURE 2.11
Compressive stress σ_x distribution after plate buckling: (a) distribution obtained by Eq. (2.39); (b) results of FEM analysis.

instead of assuming $N_y = 0$ as the boundary condition at the edges AB and CD, the stress distribution across the plate width is given by

$$\sigma_x = \sigma^e_{buc}|_{U_y=0} + \frac{E}{3-\nu}\left(\varepsilon_{x0} - \varepsilon^e_{buc}|_{U_y=0}\right)\left(\frac{2}{1+\nu} + \cos\frac{2\pi y}{b}\right) \qquad (2.44)$$

and the relation between the average stress and the compressive strain is given by

$$\bar{\sigma}_x|_{U_y=0} = \frac{\displaystyle\int_0^b \sigma_x dy}{b} = \sigma^e_{buc}|_{U_y=0} + \frac{2E}{(1+\nu)(3-\nu)}\left(\varepsilon_{x0} - \varepsilon^e_{buc}|_{U_y=0}\right) \quad (2.45)$$

Here, $\sigma^e_{buc}|_{U_y=0}$ and $\varepsilon^e_{buc}|_{U_y=0}$ are the stress and the strain at buckling of the plate under the boundary condition that the two edges AB and CD are simply supported and the displacement in the y-direction is zero; they are related to the buckling stress $\sigma^e_{buc}|_{N_y=0}$ under the boundary condition that the two edges AB and CD are simply supported and the force in the y-direction is zero by using

$$\sigma^e_{buc}|_{U_y=0} = \frac{\sigma^e_{buc}|_{N_y=0}}{1+\nu}, \qquad \varepsilon^e_{buc}|_{U_y=0} = \frac{1-\nu^2}{E}\sigma^e_{buc}|_{U_y=0} \qquad (2.46)$$

In this case, as Eq. (2.45) shows, although the stiffness of the plate depends on the Poisson ratio ν of the material, it is still close to half that before buckling (if $\nu = 1/3$, for example, the stiffness is $9/16E$, and the stiffness before buckling is $9/8E$).

Fig. 2.11(b) shows the cross-sectional distribution of compressive stress σ_x obtained by FEM simulation at buckling (denoted by O) and after buckling (denoted by □, △ and ▼). The compressive deformations corresponding to the symbols are shown in the plot of compressive force versus displacement in the inset of the figure. As the figure shows, as the compressive deformation proceeds after buckling, the stress at the center decreases below the level before buckling. Therefore, it is clear that the theoretical analysis shown in Fig. 2.11(a) can be applied to only the immediate aftermath of buckling because the large deformation after buckling is not considered in the analysis.

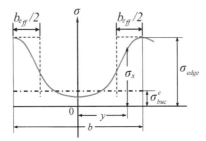

FIGURE 2.12
Concept of effective width.

The increase of force after plate buckling is usually explained using the concept of **effective width** proposed by Karman et al. [211]. Fig. 2.12 illustrates the concept of effective width based on the stress distribution before and after plate buckling under uniform compression. Although the stress distribution is uniform before buckling, the compressive stress after buckling increases further at the edges of both sides. It is considered that most of the increased force is applied to both sides of the plate. On the basis of these considerations, to make the calculations easier, Karman et al. [211] assumed that the stress was negligibly small at the center of the plate and that the total compressive force could be replaced by the stress of magnitude equal to the stress at the edge σ_{edge} uniformly distributed in the strip region of width b_{eff} (effective width) on both sides; that is, they assumed

$$P_x = \int_0^b \sigma_x t dx = \sigma_{edge} \left(t b_{eff} \right) \tag{2.47}$$

Furthermore, Karman et al. [211] assumed the stress σ_{edge} in the effective width area to be always equal to the buckling stress in a plate of width b_{eff}; because stress σ_{edge} and effective width b_{eff} satisfy the formula for the elastic buckling stress

$$\sigma_{edge} = \frac{k\pi^2 E}{12(1 - \nu^2)} \left(\frac{t}{b_{eff}} \right)^2 \tag{2.48}$$

they proposed the following formula for effective width b_{eff} as a function of the stress σ_{edge} at both edges:

$$\frac{b_{eff}}{b} = \sqrt{\frac{\sigma^e_{buc}}{\sigma_{edge}}} \tag{2.49}$$

This equation indicates that, due to the stress increase at both plate edges, the stress concentrates more at the edges and the effective width b_{eff} decreases.

It is assumed that the stress limit at both edges σ_{edge} is the yield stress of the material σ_s. Therefore, the **critical value of effective width** $b_{eff}|_{cri}$ can be obtained from Eq. (2.49) as follows:

$$\frac{b_{eff}|_{cri}}{b} = \sqrt{\frac{\sigma^e_{buc}}{\sigma_s}} \tag{2.50}$$

Therefore, for a perfect elasto-plastic material the **collapse stress** σ_{col}, namely the load-carrying capacity of a plate, can be determined by

$$\sigma_{col} = \sigma_s \times b_{eff}|_{cri} \tag{2.51}$$

The effective width concept has been extended to general formulation and been discussed by, for example, Winter [221], Wang and Rammerstorfer [212], Bambach [14], including detailed reviews by Faulkner [72], Rhodes [172] and Paik [160]. In addition, the effective width concept is used in many national design specifications and in international specifications such as Eurocode 3 [76].

In the case that the elastic buckling stress of the plate is close to the yield stress of the material σ_s, it is said that the calculation using Eq. (2.50) is on the dangerous side. Therefore, Winter [221] proposed a revised formula for the effective width $b_{eff}|_{cri}$

$$\frac{b_{eff}|_{cri}}{b} = \sqrt{\frac{\sigma^e_{buc}}{\sigma_s}} \left(1 - 0.25 \sqrt{\frac{\sigma^e_{buc}}{\sigma_s}} \right) \tag{2.52}$$

considering the experimental results. This formula is adopted as the basis of design in many countries. The coefficient in the formula was later modified and improved as shown in Eqs. (2.53) and (2.54) depending on the boundary conditions. That is, in the case that both edges, AB and CD, are simply supported, the effective width $b_{eff}|_{cri}$ is given by

$$\frac{b_{eff}|_{cri}}{b} = \sqrt{\frac{\sigma^e_{buc}}{\sigma_s}} \left(1 - 0.22 \sqrt{\frac{\sigma^e_{buc}}{\sigma_s}} \right) \tag{2.53}$$

and in the case that one of the edges, AB and CD, is simply supported and the other is free, the effective width $b_{eff}|_{cri}$ is given by

$$\frac{b_{eff}|_{cri}}{b} = 1.19 \sqrt{\frac{\sigma^e_{buc}}{\sigma_s}} \left(1 - 0.298 \sqrt{\frac{\sigma^e_{buc}}{\sigma_s}} \right) \tag{2.54}$$

Many formulas have been proposed for the effective width of a plate. For example, Timoshenko proposed [203]

$$\frac{b_{eff}|_{cri}}{b} = 0.566\frac{\sigma_{buc}^e}{\sigma_s} + 0.434 \tag{2.55}$$

and Koiter [113] proposed

$$\frac{b_{eff}|_{cri}}{b} = \left[1.2 - 0.65\left(\frac{\sigma_{buc}^e}{\sigma_s}\right)^{2/5} + 0.45\left(\frac{\sigma_{buc}^e}{\sigma_s}\right)^{4/5}\right]\left(\frac{\sigma_{buc}^e}{\sigma_s}\right)^{2/5} \tag{2.56}$$

Further, Marguerre [143] proposed the formula

$$\frac{b_{eff}|_{cri}}{b} = \sqrt[3]{\frac{\sigma_{buc}^e}{\sigma_s}} \tag{2.57}$$

based on experimental data, as well as

$$\frac{b_{eff}|_{cri}}{b} = 0.81\sqrt{\frac{\sigma_{buc}^e}{\sigma_s}} + 0.19 \tag{2.58}$$

based on theoretical analysis of the deformation behavior after plate buckling. Comparing Eqs. (2.57) and (2.58) as a function of plate thickness, Rammerstorfer [166] proposed using the former for a thin plate satisfying $\sigma_s / \sigma_{buc}^e > 7$ and the latter for a moderately thick plate satisfying $1 < \sigma_s / \sigma_{buc}^e < 7$.

The parameter λ, which is called **plate slenderness**, is often used for the effective width of a plate. This parameter is an extended concept of thickness ratio and is defined by

$$\lambda = \sqrt{\frac{\sigma_s}{\sigma_{buc}^e}} \tag{2.59}$$

Using the parameter λ, Karman's formula for the effective width (2.50) is expressed as

$$\frac{b_{eff}|_{cri}}{b} = \frac{1}{\lambda} \tag{2.60}$$

Gerard and Becker [82] proposed that the effective width $b_{eff}|_{cri}$ is always given as a function of λ by

$$\frac{b_{eff}|_{cri}}{b} = C\left(\frac{1}{\lambda}\right)^m \tag{2.61}$$

using the constants C and m.

Further, instead of λ, the parameter β given by

$$\beta = \frac{b}{t}\sqrt{\frac{\sigma_s}{E}} \tag{2.62}$$

is also used in many cases. Faulkner [72], on the basis of a wealth of experimental data, proposed the following formula for effective width $b_{eff}|_{cri}$:

$$\frac{b_{eff}|_{cri}}{b} = \frac{2.25}{\beta} - \frac{1.25}{\beta^2} \tag{2.63}$$

Note here that the relation

$$\lambda = \sqrt{\frac{\sigma_s}{\frac{k\pi^2 E}{12(1-\nu^2)}\left(\frac{t}{b}\right)^2}} = \sqrt{\frac{12(1-\nu^2)}{k\pi^2}}\beta$$

holds between the parameter β and the plate slenderness λ.

Note that the above formulas are applicable only if the length L in the compressive force direction of the plate is greater than the width b. If the length of plate becomes short, $b_{eff}|_{cri}$ becomes small because the area where stress increases owing to the constraint of both edges becomes small. Generally, if $L > b$, the value of $b_{eff}|_{cri}$ is roughly constant and barely depends on L.

2.2.3 Collapse force after plastic buckling (for materials with strain hardening)

The discussion in the previous section concerned the increase of compressive force after the elastic buckling of a plate assumed to be a perfect elasto-plastic material (without strain hardening) and is fundamentally not applicable to analysis after plastic buckling of a plate with strain hardening. However, members such as square tubes for absorbing collision energy are in most cases thick plates in which plastic buckling occurs; due to the strain hardening of a material, the compressive force increases even after plastic buckling of the plate. Therefore, compressive stress locally decreases after buckling at the center of the plate because of the out-of-plane deformation, as in elastic buckling; it increases, however, due to strain hardening near both edges, and the force of the whole plate can become even greater than the level at buckling. Note that both the decrease of stress at the center and the increase of stress at both edges are governed by Young's modulus E of elasticity in the case of elastic buckling and that, in the case of plastic buckling, the decrease of stress at the center is elastic unloading and is similarly governed by Young's modulus E; however, the increase of stress at both edges is governed by the strain hardening coefficient E_t of plasticity. Therefore, the behavior of the increase of compressive stress after plastic buckling depends on the strain hardening coefficient E_t.

Here, based on FEM analysis [44] of the axial crushing deformation of plate, the axial collapse mechanism of plates having various strain-hardening properties will be discussed.

Assuming two types of materials ($E_t/E = 0.01$ and 0.1) that obey the bilinear hardening rule and a plate thickness ratio $t/b = 0.04$ which corresponds to plastic buckling, Fig. 2.13 shows the relation between compressive force P_x and compressive deformation U_x in axial compression under the boundary condition of simple support at both edges. Here, the symbols P_{buc} and P_{col} in the figure indicate buckling force and collapse force, respectively. As the figure shows, in the case of $E_t/E = 0.01$, the force does not increase after plastic buckling; in the case of $E_t/E = 0.1$, however, the force, after decreasing temporarily, increases and reaches the collapse force P_{col}. This feature can be understood from the variation of axial compressive stress before and after buckling of the plate.

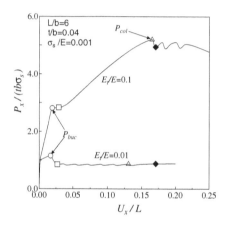

FIGURE 2.13
Two compressive force-displacement curves for plate compression [44].

Fig. 2.14 shows the distribution of axial compressive stress in the cross section over the plate width corresponding to the symbols in Fig. 2.13. As the figure shows, the axial compressive stress in the cross section of the plate is uniform up to the buckling point (symbol ○); after buckling, the compressive stress decreases at the center of the plate due to the elastic unloading caused by out-of-plane buckling whereas the axial compressive strain and hence compressive stress increase at both edges. However, since the stress increase is governed by the strain hardening coefficient E_t, if E_t/E is small (for example, in the case of $E_t/E = 0.01$ as shown in Fig. 2.13), the stress decrease rate at the center after buckling exceeds the increase rate at both edges, and therefore the compressive force decreases after plastic buckling.

2.2.3.1   Two types of collapse force

As Fig. 2.13 shows, in the case of $E_t/E = 0.1$, the compressive force continues to increase after buckling and reaches a maximum at a certain level of force (to

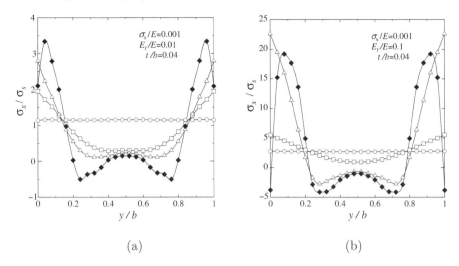

FIGURE 2.14
Compressive stress distribution for plate compression [44]: (a) $E_t/E = 0.01$;
(b) $E_t/E = 0.1$.

be called the **collapse force** as indicated by P_{col} in the figure). Corresponding
to this, as shown in Fig. 2.14(b), the compressive stress increases at both plate
edges even after buckling, but begins to decrease at the symbol $\triangle$. The stress-
strain relation of the material used in the analysis obeys the bilinear hardening
rule in which no upper limit is assumed; then the questions that arise are why
the maximum compressive stress appears in the compressive process and why
compressive collapse occurs. Here, the origins of the collapse are discussed in
reference to Figs. 2.15 to 2.18.

Fig. 2.15(b) shows the compressive stress-strain relation of a plate obtained
by FEM analysis for materials with the same initial yield stress $\sigma_s = 72.4$ MPa
and the same strain hardening coefficient $E_t/E = 0.05$, but with six different
levels (i)-(vi) of stress limit σ_B as shown in Fig. 2.15(a) (the stress limit σ_B
for (vi) is infinitely large). The figure shows that in the stress-strain relations
of (i)-(iii), the collapse stress σ_{col} increases with the stress limit σ_B. Thus, it
is supposed that in the stress-strain relations (i)-(iii), the stress limit of the
material itself is a controlling factor in the collapse behavior. On the other
hand, the collapse stress σ_{col} for the stress-strain relations (iv) and (v) does
not depend on the stress limit σ_B and agrees with that for (vi) with stress
limit $\sigma_B = \infty$. This suggests that in the stress-strain relations (iv), (v) and
(vi), there are controlling factors in collapse behavior other than the stress
limit of the material.

When the plate collapses, as the deformation behavior in Fig. 2.16 shows,
in-plane lateral deformation occurs when the out-of-plane deformation occurs
after out-of-plane buckling. The amount of in-plane lateral deformation in

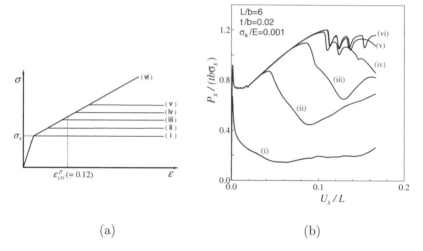

(a)   (b)

FIGURE 2.15
Relation between compressive stress and deformation in a plate obeying the
trilinear hardening rule with the stress limit σ_B [44]: (a) stress-strain curve;
(b) compressive force versus compressive deformation.

the cases of (iv)-(vi) is greater than that in the case of (ii). In addition,
the in-plane lateral deformations at collapse, appearing in the form of folds,
are almost the same between (iv)-(vi). This shows that the plate collapse is
governed by the in-plane lateral deformation at side areas, which decreases the
axial compressive stress in the plate, and that, when the deformation reaches
a certain level, the total compressive force of the plate begins to decrease.
That is, in the stress-strain relation (iv)-(vi), the controlling factor in the
collapse behavior in axial plate crushing is not the stress limitation in the
material but the limitation of in-plane lateral deformation at the side areas.
Therefore, there are two types of collapse behavior: **collapse caused by the
stress limit in the material** and **collapse caused by in-plane lateral
deformation**. The collapse stress σ_{col} of a plate under an axial compression
equals the smaller of either **the critical stress** σ_{cri} caused by the in-plane
lateral deformation or **the limit stress** σ_{lim} due to stress limit

$$\sigma_{col} = \min(\sigma_{cri}, \sigma_{lim}) \qquad (2.64)$$

In the analysis in Fig. 2.15, the boundary condition is that both plate
edges are simply supported. To further study the collapse of a plate due to the
deformation in the in-plane lateral direction, discussed here are the four types
of boundary conditions shown in Fig. 2.17: simple support (S1); fixed rotation
(C1); and S2 and C2 where the deformation in the in-plane lateral direction is
fixed under the boundary conditions of S1 and C1, respectively. Fig. 2.18(a)
shows the compressive force-deformation relation of a plate obtained for the

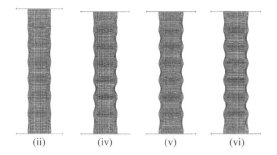

(ii) (iv) (v) (vi)

FIGURE 2.16
Collapse behavior of the plate discussed in Fig. 2.15 [44].

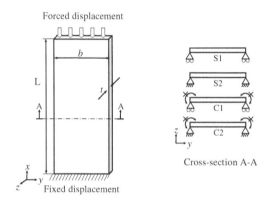

FIGURE 2.17
Four types of boundary conditions.

four boundary conditions of S1, C1, S2 and C2 assuming the plate thickness ratio $t/b = 0.03$ and a material obeying the bilinear hardening rule. The figure shows that the collapse stress appears when compression proceeds sufficiently in the cases of S1 and C1, whereas force continues to increase but collapse stress does not appear in the cases of S2 and C2. Fig. 2.18(b) shows the deformed shapes of the plate corresponding to the compression stages A, A', B and B' in Fig. 2.18(a) for the boundary conditions of S1 and S2. In the case of the boundary condition S2, the edges at the side areas remain straight due to the constraint of in-plane lateral deformation; therefore, the compressive stress locally increases there and collapse stress does not appear.

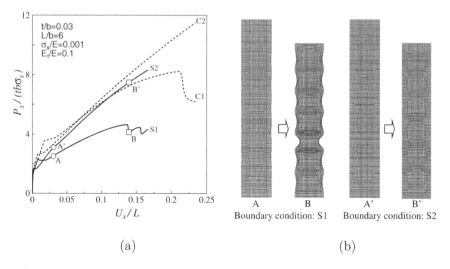

(a)   (b)

FIGURE 2.18
Influence of boundary conditions at both edges on the compressive force of a
plate [44]: (a) compression force and displacement; (b) deformed shapes.

2.2.3.2 Collapse stress due to deformation in the in-plane lateral direction

The collapse condition due to the deformation in the in-plane lateral direction can be derived from the axial **critical compressive strain** ε_{cri} at the
side area. Fig. 2.19 shows the relation between the critical compressive strain
ε_{cri} and the plate thickness ratio t/b for a material obeying either the bilinear hardening rule or the Ludwik's n-power hardening rule as defined by
Eq. (1.40). The figure shows that **the critical compressive strain in the
side area at collapse** is hardly affected by the strain-hardening property of
materials and is a function of thickness ratio; it is proportional to the plate
thickness as given by the following equation:

$$\varepsilon_{cri} = \gamma \frac{t}{b} \tag{2.65}$$

Here, the coefficient γ shows the influence of boundary conditions and is approximately given by $\gamma = 6$ in the case of S1 and $\gamma = 12$ in the case of
C1.

Substituting the critical compressive strain obtained from Eq. (2.65) into
the stress-strain constitutive equation for the material $f_\sigma(\varepsilon)$ and calculating
the critical stress $\sigma_{edge}|_{cri} = f_\sigma(\gamma t/b)$ in the plate edge area, one can predict
the critical stress due to deformation in the in-plane lateral direction using
the formula

$$\sigma_{cri} = \frac{b_{eff}}{b} \sigma_{edge}|_{cri} = \frac{b_{eff}}{b} f_\sigma(\gamma t/b) \tag{2.66}$$

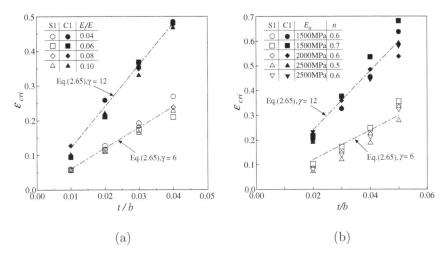

(a)   (b)

FIGURE 2.19
Critical compressive strain ε_{cri} versus plate thickness ratio t/b [44]: (a) bilinear hardening rule; (b) n-power hardening rule.

b_{eff} in Eq. (2.66) is the effective width after plastic buckling of a material with strain hardening. Although b_{eff} can be determined by the same technique as Karman et al. [211], here b_{eff} is directly evaluated based on the P_x versus U_x relation after plastic buckling obtained from FEM analysis [44]. That is, the effective width b_{eff} is calculated by the compressive force P_x, i.e.

$$b_{eff} = \frac{P_x}{t\sigma_{edge}} = \frac{P_x}{tf_\sigma(\varepsilon_{edge})} \tag{2.67}$$

Fig. 2.20 shows the log-log graph of the relation between the compressive strain at the side edge ε_{edge} ($\cong U_x/L$) and the effective width b_{eff} obtained from Eq. (2.67). As is understood from the figure, the effective width b_{eff} after plastic buckling can be evaluated by using

$$b_{eff} \cong t\sqrt{B/\varepsilon_{edge}} \tag{2.68}$$

as a function of the in-plane compressive strain at the edge. Here, the coefficient B depends on boundary conditions, and is approximately given by $B = 7$ in the case of S1 and $B = 14$ in the case of C1.

Substituting Eq. (2.68) into Eq. (2.66) yields

$$\sigma_{cri} = \sqrt{\frac{Bt}{\gamma b}} f_\sigma(\gamma t/b) \tag{2.69}$$

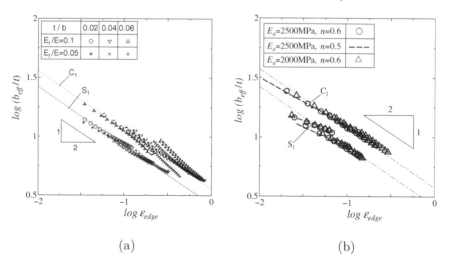

(a)   (b)

FIGURE 2.20
Relation between effective width after plastic buckling and in-plane compressive strain at the site edge [44]: (a) bilinear hardening law; (b) Ludwik's n-power hardening law.

Figs. 2.21 and 2.22 show the collapse stress obtained by FEM analysis for materials obeying the bilinear and n-power hardening rules, respectively, under the boundary conditions of S1 and C1, using symbols such as O and $\triangle$ to indicate material constants. Figs. 2.21 and 2.22 show only collapse due to in-plane lateral deformation (namely, $\sigma_{col} = \sigma_{cri}$), and, for comparison, the predicted values of critical stress calculated from Eq. (2.69) are also plotted using solid lines. The figures show that using Eq. (2.69), the collapse stress can be predicted with reasonable accuracy irrespectively of the hardening rule of the material.

2.2.3.3   Collapse stress due to stress limit

The collapse may also occur due to the stress limit before the in-plane compressive strain at both plate edges reaching the critical compressive strain ε_{cri}. Considering that the average compressive stress can be evaluated by $\sigma = \sigma_{edge}(b_{eff}/b)$ in the axial compression of a plate,

$$\frac{d\sigma}{d\varepsilon_{edge}} = \left(\frac{\sigma_{edge}}{b}\right)\frac{db_{eff}}{d\varepsilon_{edge}} + \left(\frac{b_{eff}}{b}\right)\frac{d\sigma_{edge}}{d\varepsilon_{edge}} \qquad (2.70)$$

Although the stress σ_{edge} at both edges increases due to strain hardening as compression proceeds, the effective width b_{eff} decreases as Eq. (2.68) shows. Compressive force increases after buckling if the increasing rate of force due to the increase of stress σ_{edge} at both edges is greater than the decreasing rate of force due to the decrease of effective width b_{eff}. Then, in the force-increasing

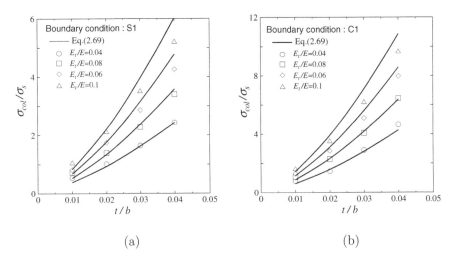

(a) (b)

FIGURE 2.21
Collapse stress due to in-plane lateral deformation (material obeying the bi-linear hardening rule) [44]: (a) boundary condition S1; (b) boundary condition C1.

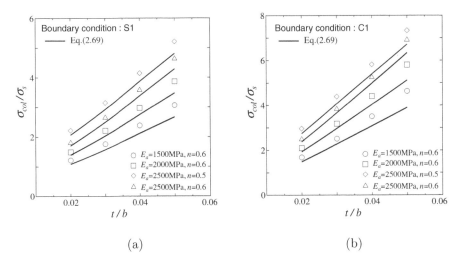

(a) (b)

FIGURE 2.22
Collapse stress due to in-plane lateral deformation (material obeying the Ludwik's n-power hardening rule) [44]: (a) boundary condition S1; (b) boundary condition C1.

stage after buckling, the maximum appears in the force-displacement curve, which is the collapse force, when the former and latter rates coincide with each other because the former (the increasing rate of force due to the increase of stress σ_{edge} at both edges) decreases and the latter (the decreasing rate of force due to the decrease of effective width b_{eff}) increases. That is, the collapse due to the stress limit appears when

$$\frac{d\sigma}{d\varepsilon_{edge}}\bigg|_{\sigma=\sigma_{lim}} = 0 \qquad (2.71)$$

Substituting Eq. (2.68) into Eq. (2.70), and considering

$$\frac{d\sigma_{edge}}{d\varepsilon_{edge}} = E_t \ , \qquad \frac{\sigma_{edge}}{\varepsilon_{edge}} = E_s$$

at the plate edge,

$$\frac{d\sigma}{d\varepsilon_{edge}}\bigg|_{\sigma=\sigma_{lim}} = \frac{t}{b}\sqrt{\frac{B}{\varepsilon_{edge}}}\left[E_t - \frac{1}{2}E_s\right] = 0 \qquad (2.72)$$

is obtained from Eq. (2.71). Therefore, the following collapse condition due to the stress limitation can be obtained:

$$E_t - \frac{1}{2}E_s = 0 \qquad (2.73)$$

In order for collapse due to stress limitation to occur, the following strain condition must hold:

$$\varepsilon_{edge} < \varepsilon_{cri} \qquad (2.74)$$

For example, in Fig. 2.15, at the bending point, where the stress reaches its limit, in the stress-strain curve $E_t = 0$ holds and $d\sigma/d\varepsilon_{edge}$ changes from positive to negative as Eq. (2.72) shows. Therefore, the collapse due to the stress limit should occur if the strain at the bending point is smaller than ε_{cri}. The critical strain corresponding to the plate thickness in Fig. 2.15 is given by $\varepsilon_{cri} = 0.12$ (dotted line in the figure) from Eq. (2.65), and so collapse occurs due to the stress limit of the material for (i)-(iii), where the bending point comes to the left of the dotted line in stress-strain relations, and collapse occurs due to in-plane deformation for (iv)-(vi), where the bending point comes to the right of the dotted line in stress-strain relations.

Thus, for predicting collapse stress due to the stress limitation of material, first, the in-plane compressive strain at the side edge, ε_{lim}, that satisfies Eq. (2.73) within the strain condition of Eq. (2.74) is evaluated.

$$\varepsilon_{lim} = \varepsilon_{edge}\bigg|_{\text{with } E_t-E_s/2=0}$$

Then, the collapse stress is given by

$$\sigma_{lim} = \frac{t f_\sigma(\varepsilon_{lim})}{b}\sqrt{\frac{B}{\varepsilon_{lim}}} \qquad (2.75)$$

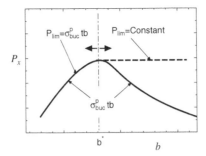

FIGURE 2.23
Evaluation of strength after plastic buckling proposed by Langseth and Hopperstad [119].

Langseth and Hopperstad [119] proposed a modification of the Stowell theory to take post-buckling strength into account. Their basic idea is shown in Fig. 2.23, which shows the plastic buckling force ($\sigma^p_{buc}tb$) versus plate width b curve by solid line. Here, the plastic buckling stress σ^p_{buc} is obtained from Eqs. (2.28) and (2.29). As shown in Fig. 2.23, the plastic buckling force takes the maximum at $b = b^*$. It is assumed that up to the critical value b^* corresponding to the maximum force $\sigma^p_{buc}bt$ the plate is fully effective with a capacity expressed as $\sigma^p_{buc}bt$; however, beyond b^* at collapse the load is taken by two edge-strips, thus neglecting the load-carrying capacity of the mid section of the plate, the failure load for the plate is then constant and is equal to the load-carrying capacity of the plate with width of b^*. Therefore, the corresponding collapse stress is given as follows:

$$\sigma_{lim} = \begin{cases} \sigma^p_{buc} & b \leq b^* \\ \sigma^p_{buc}\big|_{\text{for } b = b^*} \times \left(\dfrac{b^*}{b}\right) & b > b^* \end{cases} \tag{2.76}$$

Fig. 2.24 summarizes the collapse stress under the boundary conditions of S1 and C1 obtained from FEM analysis of the three types of materials having different material constants and obeying the n-power hardening rule as defined by Eq. (1.40). The figure only shows the results for collapse caused by the stress limit (namely, $\sigma_{col} = \sigma_{lim}$), but, for comparison, the values predicted by Eq. (2.75) and Eq. (2.76), respectively, are also shown. The figure shows that the collapse stress is predicted with good accuracy by Eq. (2.75).

The illustrative curves in Fig. 2.25 are obtained based on the variation of $E_t - E_s/2$ after the plastic buckling of a plate. Fig. 2.25(a) shows $E_s/2$ with a heavy line and 5 kinds of E_t curves with thin lines to discuss the typical plus and minus variations of $E_t - E_s/2$. From the variation of the 5 cases of $E_t - E_s/2$ shown in Fig. 2.25(a), it is seen that there are five plots of relations

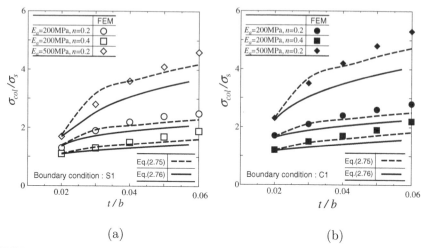

(a)   (b)

FIGURE 2.24
Collapse stress of plates due to stress limit (for materials obeying the Ludwik's
n-power hardening rule): (a) for boundary conditions of S1; (b) for boundary
conditions of C1.

between compressive force and strain shown in Fig. 2.25(b). Here the following
matters are taken into consideration:

(1) When plastic buckling occurs, the force after buckling decreases if $E_t - E_s/2 < 0$, but increases if $E_t - E_s/2 > 0$.

(2) If there is a crossing point where $E_t - E_s/2$ changes from negative to positive in the range between the plastic buckling strain ε_{buc} and the critical compressive strain ε_{cri}, the compressive force changes from decreasing to increasing when passing through the crossing point. Further, if there is a crossing point where $E_t - E_s/2$ changes from positive to negative, collapse occurs due to the stress limit.

(3) As the compressive force increases and reaches the critical strain ε_{cri}, collapse occurs due to in-plane lateral deformation.

Accordingly, the following five types of relations ($a1$, $a2$, $b1$, $b2$, and c) are derived between compressive force and strain from Fig. 2.25(a), as the rough sketch in Fig. 2.25(b) shows.

(a) The type where force continually increases, after buckling, to reach the collapse force (type a in the figure); this type is further classified into two subtypes: collapse due to in-plane lateral deformation (type $a1$ in the figure) and collapse due to the stress limit (type $a2$ in the figure)).

(b) The type where force decreases first after buckling and then, conversely, increases to reach collapse force (type b in the figure); this type is further

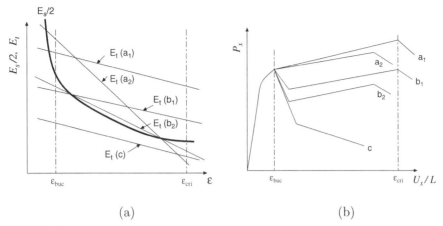

(a)   (b)

FIGURE 2.25
Schematic diagram of collapse behavior corresponding to collapse condition in axial compressive deformation of a plate: (a) illustration of $E_t - E_s/2$; (b) five types of relations P_x and U_x.

classified into two subtypes: collapse due to in-plane lateral deformation (type $b1$ in the figure) and collapse due to the stress limit (type $b2$ in the figure)).

(c) The type where force continually decreases after buckling (type c in the figure).

2.3 Buckling and collapse force of square tubes

In the axial compressive deformation of a square tube, just as in the simple case of a plate, there are two cases where the maximum force in the initial crushing process corresponds to either the buckling force or the collapse force; both can be evaluated simply from the buckling force and the collapse force of the plate constituting the square tube.

Fig. 2.26 shows the elastic buckling stress of square tubes, as simulated by FEM analysis. The figure also shows the values from the theoretical formula, Eq. (2.12), for the elastic buckling stress. The figure shows that the elastic buckling stress of square tubes also relates to the compressive deformation mode—extensional mode or inextensional mode—and can be evaluated through the buckling stress of a plate under the condition of simple support or fixed support, respectively, at both edges.

In the inextensional mode deformation of a square tube, in the corner area rotation occurs but a moment does not appear; it is reasonable to assume

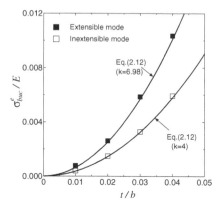

FIGURE 2.26
Elastic buckling stress of square tubes.

that the boundary condition for each plate composing the square tube is, approximately, simple support. On the other hand, in the extensional mode deformation of a square tube, a moment appears in the corner area that does not rotate; it is assumed that each plate of a square tube is, approximately, fixed at both ends. However, to be exact, the boundary condition at both edges of a plate in the axial compression of a square tube is different from both S1 and C1. Studies by Li and Reid [123, 124] showed that, owing to the deformation of the corner area of a square tube, the displacement in the out-of-plane direction at the plate edge is non zero because it couples with the lateral displacement in the in-plane direction of the adjacent plates such that the symmetry or asymmetry condition of displacements about the bisector line of the corner is satisfied. On this basis, Li and Reid [123, 124] pointed out that the buckling stress of a square tube is lower than that of the plate under the boundary condition S1 or C1.

Fig. 2.27 compares the plastic buckling stress of several polygonal tubes made of identical plates (triangular, square, and hexagonal tubes; square and hexagonal tubes with partition walls) in the extensible mode obtained by FEM analysis. As the figure shows, because identical plates are used in all tubes, the buckling stress should be approximately the same. However, the buckling stress slightly decreases with the number of the sides of the tubes. This is because the boundary constraint at both edges of a plate decreases as the number of the sides of the polygonal tubes increases.

Fig. 2.28 compares the values of plastic buckling stress obtained by FEM analysis for square tubes with those obtained by using Eq. (2.35) for corresponding plates, assuming that the material obeys the bilinear hardening rule with the strain hardening coefficients $E_t/E = 0.01$ and 0.1. In the analysis, the crushing mode of the square tube is inextensible for $E_t/E = 0.01$ and

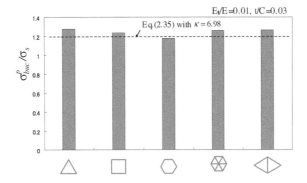

FIGURE 2.27
Plastic buckling stress of several polygonal tubes made of identical plates.

extensible for $E_t/E = 0.1$. As the figure shows, the plastic buckling stress of the square tube with $E_t/E = 0.01$ and 0.1 can be approximately evaluated as the buckling stress under the boundary conditions S1 and C1, respectively, for the corresponding plate.

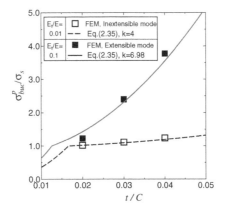

FIGURE 2.28
Plastic buckling stress of square tubes: (a) $E_t/E = 0.01$; (b) $E_t/E = 0.1$.

Fig. 2.29 shows the collapse stress of square tubes in the extensible and inextensible modes obtained by FEM analysis for materials obeying the n-power hardening rule and the bilinear hardening rule with $E_t/E = 0.05$. In the FEM analysis, to induce an inextensible deformation mode in a square tube, an infinitesimal geometric imperfection was added near the bottom of tube.

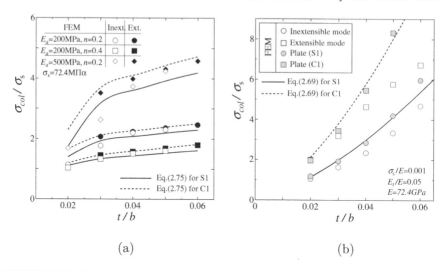

(a)   (b)

FIGURE 2.29
Collapse stress of square tubes: (a) material obeying the n-power hardening rule [44]; (b) material obeying the bilinear hardening rule with $E_t/E = 0.05$.

In Fig. 2.29(a), because the collapse of the tube is due to the stress limit, the figure also shows the collapse stress of the identical plate obtained using Eq. (2.75). The collapse stress of the square tube depends on the deformation mode. The collapse stress of square tubes in the inextensible mode and the extensible mode correspond to plates under the boundary condition S1 and C1, respectively, and then can be evaluated by Eq. (2.75). Further, in Fig. 2.29(b), since the collapse of the tube is due to the deformation in the in-plane lateral direction, the figure also shows the collapse stress of the identical plate obtained by Eq. (2.69). It is seen from the figure that although the collapse stress of the square tube corresponds to that of plates under the respective boundary conditions, the collapse stress of the square tube is smaller than that of plate, especially in cases of large thickness and of extensible mode. The reason can be understood from the fact that since the out-of-plane deformation of corner parts of square tube becomes larger with increase of the in-plane deformation in the case of the plastic buckling, the boundary conditions for a plate of a square tube are quite different from those of C1 or S1. In relation to this, Graves-Smith and Sridharan [84] pointed out that, in the axial crushing of a square tube, the "**crinkly buckling**" of the corner area which results from the displacement in the out-of-plane direction at both edges of each plate plays an important role. For comparison, Fig. 2.29(b) also shows the collapse stress of the plate analyzed by FEM.

2.4   Average force in the axial crushing of square tubes

To theoretically predict the average force in the axial crushing of a square tube, it is necessary to investigate deformation models of the crushing after buckling. As discussed in the preceding section, there are various modes of square tubes. On the other hand, Abramowicz and Jones [4] focused on the deformation in the corner area of a square tube; they pointed out that there are basically two types of folding mechanism [91, 216] in the corner area shown in Fig. 2.30: inextensional folding mechanism and extensional folding mechanism, and that the average crushing force corresponding to the various deformation modes of square tubes can be calculated by appropriately combining the two types of deformation mode.

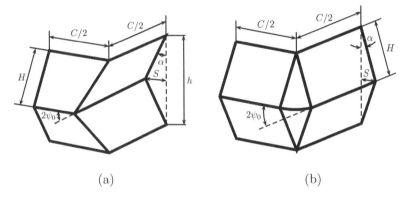

(a)   (b)

FIGURE 2.30
Basic types of collapse mechanism in the corner area [216]: (a) inextensional folding mechanism; (b) extensional folding mechanism.

The inextensional folding mechanism corresponds to the **inextensional deformation mode** of a square tube. The most important characteristics of the inextensional mode in the crushing of a square tube are that the crests and valleys of the buckles appear alternately in the adjacent planes. Fig. 2.31 shows the origami model for a square tube to help understand the mechanism of the inextensional deformation of a square tube. As known from the **origami model**, while the square tube is compressed, the horizontal hinges moving outward and moving inward are created alternately in the axial direction, and so the lines of intersection of tube sides tilt to create inclined hinges. The buckles are created through the formation of these horizontal and inclined hinges and the rotation about those hinges. In creating the buckle, in the cross section passing through the buckle top, the two horizontal hinges facing each other and moving outward become shorter while the two adjacent horizontal hinges moving inward become longer; as a result, the total circum-

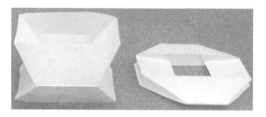

FIGURE 2.31
Origami model of the inextensional mode.

ferential length is almost unchanged. Therefore, circumferential expansion is not necessary, and the deformation mode is called inextensional.

The extensional folding mechanism corresponds to the **extensional deformation mode** of a square tube. This is similar to the axisymmetric deformation of a thin-walled circular tube; and the cross section in Fig. 2.4(a) shows that either the hinges moving outward or the hinges moving inward are created at the same position of the axial direction on each side plane of a square tube. Therefore, circumferential expansion occurs, and the deformation is called extensional.

In the following, according to the research given by Wierzbicki [214], Wierzbicki and Abramowicz [216], Abramowicz and Jones [5], Abramowicz and Wierzbicki [7], etc., the strain energy needed for creating the buckles of the two basic types of crushing in the corner area of a square tube is first calculated and then the analysis of average crushing force of various square tubes is discussed based on the formulas derived for strain energy.

2.4.1   Inextensional folding mechanism

Fig. 2.32(a) shows the origami model of the inextensional folding mechanism. As compared with Fig. 2.31, in Fig. 2.32(a) BA is a crest of the buckle, and B_1A_1 and B_2A_2 are the intermediate lines between the crest and the adjacent valleys above and below the crest in the axial direction. Similarly, BC in the figure is a valley of the buckle, and B_1C_1 and B_2C_2 are the intermediate lines between this valley and the adjacent crests above and below.

The angle between the horizontal hinges AB and BC is always constant in the deformation process. Assuming the external angle $2\psi_0$ shown in the figure, ψ_0 is determined by the number n of tube sides and is given by

$$\psi_0 = \frac{\pi}{n} \tag{2.77}$$

As a parameter of the deformation process of buckles, the angle α between the side surface (face BCC_1B_1 in the figure) and the axial direction is used. In the crushing process, the angle α increases from 0 to $\pi/2$. For an assumed

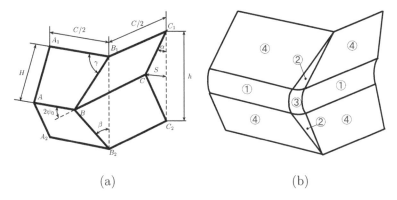

FIGURE 2.32
Geometry of inextensional folding mechanism: (a) origami model; (b) quasi-inextensional model.

angle α, the angle β (angle $\angle BB_1B_2$ in the figure) satisfies

$$\tan \beta = \frac{\tan \alpha}{\sin \psi_0} \tag{2.78}$$

and the angle γ (between inclined hinge BB_1 and horizontal line A_1B_1) satisfies

$$\tan \gamma \sin \alpha = \tan \psi_0 \tag{2.79}$$

As known from Eqs. (2.78) and (2.79), during the deformation of the buckle in which the angle α increases from 0 to $\pi/2$, the angle γ decreases from $\pi/2$ to ψ_0, and the angle β increases from 0 to $\pi/2$.

In the origami model of Fig. 2.32(a), it is here assumed that the plastic deformation concentrates on the lines of the plastic hinges. In such a model, there are no mechanical problems if the plastic hinges do not travel. However, the inclined hinges BB_1 and BB_2 should travel as the buckles are created. As later shown in Eq. (2.87), the energy needed for the travelling of a hinge is proportional to the curvature of the hinge. This means that in the origami model, the energy needed for the travelling becomes infinite because the curvature radius of the inclined hinge is assumed to be zero, indicating that the model is impractical from a mechanical viewpoint. Fig. 2.32(b) is a model proposed to solve the problem; this is called the quasi-inextensional model because in-plane deformation can occur in the area denoted by ③ [216].

In the quasi-inextensional model shown in Fig. 2.32(b), the plastic deformation, which is concentrated on the plastic hinges in the origami model shown in Fig. 2.32(a), is assumed to be distributed over an extensive zone. That is, the hinges AB and BC of the origami model are replaced by zone ①, the inclined hinges BB_1 and BB_2 are replaced by zone ②, and the intersection B of horizontal and inclined hinges is replaced by zone ③; in summary,

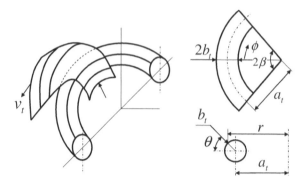

FIGURE 2.33
Toroidal curved surface.

the model is composed of two parts of zone ①, two parts of zone ②, a part of zone ③, and four parts of zone ④. Zone ④, being a trapezoidal plate, moves like a rigid body during the deformation. Zone ① is a part of circular tube surface bounded by two straight lines, which move during the deformation in the opposite directions so that the zone becomes wider. Zone ② is a part of a conical surface bounded by two straight lines, and the bottom part of this moves with the adjacent zone ③. Meanwhile, zone ③ is a toroidal curved surface connecting the two parts of zone ① and the two parts of zone ②, and is bent in two vertical directions; in zone ③, in-plane strain is created during the deformation. The zone of the toroidal curved surface, zone ③, is part of the surface of a circular tube of radius b_t, which is bent with radius a_t as shown in Fig. 2.33 [216]. Using the (θ, ϕ) coordinate (see Fig. 2.33), this is the region of (θ, ϕ) defined by

$$\frac{\pi}{2} - \psi \le \theta \le \frac{\pi}{2} + \psi, \qquad -\beta \le \phi \le \beta$$

where ψ depends on the coordinate ϕ and is given by

$$\psi = \psi_0 + \left(\frac{\pi - 2\psi_0}{\pi}\right)|\phi|$$

Note that if $\phi = 0$, $\pi/2 - \psi_0 \le \theta \le \pi/2 + \psi_0$ holds.

The average force in the axial crushing of a square tube is calculated from the balance with the strain energy needed to create crushing buckles. As known from the mechanical model discussed above, the strain energy needed to create crushing buckles in the quasi-inextensional model shown in Fig. 2.32(b) can be evaluated from (1) the strain energy of expansion due to the extensional deformation in the toroidal zone, (2) the rotation about the horizontal hinges and (3) the rotation about the inclined travelling hinges. The strain energy of (1), (2) and (3) shall be determined in the following.

(1) Energy dissipation in the toroidal zone

In the compressive deformation of a square tube, the horizontal hinge BC moves to the inside of the square tube. The movement velocity is $\dot{S}$. Thus, the movement velocity V_t of each point in the toroidal area (see Fig. 2.33) is given by

$$V_t = \frac{\dot{S}}{\tan \psi_0}, \quad S = H \sin \alpha \tag{2.80}$$

Strain rate $\dot{\varepsilon}_\phi$ appears in the circumferential direction, originating from the horizontal component $V_t \sin \theta$ of the movement velocity V_t. The strain rate $\dot{\varepsilon}_\phi$ at the radius $r = b_t \cos \theta + a_t$ is given by

$$\dot{\varepsilon}_\phi = \frac{V_t \sin \theta}{r} = \frac{H \cos \alpha}{\tan \psi_0} \frac{\sin \theta}{b_t \cos \theta + a_t} \dot{\alpha} \tag{2.81}$$

Calculating the expansion energy corresponding to $\dot{\varepsilon}_\phi$ in the toroidal area yields[2]

$$\dot{E}_1 = \int_{-\beta}^{\beta} \int_{\pi/2-\psi}^{\pi/2+\psi} \sigma_s t \dot{\varepsilon}_\phi (r d\phi b_t d\theta)$$

$$= 4\sigma_s t b_t H \frac{\pi}{(\pi - 2\psi_0) \tan \psi_0} \cos \alpha \left[\cos \psi_0 - \cos \left(\psi_0 + \frac{\pi - 2\psi_0}{\pi} \beta \right) \right] \dot{\alpha} \tag{2.82}$$

Further, by integrating with respect to α, the expansion energy is given by

$$E_1 = 4\sigma_s t b_t H I_1(\psi_0) = 16 M_0 \frac{H b_t}{t} I_1(\psi_0) \tag{2.83}$$

until the buckle collapses completely, where

$$I_1(\psi_0) = \frac{\pi}{(\pi - 2\psi_0) \tan \psi_0} \int_0^{\pi/2} \cos \alpha \left[\cos \psi_0 - \cos \left(\psi_0 + \frac{\pi - 2\psi_0}{\pi} \beta \right) \right] d\alpha \tag{2.84}$$

In particular, $I_1(\pi/4) = 0.58$, $I_1(\pi/6) = 1.04$, and $I_1(\pi/8) = 1.51$ hold.

(2) Energy dissipation in the horizontal hinges

When the angle $d\alpha$ increases, the angle of rotation about the horizontal hinges BA and BC is given by $2d\alpha$; and the incremental energy $\dot{E}_2$ needed for rotation about the horizontal hinges is given by

$$\dot{E}_2 = 2 M_0 C \dot{\alpha} \tag{2.85}$$

By integrating Eq. (2.85) with respect to α, the energy until the buckle is completely crushed can be obtained:

$$E_2 = \int_0^{\pi/2} 2 M_0 C d\alpha = \pi M_0 C \tag{2.86}$$

where $M_0 = \sigma_0 t^2 / 4$, t is the plate thickness.

---

[2]Though there is a non-zero curvature change in circumferential direction, the corresponding bending energy can be shown to be zero for small b_t/a_t.

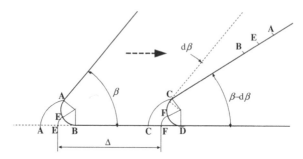

FIGURE 2.34
Travelling of a plastic hinge.

(3) Energy dissipation in the inclined travelling hinges

First, consider a strip of width b_s with a travelling hinge as shown in Fig. 2.34 in order to discuss the energy dissipation in a travelling hinge. Here the center of a hinge travels from point E (center of circular arc AB) to point F (center of circular arc CD), which are separated by a distance Δ. The angle changes from β to $\beta - d\beta$, while the curvature radius ρ remains constant.

As Fig. 2.34 shows, the energy associated with the travelling of the hinge is calculated from the following three terms.

(a) Energy dW_1 needed for unbending part AB to a planar state:

$$dW_1 = b_s \frac{|AB|}{\rho} M_0$$

where M_0 is the fully plastic bending moment.

(b) Energy dW_2 needed for the initial bending part BC and the subsequent reverse bending to a planar state:

$$dW_2 = b_s \frac{|BC|}{\rho} M_0 \times 2$$

(c) Energy dW_3 needed for bending part CD:

$$dW_3 = b_s \frac{|CD|}{\rho} M_0$$

In summary, the energy dissipation dW in the travelling of hinge is calculated by using

$$dW = dW_1 + dW_2 + dW_3 = 2b_s \frac{\Delta}{\rho} M_0 = \frac{2dA}{\rho} M_0 \qquad (2.87)$$

where $dA = b_s \Delta$ is the area swept by the travelling hinge.

Therefore for zone ②, assuming distance s from the vertex of the circular cone, Eq. (2.87) shows that the velocity of energy dissipation in the travelling hinges BB_1 and BB_2 can be evaluated by

$$\dot{E}_3 = 2 \int_0^{|BB_1|} \frac{2v(s)ds}{\rho} M_0 \tag{2.88}$$

where $v(s)$ is the travelling velocity of point s of the conic surface, and ρ is the cone radius of the point s as given by

$$v(s) = \frac{s}{|BB_1|} V_t, \quad \rho = \frac{s}{|BB_1|} b_t, \quad |BB_1| = \frac{H}{\sin \gamma} \tag{2.89}$$

Substituting Eq. (2.89) into Eq. (2.88) yields

$$\dot{E}_3 = \frac{4V_t H}{b_t \sin \gamma} M_0 = 4M_0 \frac{H^2 \cos \alpha}{b_t \sin \gamma \tan(\psi_0)} \dot{\alpha} \tag{2.90}$$

and integrating this in the range $\alpha = 0{-}\pi/2$ yields

$$E_3 = \int_0^{\pi/2} 4M_0 \frac{H^2 \cos \alpha}{b_t \sin \gamma \tan(\psi_0)} d\alpha = 4M_0 \frac{H^2}{b_t} I_3(\psi_0) \tag{2.91}$$

where

$$I_3(\psi_0) = \frac{1}{\tan(\psi_0)} \int_0^{\pi/2} \frac{\cos \alpha}{\sin \gamma} d\alpha \tag{2.92}$$

In particular, $I_3(\pi/4) = 1.15$, $I_3(\pi/6) = 2.39$ and $I_3(\pi/8) = 3.96$.

Taking together the energies E_1, E_2 and E_3 in the terms (1), (2) and (3), respectively, needed for the deformation of the buckle,[3] in inextensional folding mechanism the strain energy E_{inex} required to crush the corner area with the outer angle $2\psi_0$ can be evaluated as

$$E_{inex} = M_0 \left(16 \frac{b_t H}{t} I_1(\psi_0) + 2\pi C + 4 \frac{H^2}{b_t} I_3(\psi_0) \right) \tag{2.93}$$

In order to consider the strain hardening of a material, here used is the energy equivalent flow stress σ_0, which is larger than the initial yield stress of the material σ_s but is smaller than the tensile strength σ_u. In this way, to consider that the deformation, and therefore the flow stress, is different with the three terms E_1, E_2, and E_3, there is also a proposal which sets those flow stresses to $\sigma_0^{(1)}$, $\sigma_0^{(2)}$ and $\sigma_0^{(3)}$, respectively [213]. Based on the stress-strain relation $f_\sigma(\varepsilon)$, the values of flow stresses $\sigma_0^{(1)}$, $\sigma_0^{(2)}$ and $\sigma_0^{(3)}$ are derived by using

$$\sigma_0^N = \frac{1}{\varepsilon_f} \int_0^{\varepsilon_f} f_\sigma(\varepsilon) d\varepsilon \tag{2.94}$$

---

[3]Note that, in studies by Wierzbicki and Abramowicz [216] and Abramowicz and Jones [5] the strain energy associated with the rotation about a horizontal hinge is assumed to be twice that of $\dot{E}_2$, $2\dot{E}_2$.

for extensional deformation, and by using

$$\sigma_0^M = \frac{2}{\varepsilon_f^2} \int_0^{\varepsilon_f} f_\sigma(\varepsilon)\varepsilon d\varepsilon \tag{2.95}$$

for bending deformation. Here, ε_f is the maximum strain in each term of deformation. White et al. [213] proposed an expression for the maximum strain $\varepsilon_f|_1$ due to the extensional deformation in the toroidal zone corresponding to $\sigma_0^{(1)}$,

$$\varepsilon_f|_1 = \frac{t}{2b} \tag{2.96}$$

an expression for the maximum strain $\varepsilon_f|_2$ due to the bending deformation around the horizontal hinge corresponding to $\sigma_0^{(2)}$,

$$\varepsilon_f|_2 = 0.93\frac{t}{H} \tag{2.97}$$

and an expression for the maximum strain $\varepsilon_f|_3$ due to the bending deformation around the inclined travelling hinge corresponding to $\sigma_0^{(3)}$,

$$\varepsilon_f|_3 = \frac{t}{2b} \tag{2.98}$$

Meanwhile, in a theoretical analysis of the average compressive force of honeycomb structures, Wierzbicki [214] assumed $\sigma_0^{(1)} = \sigma_0^{(2)} = \sigma_0^{(3)} = \sigma_0$ based on the finding that E_1, E_2 and E_3 are equally weighed in the inextensional compressive deformation at the corner area, and derived σ_0 from the average strain ε_{ave} in the bending deformation around the inclined hinge by using $\sigma_0 = f_\sigma(\varepsilon_{ave})$. Here, the average strain ε_{ave} is taken to be

$$\varepsilon_{ave} = \frac{t}{4b} \tag{2.99}$$

2.4.2   Extensional folding mechanism

The characteristic feature of this deformation mode is that either the crests or valleys of the buckles appear on the side surfaces at the same position in the axial direction. Then, expansion or shrinkage strain appears in the circumferential direction. Therefore the deformation is called extensional. Fig. 2.35 shows the corner area model of extensional folding mechanism. It is seen from comparison of Fig. 2.35 with Fig. 2.4(a) that AB and CD in Fig. 2.35 are both the crests of the buckle, and A_1B_1, B_1C_1 and A_2B_2 are located at the adjacent valleys above and below in the axial direction. Here, for simplicity, it is assumed that the lengths of the line segments at the valley positions such as A_1B_1, B_1C_1, A_2B_2 and B_2C_2 are unchanged.

Similarly to the analysis of inextensional mode, the inclined angle α is defined in the corner area and is used as a parameter for describing the deformation process.

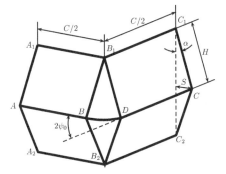

FIGURE 2.35
Geometry of extensional folding mechanism [5].

As suggested by the mechanical model discussed above, the energy needed for the deformation of the buckle in the extensional folding model can be evaluated through the sum of (1) the strain energy due to expansion in the circumferential direction, (2) the strain energy due to rotation about the horizontal hinge, and (3) the strain energy due to rotation between side planes. The strain energy components will be analyzed in the following. This idealized mechanism and the subsequent analysis is given by Abramowicz and Jones [5].

(1) Strain energy due to expansion in the circumferential direction

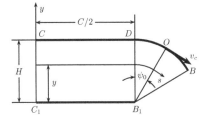

FIGURE 2.36
View of buckle development.

Fig. 2.36 shows a view of the development of a part of a buckle, $BDCC_1B_1$, shown in Fig. 2.35. Here, the point O is the midpoint of the circular arc BD. During deformation, with respect to the velocity distribution along the line segment B_1O shown in the figure, Abramowicz and Jones [5] assume that the velocity is zero at the point B_1, is v_c at the point O, and linearly increases from the point B_1 to the point O in accordance with

$$v = \frac{v_c ys}{H(C/2 + \psi_0 y)} \tag{2.100}$$

using the (y, s) coordinate in the figure where

$$\psi_0 = \frac{\pi}{n}$$

As also shown in Fig. 2.36, the velocity in the radius direction at both D and O is given by $\dot{S} = H \cos \alpha \dot{\alpha}$, and so v_c is calculated using

$$v_c = \dot{S} \psi_0 = H \psi_0 \cos \alpha \dot{\alpha} \tag{2.101}$$

The strain rate $\dot{\varepsilon}$ is evaluated from Eq. (2.100) by

$$\dot{\varepsilon} = \frac{v_c y}{H(C/2 + \psi_0 y)} \tag{2.102}$$

Thus, by considering that the buckle shown in Fig. 2.35 corresponds to the four in Fig. 2.36, the incremental energy $\dot{E}_4$ due to expansion is obtained by integrating over the area of the buckle as

$$\dot{E}_4 = 4 \int_0^H \sigma_s t \dot{\varepsilon}(C/2 + \psi_0 y) dy = 2\sigma_s t v_c H = 8 M_0 \frac{H^2}{t} \psi_0 \cos \alpha \dot{\alpha} \tag{2.103}$$

Further, the energy E_4 until the buckle is completely crushed is given by

$$E_4 = \int_0^{\pi/2} 8 M_0 \frac{H^2}{t} \psi_0 \cos \alpha d\alpha = 8 M_0 \frac{\pi H^2}{nt} \tag{2.104}$$

(2) Strain energy due to rotation about horizontal hinges

For an increment $d\alpha$ of the inclined angle α, the rotation angle around the middle horizontal hinge is $2d\alpha$, and the rotation angle is $d\alpha$ either around the upper or lower horizontal hinge. Therefore, the incremental bending energy $\dot{E}_5$ due to the rotation about the horizontal hinge is given by

$$\dot{E}_5 = (2C \times \dot{\alpha} + C \times 2\dot{\alpha}) \times M_0 = 4C M_0 \dot{\alpha} \tag{2.105}$$

and E_5 is given by

$$E_5 = \int_0^{\pi/2} 4C M_0 d\alpha = 2\pi C M_0 \tag{2.106}$$

(3) Strain energy due to rotation between adjacent side plates

In the initial configuration, the angle between two adjacent side plates is $\pi - 2\psi_0$, as shown in Fig. 2.35. When the deformation progresses this angle diminishes and is zero in the final configuration. The accompanying strain energy E_6 until the buckle is crushed is given by

$$E_6 = \int_{2\pi/n}^0 2 M_0 H(-d\phi) = \frac{4\pi M_0 H}{n} \tag{2.107}$$

Summing up the energy components E_4, E_5 and E_6, as obtained in the aforementioned terms (1), (2) and (3), for the extensional folding mechanism, the strain energy E_{ex} needed for crushing the corner area with external angle $2\psi_0$ is evaluated by

$$E_{ex} = M_0 \left(8\frac{H^2}{t}\psi_0 + 2\pi C + 4H\psi_0 \right) \tag{2.108}$$

2.4.3 Average crushing force

The average crushing force in the axial crushing of various polygonal tubes can be obtained by appropriately combining Eqs. (2.93) and (2.108).

2.4.3.1 Square, regular hexagonal and regular octagonal tubes

Based on Eq. (2.93), the **average crushing force** of the inextensional mode of a regular polygonal tube with side number being n is expressed as

$$\frac{P_{ave}}{nM_0} = A_1\frac{b}{t} + A_2\frac{C}{H} + A_3\frac{H}{b} \tag{2.109}$$

where

$$A_1 = 8I_1(\psi_0), \quad A_2 = \pi, \quad A_3 = 2I_3(\psi_0) \tag{2.110}$$

In Eq. (2.109), H and b are unknown parameters, which can be obtained from the following condition that minimizes P_{ave}:

$$\frac{\partial P_{ave}}{\partial H} = 0, \quad \frac{\partial P_{ave}}{\partial b} = 0 \tag{2.111}$$

Then the parameters are derived as follows:

$$H = \sqrt[3]{\frac{A_2^2}{A_3 A_1}C^2 t}, \quad b = \sqrt[3]{\frac{A_2 A_3}{A_1^2}C t^2} \tag{2.112}$$

By substituting Eq. (2.112) into Eq. (2.109), the average crushing force is obtained:

$$\frac{P_{ave}}{nM_0} = 3\sqrt[3]{A_1 A_2 A_3}\sqrt[3]{\frac{C}{t}} \tag{2.113}$$

Note that, in the total deformation strain energy, the fractional energy components are the same between the energy dissipated in the toroidal zone, the energy dissipated in the horizontal hinge and the energy dissipated in the inclined travelling hinge.

Further, substituting Eq. (2.110) into Eq. (2.113), the average crushing force can be rewritten as

$$P_{ave} = \frac{3n}{2}\sqrt[3]{2\pi I_1(\psi_0)I_3(\psi_0)} \times \sigma_s t^{5/3}C^{1/3} \tag{2.114}$$

indicating that the average crushing force is proportional to the 5/3 power of the wall thickness ($t^{5/3}$) of a tube. If the energy needed in the crushing process is bending energy only, the average crushing force is proportional to the square of wall thickness (t^2), and if extensional energy only, it is linearly proportional to wall thickness (t). Therefore, from Eq. (2.114), the ratio of the bending energy and the extensional energy needed is 2 : 1 in the axial crushing of a tube.

Eq. (2.114) shows that the average crushing stress σ_{ave} ($= P_{ave}/ntC$) is proportional to the 2/3 power of t/C.

$$\sigma_{ave} = \frac{3}{2}\sqrt[3]{2\pi I_1(\psi_0) I_3(\psi_0)} \times \sigma_s \left(\frac{t}{C}\right)^{2/3} \tag{2.115}$$

The average force of a tube is sometimes expressed by using **structural effectiveness** η ($= \sigma_{ave}/\sigma_s$) and **relative density**, namely, the **solidity ratio** ϕ (defined as the ratio of the actual cross section A of the tube and the area A_s occupied by the tube: $\phi = A/A_s$, $A = ntC$, $A_s = nC^2/(4\tan\psi_0)$). By using η and ϕ, the average crushing force of square, hexagonal and octagonal tubes in the inextensional mode can be expressed, respectively, as

$$\eta = 0.958\phi^{2/3} \qquad \text{(for square tube)} \tag{2.116}$$

$$\eta = 2.14\phi^{2/3} \qquad \text{(for hexagonal tube)} \tag{2.117}$$

$$\eta = 3.59\phi^{2/3} \qquad \text{(for octagonal tube)} \tag{2.118}$$

The formulas for the average crushing force of square, regular hexagonal and regular octagonal tubes in the inextensional mode obtained from the above analysis are summarized in Table 2.2.

TABLE 2.2
Evaluation formulas for the average crushing force of square, regular hexagonal and regular octagonal tubes in the inextensional mode.

	Square tubes	Hexagonal tubes	Octagonal tubes
H	$0.976\sqrt[3]{C^2 t}$	$0.629\sqrt[3]{C^2 t}$	$0.469\sqrt[3]{C^2 t}$
$\dfrac{P_{ave}}{M_0}$	$38.62(C/t)^{1/3}$	$89.92(C/t)^{1/3}$	$160.6(C/t)^{1/3}$
η	$0.958\phi^{2/3}$	$2.14\phi^{2/3}$	$3.59\phi^{2/3}$

Similarly, based on Eq. (2.108), the **average crushing force** of a polygonal tube in the extensional mode can be obtained as follows:

$$\frac{P_{ave}}{nM_0} = A_4\frac{H}{t} + A_5\frac{C}{H} + A_6 \tag{2.119}$$

where

$$A_4 = \frac{4\pi}{n}, \quad A_5 = \pi, \quad A_6 = \frac{2\pi}{n} \tag{2.120}$$

In Eq. (2.119), the unknown parameter H can be obtained by assuming

$$\frac{\partial P_{ave}}{\partial H} = 0 \tag{2.121}$$

so that P_{ave} is minimized:

$$H = \sqrt{\frac{A_5}{A_4}Ct} = \sqrt{\frac{n}{4}Ct} \tag{2.122}$$

Substituting Eq. (2.122) into Eq. (2.119) yields

$$\frac{P_{ave}}{nM_0} = 2\sqrt{A_4 A_5}\sqrt{\frac{C}{t}} + A_6 = \frac{4\pi}{\sqrt{n}}\sqrt{\frac{C}{t}} + \frac{2\pi}{n} \tag{2.123}$$

Therefore, using η and ϕ, the average crushing force of square, regular hexagonal and regular octagonal tubes in the extensional mode can be given by the following formulas:

$$\eta = 0.098\phi + 0.785\sqrt{\phi} \qquad \text{(for square tube)} \tag{2.124}$$

$$\eta = 0.114\phi + 0.844\sqrt{\phi} \qquad \text{(for hexagonal tube)} \tag{2.125}$$

$$\eta = 0.119\phi + 0.863\sqrt{\phi} \qquad \text{(for octagonal tube)} \tag{2.126}$$

Table 2.3 summarizes the formulas for the average crushing force of square, regular hexagonal and regular octagonal tubes in the extensional mode obtained in the above analysis.

TABLE 2.3
Formulas for the average crushing force of square, regular hexagonal and regular octagonal tubes in the extensional mode.

	Square tubes	Hexagonal tubes	Octagonal tubes
H	$\sqrt{Ct}$	$\sqrt{6Ct}/2$	$\sqrt{2Ct}$
$\dfrac{P_{ave}}{M_0}$	$8\pi(C/t)^{1/2} + 2\pi$	$4\sqrt{6}\pi(C/t)^{1/2} + 2\pi$	$8\sqrt{2}\pi(C/t)^{1/2} + 2\pi$
η	$0.098\phi + 0.785\sqrt{\phi}$	$0.114\phi + 0.844\sqrt{\phi}$	$0.119\phi + 0.863\sqrt{\phi}$

With respect to the crushing deformation of polygonal tubes, generalizing the mechanisms of inextensional and extensional corner area crushing model and combining the deformation characteristics, Abramowicz and Wierzbicki [7] further proposed a mixed model of inextensional and extensional folding mechanism. In this generalized mixed folding mechanism, the deformations are divided into two phases. Just as in the inextensional and extensional corner area crushing models, the angle α, which changes from 0 to $\cong 90°$ as the crushing progresses, is chosen as a timelike parameter to represent the compressive process. In the mixed model, at the first phase where α changes from

0 to $\bar{\alpha}$ (switch angle determined by the geometry of the tube), the deformation is the same as in the inextensional folding mechanism; after that, at the second phase where α changes from $\bar{\alpha}$ to $\cong 90°$, the deformation is the same as in the extensional folding mechanism. Summarizing the two-phase deformation, the average crushing force is expressed as follows:

$$\frac{P_{ave}}{M_0} = \left(A_1 \frac{b}{t} + (A_2 + A_5) \frac{C}{H} + A_3 \frac{H}{b} + A_4 \frac{H}{t} + A_6 \right) \frac{2H}{\delta_e} \qquad (2.127)$$

where δ_e is the effective crushing distance of a single buckle in the axial compression of a polygonal tube.

To determine A_1, A_2 c in Eq. (2.127), first, it is necessary to extract the parameter $\bar{\alpha}$ to decide between the first and the second deformation phases. Although $\bar{\alpha}$ can be extracted by minimizing the crushing force, the calculation is very complex. Abramowicz and Wierzbicki [7] showed that in the case of a hexagonal tube, $\bar{\alpha}$ is about $6°$–$40°$ depending on thickness C/t, and in the case of a square tube with $C/t > 20$, the process of buckle formation is always inextensional and $\bar{\alpha} \cong 90°$ holds.

Based on the analytical results of the generalized mixed folding mechanism, Abramowicz and Wierzbicki [7] proposed an improved formula for the average force of a square tube,

$$\frac{P_{ave}}{M_0} = 48.64(C/t)^{0.37} \qquad (2.128)$$

or

$$\eta = 1.269\phi^{0.63} \qquad (2.129)$$

using the **structural effectiveness** η and the **solidity ratio** ϕ. They also proposed an improved formula for the average force of a regular hexagonal tube:

$$\frac{P_{ave}}{M_0} = 80.92(C/t)^{0.4} \qquad (2.130)$$

or

$$\eta = 2.041\phi^{0.6} \qquad (2.131)$$

Figs. 2.37 and 2.38 compare the average force of polygonal tubes obtained from the inextensional and extensional analytical models (Ex-mode and Inex-mode in the figure), the average force obtained by FEM numerical simulation, and the experimental results of other researchers [63, 132, 130, 148, 4, 5, 138, 182]. But considering that the **effective crushing distance** δ_e for creating the buckle is given by

$$\frac{\delta_e}{2H} \cong 0.73 \qquad (2.132)$$

for the inextensional mode and by

$$\frac{\delta_e}{2H} \cong 0.77 \qquad (2.133)$$

for the extensional mode, the average forces of inextensional and extensional

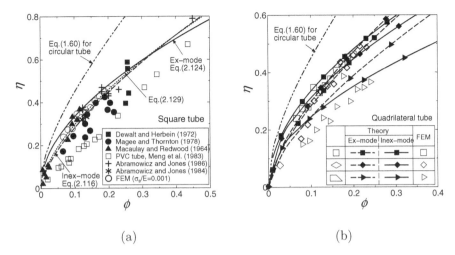

FIGURE 2.37
Average force in axial crushing of quadrangular tubes: (a) square cross section; (b) non-square cross section.

modes in the figure use the analytical values of the theoretical model divided by the coefficients 0.73 and 0.77, respectively.

For rhomboidal and trapezoidal tubes shown in Fig. 2.37(b) and for irregular pentagonal tubes shown in Fig. 2.38(a), the analytical values of corresponding models of inextensional and extensional modes are obtained by applying the theoretical methods in Sections 2.4.1 and 2.4.2 to the problem of axial crushing of arbitrary tubes with n-gonal cross sections, and are given by

$$\frac{P_{ave}}{M_0} = 3\sqrt[3]{B_1 B_2 B_3}\sqrt[3]{\frac{\bar{C}}{t}} \qquad (2.134)$$

for the inextensional mode where

$$
\begin{aligned}
B_1 &= 8\sum_{k=1}^{n} I_1(\psi_k) \\
B_2 &= \pi\sum_{k=1}^{n} \frac{(C_k + C_{k+1})/2}{\bar{C}} \\
B_3 &= 2\sum_{k=1}^{n} I_3(\psi_k)
\end{aligned} \qquad (2.135)
$$

For the extensional mode, the average compressional force is

$$\frac{P_{ave}}{M_0} = 2\sqrt{B_4 B_5}\sqrt{\frac{\bar{C}}{t}} + B_6 \qquad (2.136)$$

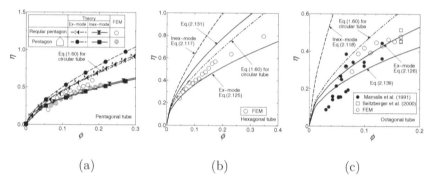

(a)   (b)   (c)

FIGURE 2.38
Average force in axial crushing of various polygonal tubes: (a) pentagonal
tube; (b) hexagonal tube; (c) octagonal tube.

where

$$B_4 = 4\sum_{k=1}^{n} \psi_k$$

$$B_5 = \pi\sum_{k=1}^{n} \frac{(C_k + C_{k+1})/2}{\bar{C}} \qquad (2.137)$$

$$B_6 = 2\sum_{k=1}^{n} \psi_k$$

In the above formulas, C_k is the length of the side facing the kth corner,
ψ_k is half the external angle for the kth corner (see Fig. 2.32(a)), and $\bar{C}$ is the
average length of the sides of the cross section of an n-gon:

$$\bar{C} = \frac{\sum_{k=1}^{n} C_k}{n} \qquad (2.138)$$

Also, Fig. 2.37(a) shows the values of average crushing force of square tubes
calculated by Eq. (2.129); Fig. 2.38(b) shows the values of average crushing
force of hexagonal tubes calculated by Eq. (2.131); and Fig. 2.38(c) shows the
values of average crushing force of octagonal tubes calculated by the following
formula proposed by Mamalis et al. [138]:

$$\eta = 0.952\sqrt{\phi} \qquad (2.139)$$

Figs. 2.37 and 2.38 indicate the following.

(1) As shown in Fig. 2.37(a), the theoretical analytical values of inextensional
and extensional modes for square tubes are approximately identical, and
roughly equal to the analytical values calculated by using Eq. (2.129) based

on the mixed model. The values roughly agree with the FEM numerical analysis, but are greater than the experimental results shown in the figure.

(2) As shown in Fig. 2.37(b), both the theoretical analysis and the FEM analysis give roughly the same crushing force for square and rhomboidal cross sections. On the other hand, for the trapezoidal (irregular quadrilateral) cross section, both the theoretical analysis and the FEM analysis give smaller values of crushing force compared with the square tube.

(3) As Fig. 2.38 shows, in the regular pentagonal, regular hexagonal and regular octagonal tubes, the average force obtained from the extensional mode formula agrees relatively well with the FEM analysis and the experimental results. The average force obtained from the inextensional mode formula is greater than that of extensional mode, and the difference increases with the number of sides in the cross section of the polygonal tube. However, in the non-equal sides of polygonal tubes, for example, the trapezoidal tube in Fig. 2.37(b) and the irregular pentagonal tube in Fig. 2.38(a), the average force obtained from the inextensional mode formula is lower than that from the extensional mode formula, and agrees relatively well with the FEM analysis.

(4) In Figs. 2.37 and 2.38, for comparison, the average force for the axisymmetric deformation of circular tubes extracted using the formula of Abramowicz and Jones [5], Eq. (1.60), is also shown. If the number n of tube sides is greater than 5, as also shown in Fig. 2.6, the deformation mode resembles that of the crushing a circular tube rather than the inextensional or extensional mode. Consequently, the results of FEM numerical analysis shown in Fig. 2.38(b) and (c) are between the results from Eq. (1.60) for the axisymmetric deformation of a circular tube and Eqs. (2.125) and (2.126) for a polygonal tube.

2.4.3.2 Honeycomb structures

Fig. 2.39 shows the honeycomb material with corrugated plate joints; the cross section of each cell of the honeycomb is a regular hexagon of length C where the thickness of the diagonal wall is t and the thickness of the bonded wall is twice that, $2t$. On the axial crushing of the honeycomb (in the height direction of the wall), from the cyclic nature of the problem, Wierzbicki [214] applied the inextensional folding mechanism, by the way shown in Fig. 2.40(b), to the deformation of the single unit shown in Fig. 2.40(a) (the part with length $C/2$ and height $2H$, which equals the half-wavelength of the compressed buckle, from the two diagonal walls and a bonded wall with common lines of intersection). That is, in axial compression, because cracks appear along the jointed surface in the bonded wall and extend as the deformation proceeds, the deformation of the unit can be approximated by the inextensional folding mechanism of the two corners with angle $120°$ composed of plates of thickness t. Therefore, the strain energy needed for the deformation is evaluated as twice

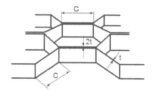

FIGURE 2.39
Honeycomb materials with corrugated plate joints.

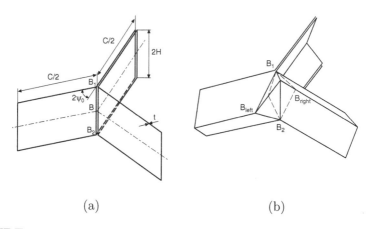

(a)   (b)

FIGURE 2.40
Deformation mechanism of honeycomb structures proposed by Wierzbicki
[214]: (a) one unit; (b) deformation mechanism.

the value of Eq. (2.93). However, considering the rotation before cracks appear
in the bonded wall, the strain energy E_2 needed for the rotation about the
horizontal hinge is given by

$$
\begin{aligned}
E_2 &= 2 \times \left[4 \left(\frac{\sigma_s t^2}{4} \right) \frac{C}{2} + 2 \left(\frac{\sigma_s (2t)^2}{4} \right) \times \frac{C}{2} \right] \times \frac{\pi}{2} \\
&= 6\pi M_0 C
\end{aligned}
\tag{2.140}
$$

As a result, on the single unit shown in Fig. 2.40(a), the strain energy
E_{1unit} needed for crushing is given by

$$
E_{1unit} = M_0 \left(2 \times 16 \frac{bH}{t} I_1(\pi/6) + 6\pi C + 2 \times 4 \frac{H^2}{b} I_3(\pi/6) \right)
\tag{2.141}
$$

Thus, the average crushing force is given by

$$
\frac{P_{ave}}{M_0} = \left(16 \times 1.05 \frac{b}{t} + 3\pi \frac{C}{H} + 4 \times 2.39 \frac{H}{b} \right)
\tag{2.142}
$$

The requirement of minimizing P_{ave} gives

$$H = 0.821\sqrt[3]{C^2 t}\,, \quad b = 0.683\sqrt[3]{Ct^2} \tag{2.143}$$

and therefore

$$\frac{P_{ave}}{M_0} = 34.45\sqrt[3]{\frac{C}{t}} \tag{2.144}$$

Here, by recalling that the structural effectiveness η and the solidity ratio ϕ are given by

$$\eta = \frac{P_{ave}}{2tC\sigma_0}\,, \quad \phi = \frac{8}{3\sqrt{3}}\left(\frac{t}{c}\right) \tag{2.145}$$

$$\eta = 3.23\phi^{2/3} \tag{2.146}$$

is obtained from Eq. (2.144).

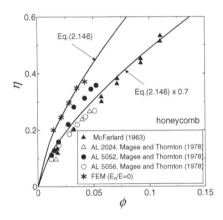

FIGURE 2.41
Average crushing force of honeycomb structure under axial compression.

With respect to the average crushing force of the honeycomb structure, Fig. 2.41 compares the average crushing force obtained from the theoretical formula, Eq. (2.146), with the experimental results [147, 132] and the FEM numerical analyses. As the figure shows, the calculation using Eq. (2.146) agrees well with the FEM numerical analysis. On the other hand, Eq. (2.146) gives similar but larger results compared with the experiment. This relates to the strain hardening of the material used in the experiment. That is, in the experiment the structural effectiveness η is calculated by using the material tensile strength σ_u ($\eta = \sigma_{ave}/\sigma_u$); however, in the actual deformation of the material after yield, as Wierzbicki [214] pointed out, the average flow stress σ_0 involved in the crushing of the honeycomb structure should be roughly close to $\sigma_0 = 0.7\sigma_u$. To take this into consideration, it is reasonable to apply a correction factor of 0.7 to Eq. (2.146). As shown in the figure, by multiplying

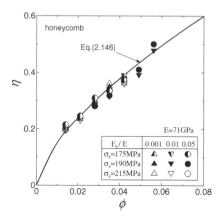

FIGURE 2.42
Average crushing force of the honeycomb structure considering strain hardening.

the correction factor 0.7, the prediction agrees well with the experimental data.

Fig. 2.42 compares the average force obtained from FEM numerical analysis and its counterpart obtained from the theoretical formula, Eq. (2.146), for the materials obeying the bilinear hardening rule with various strain hardening coefficients. For the values obtained from FEM numerical analyses, the flow stress σ_0 is used to derive $\eta = \sigma_{ave}/\sigma_0$. Note that the flow stress σ_0 is calculated from the stress-strain relation of the material ($\sigma_0 = f_\sigma(\varepsilon_{ave})$) based on the method of Wierzbicki [214], using the average strain ε_{ave} calculated by Eq. (2.99). As the figure shows, the influence of strain hardening of the material can be evaluated accurately by using the method of Wierzbicki [214].

Note that in the numerical analyses of FEM shown in Fig. 2.41 and 2.42 the plate of thickness $2t$ in the three-fold point corner does not split into two plates during axial collapse, while their results are in good agreement with Eq. (2.146) based on Wierzbicki's model. In Wierzbicki's model [214], as shown in Fig. 2.40, it is assumed the honeycomb is manufactured by gluing together two plates of thickness t, and in the crushing process the glued plates of thickness $2t$ split into two plates having thickness t, and so the three-fold point corner can be analyzed as two two-fold point corners. Factually, the three-fold point corners in a thin-walled member do not necessarily become two-fold point corners by splitting. For example, in the case of a three-fold point corner consisting of single plates manufactured by a method different from gluing, such as powder processing, single plates cannot split into two plates. Therefore, here the deformation mechanism for the three-fold point corner that does not split during axial collapse is investigated using FEM simulation

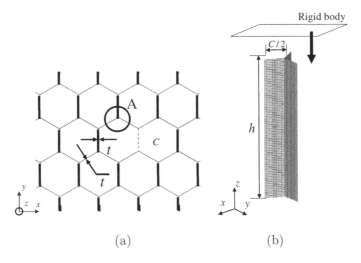

(a)   (b)

FIGURE 2.43
Three-fold point corner [54]: (a) honeycomb; (b) one corner model.

analysis [54]. In numerical analysis the one-corner model was adopted, as shown in Fig. 2.43(b), which corresponds to circle A in Fig. 2.43(a).

Fig. 2.44 shows the axial deformed shapes of plate-1 with thickness t, plate-2 with thickness t, and another plate with thickness $2t$ from directions a, b and c, respectively, as shown in the upper figure, for a honeycomb with $t' = 2t$ at 10% compression. The "S" shaped line in the figures corresponds to the edges of the cell walls constituting the three-fold point corner, i.e., the deformed shapes of the centerlines of each cell wall. The shapes of the folds are different from each other as shown in Fig. 2.45, which is obtained by overlapping these shapes.

Fig. 2.46 shows cross-sectional views of this three-fold point corner during axial compression processes A, B and C (marked on the load-displacement curve shown in the left-hand side of the figure). The axial positions selected are cross sections (a) I-I, (b) II-II and (c) III-III in Fig. 2.45. The thin lines in Fig. 2.46 indicate the reference position of the cell walls before axial compression, and the dashed lines on deformation stage C in Fig. 2.46 show the deformed cross sections based on the theoretical model proposed by Wierzbicki [214] shown in Fig. 2.40. Wierzbicki's model assumes that the glued plate having thickness $2t$ splits into two plates having thickness t around the corner part, and two traveling hinges, shown as points S, T in Fig. 2.46(a) and (c), then occur. In Wierzbicki's model, two traveling hinges accompany the same generated folds in the three cell walls. One of the traveling hinges exists in the plate with thickness t, and the other one exists in the plate produced by splitting of the plate with thickness $2t$. The phase of these two hinges is the same. However, for the three-fold point corner analyzed here, the splitting of

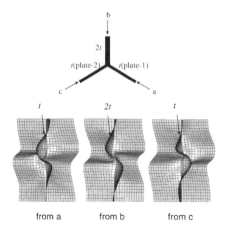

FIGURE 2.44
Axial deformed shapes of folds from direction a, b and c [54].

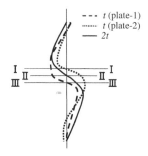

FIGURE 2.45
Comparison of axial folds shape of three plates constituting three-fold point corner [54].

the plate is not considered. Thus, two traveling hinges are generated in two plates of thickness t, respectively, but the phase of these two hinges is different. As shown in cross section I-I of Fig. 2.45, the amplitude of the fold of plate-1 becomes the maximum, and the amplitude of the fold of the other plate with thickness t (plate-2) becomes almost zero. In contrast, the amplitudes of the folds in cross section III-III are just the opposite; that is, plate-2 shows the maximum amplitude of the fold. Therefore, the estimation of the average load proposed by Wierzbicki [214] is applicable to the present three-fold point corner, in which the plate with thickness $2t$ does not split, because for the deformation of buckle the energy dissipated in the toroidal zone E_1 and the energy dissipated in the inclined travelling hinges E_3 are the same as those in Wierzbicki's model, only the energy dissipated in the horizontal hinges E_2

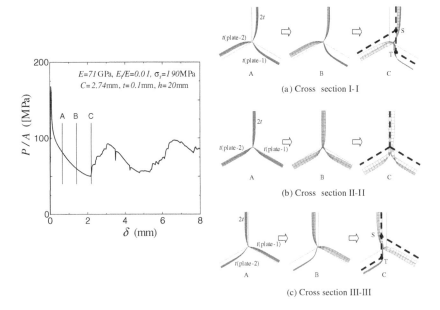

(a) Cross section I-I

(b) Cross section II-II

(c) Cross section III-III

FIGURE 2.46
Deformed cross section of one corner model [54].

needs to re-calculate. That is, instead of Eq. (2.140) the strain energy E_2 due to rotation around the horizontal hinge is given by

$$E_2 = \pi C (2M_0 + M_0') \qquad (2.147)$$

where $M_0 = \sigma_0 t^2/4$, and $M_0' = \sigma_0 (t')^2/4$ corresponding to plate of thickness t'. Here, t' is greater then t, but does not necessarily equal $2t$.

As a result, the average crushing force of Eq. (2.142) becomes

$$\frac{P_{ave}}{M_0} = \left(16 \times 1.05 \frac{b}{t} + \kappa_c \pi \frac{C}{H} + 4 \times 2.39 \frac{H}{b} \right) \qquad (2.148)$$

where

$$\kappa_c = 1 + \frac{1}{2} \left(\frac{t'}{t} \right)^2 \qquad (2.149)$$

And finally, the average load P_{ave} can be calculated by

$$\frac{P_{ave}}{M_0} = 23.88 \sqrt[3]{\kappa_c} \sqrt[3]{\frac{C}{t}} \qquad (2.150)$$

Note that Eq. (2.150) results in Eq. (2.144) of Wierzbicki's analysis in the case of $t'/t = 2$, whereas it can be also applied to the case of $t'/t \neq 2$.

Fig. 2.47 compares the values for the average load obtained by FEM with the predicted values obtained from Eq. (2.150) for honeycomb with $t'/t = 1.5$. It is seen from this figure that the P_{ave} values obtained from Eq. (2.150) are in reasonably good agreement with those obtained by FEM. Therefore it is reasonable that the average load obtained from FEM shown in Fig. 2.41 and 2.42 is in good agreement with Eq. (2.146) based on Wierzbicki's model, although in the FEM analysis the plate of thickness $2t$ does not split into two plates during axial collapse.

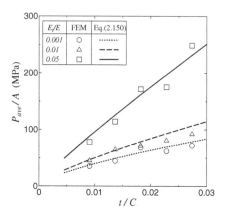

FIGURE 2.47
Comparison of average load obtained from Eq. (2.150) and FEM for honeycomb with $t'/t = 1.5$ [54].

2.4.3.3   Top-hat and double-hat square tubes

Fig. 2.48 shows the cross-sectional view of a top-hat tube and a double-hat tube. In the cross section of the top-hat tube, some corners are shared by three plates. Based on the fact that the deformation of a three-fold corner can be expressed by the deformation of two inextensional folding corners similarly to the above-mentioned analysis of the honeycomb structure, the average crushing force is expressed as [213]

$$P_{ave} = \frac{t^2}{4} \left\{ 4 \times \sigma_0^{(1)} 8 I_1 \frac{b}{t} + \sigma_0^{(2)} \pi \frac{L}{H} + 4 \times 2\sigma_0^{(3)} I_3 \frac{H}{b} \right\} \frac{2H}{\delta_e} \qquad (2.151)$$

Here, the coefficient 4 in the first and third terms in the curly bracket $\{\cdots\}$ corresponds to the number of corners, L in the second term corresponds to

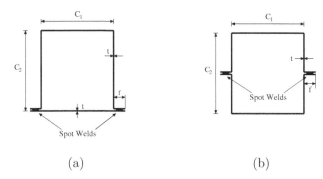

FIGURE 2.48
Top-hat and double-hat square tubes: (a) cross section of a top-hat tube; (b) cross section of a double-hat tube.

the length of the horizontal hinge where $L = 2C_1 + 2C_2 + 4f$,[4] and $\sigma_0^{(i)}$ is the energy equivalent flow stress in the ith region of plastic flow.

With respect to the material without strain hardening, by assuming $\sigma_0^{(1)} = \sigma_0^{(2)} = \sigma_0^{(3)} = \sigma_s$,

$$\frac{H}{t} = 0.387 \left(\frac{L}{t}\right)^{2/3} \quad , \quad \frac{b}{t} = 0.439 \left(\frac{L}{t}\right)^{1/3} \tag{2.152}$$

is obtained from Eq. (2.151), and therefore the average force is given by

$$\frac{P_{ave}}{M_0} = 33.33 \left(\frac{L}{t}\right)^{1/3} \tag{2.153}$$

Similarly, the average crushing force of the double-hat square tube is given by

$$P_{ave} = \frac{t^2}{4} \left\{ 8 \times \sigma_0^{(1)} 8 I_1 \frac{b}{t} + \sigma_0^{(2)} \pi \frac{L}{H} + 8 \times 2\sigma_0^{(3)} I_3 \frac{H}{b} \right\} \frac{2H}{\delta_e} \tag{2.154}$$

where $L = 2C_1 + 2C_2 + 4f$.

Further, if the material is perfectly elasto-plastic,

$$\frac{H}{t} = 0.244 \left(\frac{L}{t}\right)^{2/3} \quad , \quad \frac{b}{t} = 0.348 \left(\frac{L}{t}\right)^{1/3} \tag{2.155}$$

is derived from Eq. (2.154) and the average force is given by

$$\frac{P_{ave}}{M_0} = 52.91 \left(\frac{L}{t}\right)^{1/3} \tag{2.156}$$

---

[4]The bending around the horizontal hinge of thickness $2t$ at the flange is approximated by the bending of two plates of thickness t because the length of the flange f is usually small.

2.5 Influence of partition walls on the axial crushing of polygonal tubes

To improve the energy-absorbing capacity of thin-walled polygonal tubes widely used as collision energy absorbers, partition walls are often placed inside the polygonal tubes (e.g., [222, 68]). In this section, the influence of partition walls on the axial crushing of polygonal tubes is discussed, based on the results of elasto-plastic FEM analysis by Chen et al. [45, 46] on the axial crushing of thin-walled polygonal tubes with partition walls. Here, the thicknesses of the outer wall and the partition wall are assumed to be t and t', respectively.

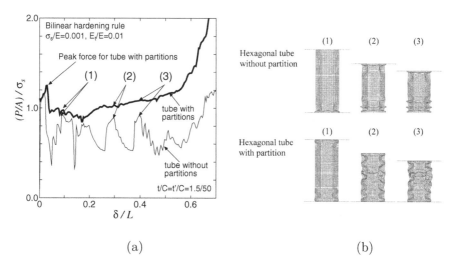

(a) (b)

FIGURE 2.49
Axial crushing deformation of regular hexagonal tubes with and without partition walls [45]: (a) crushing force-displacement relation; (b) deformation behavior under axial compression.

Fig. 2.49(a) shows the relation between the crushing force P/A (A is actual cross-sectional area of the tube) and the axial crushing distance δ in the axial crushing of regular hexagonal tubes with and without partition walls, from which it is seen that the crushing force-displacement curve behaves quite differently with and without partition walls. Without partition walls, the crushing force increases and decreases repeatedly with a certain period just as in circular tubes; with partition walls, the crushing force increases without fluctuating greatly, passing over the peaks in the former curve. The difference in the force-displacement curve is attributed to the difference in the crushing deformation mode between the two. Fig. 2.49(b) shows the deformation behavior of the tubes with and without partition walls corresponding to the symbols (1), (2) and (3) in Fig. 2.49(a). As shown in Fig. 2.49(b), the process of buckle formation is clearly different. In the tube without partition walls, it is found that the progressive crushing mode where the buckles are formed in sequence one after another is just as in circular tubes; in the tube with partition walls, it is found that the buckles are first formed over the entire length in the axial direction of the tube and then all of them are collapsed and crushed at the same time in the simultaneous crushing mode.

In the axial crushing of a polygonal tube with partition walls, the compressive stress locally decreases in the outer wall during buckle formation; on the other hand, buckles are hardly formed in the central area where the partition walls intersect with each other, and so a greater crushing force is required to keep the compressive deformation going. Therefore, in the axial crushing of a polygonal tube with partition walls, the sequential process, in which another buckle appears after a buckle is crushed, does not occur; rather, the buckles are created over the entire length in the axial direction and then simultaneously crushed, and the crushing stress does not appreciably fluctuate after the peak force.

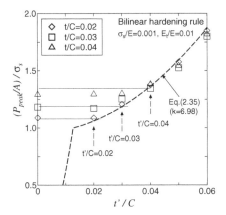

FIGURE 2.50
Peak force in the axial crushing deformation of regular hexagonal tubes with partition walls [46].

Fig. 2.50 shows the peak force as a function of partition wall thickness t'/C for the regular hexagonal tube with partition walls and with three values of outer wall thickness $t/C = 0.02$, 0.03 and 0.04. As the figure shows, the peak force increases with thickness t'/C; however, in the range $t' < t$, the influence of partition wall thickness is weak, and the peak force depends mainly on the outer wall thickness t/C. On the other hand, in the range $t' > t$, the peak force is independent of the outer wall thickness t/C and determined mainly by the partition wall thickness t'/C. Further, the broken lines in Fig. 2.50 show the buckling force derived using Eq. (2.35) for the problem where the plate, which has the same size as the plate composing the regular hexagonal tube, is fixed at two sides and a uniform compressive stress is applied across the top and bottom sides. As the figure shows, the peak force of a hexagonal tube in the ranges $t' < t$ and $t' > t$ is roughly equal to the buckling force of the plate of the outer wall and the partition wall, respectively.

The following is found from the deformation behavior of the respective wall plates before and after buckling of the polygonal tube [45]. That is, in the axial compressive deformation of a polygonal tube with partition walls, the buckling starts from the outer wall. If the partition wall is thin, although the partition walls also buckle after the outer walls buckle, the buckling force of the outer wall becomes the peak force because the in-plane compressive force of the partition wall is lower than that of the outer wall. On the other hand, if the partition wall is thick, the compressive force corresponding to the buckling of the partition wall is higher than that of the outer wall, and the buckling force of the partition wall becomes the peak force.

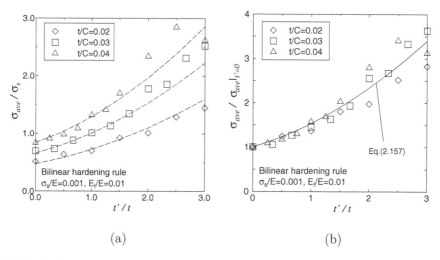

(a)   (b)

FIGURE 2.51
Average stress in the axial crushing deformation of regular hexagonal tubes with partition walls [46]: (a) average stress; (b) ratio of average stress of a regular hexagonal tube with partition walls to that of a hexagonal tube without partition walls.

Fig. 2.51(a) shows the average crushing stress σ_{ave} ($\sigma_{ave} = P_{ave}/A$) as a function of partition wall thickness t'/C for regular hexagonal tubes with partition walls and three outer wall thicknesses $t/C = 0.02$, 0.03 and 0.04. As the figure shows, the average crushing stress is related to both the outer walls and partition walls, increasing with the thickness of each; nonetheless, the influence is greater in the partition wall than in the outer wall. Fig. 2.51(b) re-plots the average crushing stress in Fig. 2.51(a), where the vertical axis shows the ratio of the average crushing stress σ_{ave} against that of a hexagonal tube without partition walls $\sigma_{ave}\big|_{t'=0}$. As shown in the figure, as the thickness ratio between partition and outer walls increases, the ratio of average crushing stress also increases between the regular hexagonal tubes with and without partition walls. Further, although the value also depends on the outer wall thickness, all calculated points within the thickness range examined fall in a narrow band, and the ratio of average stress can be approximately evaluated as a function of the thickness ratio of partition and outer walls as the following equation shows:

$$\frac{\sigma_{ave}}{\sigma_{ave}\big|_{t'=0}} \cong 1 + 0.41 \left(\frac{t'}{t}\right) + 0.13 \left(\frac{t'}{t}\right)^2 \tag{2.157}$$

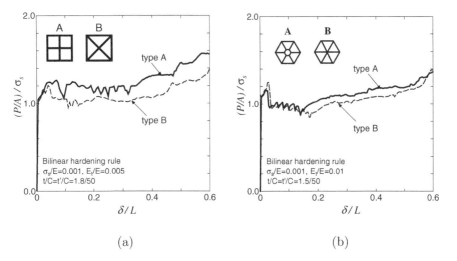

(a)   (b)

FIGURE 2.52
Role of partition walls in the axial compression deformation of polygonal tubes [45]: (a) square tubes; (b) hexagonal tubes.

The increase of crushing force by adding partition walls mainly relates to the compressive deformation behavior of both the three-fold point corner where the partition and outer walls intersect and the multi-fold point corner where the partition walls intersect. At these intersections, in the axial compressive deformation of the polygonal tubes, the crushing force locally increases even after the buckles are created in the wall plates. Thus, the vibration of the crushing force is suppressed, and the average crushing force increases. Therefore, having many intersections could be an effective method for improving the collision energy absorption efficiency of thin-walled materials. Fig. 2.52 generally verifies this. Fig. 2.52(a) compares the crushing stress in the axial compression of two types of square tubes with partition walls. As the figure shows, the crushing stress is higher in type A, which has more intersections than type B. Further, Fig. 2.52(b) shows the crushing stress in the axial compression where a small circular tube is placed at the center of a hexagonal tube with partition walls in order to increase the cross-sectional area of the central part. The figure indicates that the crushing stress increases by increasing the cross-sectional area of the central part where crushing stress is locally high.

3

Axial compression of corrugated tubes

The geometry of tubes is an important factor in their energy absorption capacity. Many studies have examined the axial crushing of tubes with various section configurations in order to improve the performance of energy absorption devices. Farley [71] investigated the effect of tube geometry on energy absorption capacity. Sigalas et al. [183] investigated the role played by the chamfer of tubes. Mahdi et al. [133, 134] experimentally studied the behavior of cone-cylinder-cone tubes. Many studies have shown that one way in particular to create a high-performance energy absorption device is to introduce corrugations along the axis or circumference of a tube (e.g., [185, 67, 1]). This chapter discusses how introducing corrugations affects the force-displacement and energy absorption characteristics of tubes. In Section 3.1, axially grooved tubes are investigated to provide a basis for comparison. Then, axial corrugations and circumferential corrugations will be discussed in Sections 3.2 and 3.3, respectively.

3.1 Axially grooved circular tubes

If a circular tube with a uniform cross section is axially crushed, a large initial peak force occurs due to buckling, and fluctuation occurs with a force smaller than the initial peak force due to the folding of local buckles. A method for suppressing such a high initial peak force and the subsequent force fluctuation is to introduce grooves on the tube surface, about which many studies have been reported (e.g., [139, 140, 61, 94]). Here, based on a finite-element method (FEM) numerical analysis of the axial crushing deformation of a thin-walled tube in which inner and outer surfaces are periodically grooved [209] as shown in Fig. 3.1, the effects of grooves on deformation mode and compressive force are discussed. Note that the materials used in the numerical analysis are assumed to obey the bilinear hardening rule.

The deformation modes of the tubes that exhibit axisymmetric deformation are roughly classified into the following two categories when they have grooves:

- Deformation modes where the fold half-wavelength λ is equal to the grooves interval λ_g ($\lambda = \lambda_g$).

157

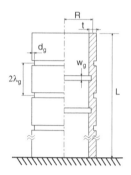

FIGURE 3.1
Geometry of circular tube with a grooved surface.

- Deformation modes where the fold half-wavelength λ is not equal to the grooves interval λ_g ($\lambda \neq \lambda_g$).

The circular tubes of the first type ($\lambda = \lambda_g$) show two deformation modes: the **progressive crushing mode** (called the P-mode) where the folds appearing in the crushing process are continuously folded just as in a tube without grooves, and the **simultaneous crushing mode** (called the S-mode) where all folds are simultaneously folded. The S-mode is not observed in circular tubes without grooves. Fig. 3.2 shows the S-mode deformation behavior of a grooved wall during crushing.

Fig. 3.3 shows a chart for classifying the deformation modes of a circular tube through the two parameters: groove interval λ_g and groove depth d_g. In the mode chart, the vertical axis shows the dimensionless groove interval obtained by normalizing λ_g by the fold half-wavelength λ_0 in a similar tube without grooves. The symbols $\bigcirc$ and $\triangle$ indicate the deformation mode '$\lambda \neq \lambda_g$' ('$\lambda < \lambda_g$' and '$\lambda > \lambda_g$', respectively), and the symbols $\blacksquare$ and $\blacktriangle$ indicate the P- and S-mode deformations, respectively. As the figure shows, λ_g must be close to λ_0 to have the deformation mode '$\lambda = \lambda_g$' such that the folds are folded along the grooves; and the tolerance for the groove interval λ_g becomes wider as the grooves become deeper.

Fig. 3.4 shows the relationship between crushing force P and displacement δ in the four representative circular tubes that exhibit S- and P-mode deformations ($\lambda = \lambda_g$), '$\lambda < \lambda_g$' deformation and '$\lambda > \lambda_g$' deformation. For comparison, the figure also shows the result for a similar circular tube without grooves ($d_g = 0$) using the symbol $\triangle$. The figure shows that force fluctuation does not appear and the force increases monotonically in the tube exhibiting S-mode deformation, and that fluctuation in crushing force P appears as deformation proceeds in all other tubes. In addition, if the groove interval λ_g and fold half-wavelength λ are not in agreement (curves '$\lambda_g/L = 6/108, d_g/t = 0.125/2$' and '$\lambda_g/L = 21/84, d_g/t = 0.375/2$' in the figure), the crushing force of the

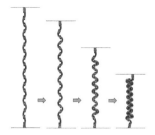

FIGURE 3.2
S-mode deformation behavior of a wall during crushing of a circular tube with grooves [209].

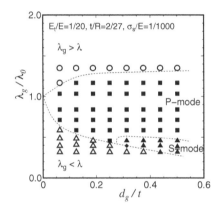

FIGURE 3.3
Classification chart of deformation mode [209].

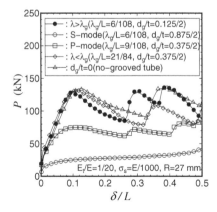

FIGURE 3.4
Crushing forces for various different grooved circular tubes [209].

tube is roughly equal to that of a tube without grooves ($d_g = 0$). If $\lambda = \lambda_g$, however, both the force fluctuation and the crushing force decrease. In particular, in the S-mode, the crushing force greatly decreases compared with the case of a similar tube without grooves. Therefore, peak force and crushing force fluctuation can be reduced by adding grooves to a tube, but the drawback is the decrease of crushing force.

3.2   Axially corrugated tubes

As discussed in the preceding section, during the crushing of a tube, the grooves suppress the fluctuation of crushing force, but the crushing force decreases greatly. To suppress the force fluctuation while keeping the crushing force high, it has been proposed to use axial corrugated circular or rectangular tubes [185, 30, 189, 67]. For example, Singace and El-Sobky [185] conducted a quasi-static crushing test of axially corrugated, thin-walled tubes made of an aluminum alloy and PVC; they showed that the use of axial corrugations decreases the fluctuation of crushing force and proposed that axial corrugations are good for controlling the force of an impact energy absorber.

3.2.1   Axially corrugated circular tubes

Fig. 3.5 shows a corrugated circular tube with wavelength $2\lambda_c$ and amplitude a_c.

Table 3.1 classifies the deformation modes of corrugated circular tubes under axial crushing force [42], in a similar way to grooved tubes. First the deformation modes can be classified according to whether axial folds appear along the corrugations: the one mode in which deformation occurs axially along the corrugations and the other mode in which deformation does not occur axially along the corrugations. Further, the deformation in which folds are axially formed along the corrugations can be classified into axisymmetric and non-axisymmetric modes from the circumferential shape of the folds. For the case where the folds are axially formed along the corrugations of a tube, Fig. 3.6 compares the circumferentially axisymmetric and non-axisymmetric deformation behavior.

The axisymmetric modes of corrugated circular tubes can be further classified into the P- and S-modes depending on the folding process. Fig. 3.7 shows the deformation behavior during the crushing of the two types of circular tubes made with the same material, tube radius and wall thickness but with different corrugation shapes: ' $2\lambda_c = 27.8$ mm, $a_c = 0.5$ mm', and ' $2\lambda_c = 16.7$ mm, $a_c = 1.0$ mm' (called tubes A and B here). When the two types of tubes A and B are compared, both deformation modes are axisymmetric but the process for forming folds is clearly different. In tube A, folds are formed one by one just as in a smooth tube without corrugations, but in tube B, the

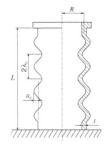

FIGURE 3.5
Geometry of circular tube with corrugated surface.

TABLE 3.1
Classification of deformation modes for corrugated tubes.

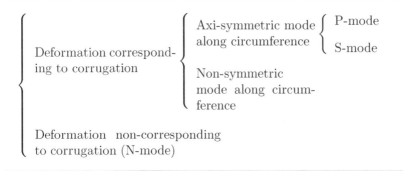

folds are simultaneously formed and folded along the full length of the tube in the crushing process. Here, as in the case of a grooved tube, the deformation of tube A is called the "**progressive crushing mode**" or the P-mode, and the deformation of tube B is called the "**simultaneous crushing mode**" or the S-mode. The two deformation modes produce different curves relating crushing force and crushing distance. Fig. 3.8 shows the relationship between crushing stress and crushing distance where the vertical axis shows the crushing force P divided by the average cross section $2\pi Rt$ (called crushing stress in the following) and the horizontal axis shows the crushing distance δ. The curve for tube A shows a smaller force variation than that of a smooth tube, but still the force fluctuates periodically. In contrast, the curve for tube B continues increasing without force fluctuation.

Experiments have also confirmed such compression response in axial crushing of corrugated circular tubes. Figs. 3.9(a) and (b) show experimental results of the compressive load-displacement curves corresponding to each mode

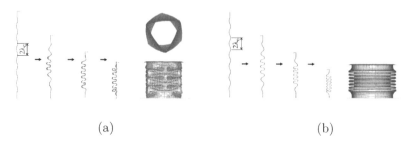

(a)   (b)

FIGURE 3.6
Deformation corresponding to corrugation [42]: (a) non-axisymmetric deformation and (b) axisymmetric deformation.

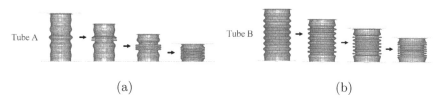

(a)   (b)

FIGURE 3.7
Axisymmetric mode of deformation [42]: (a) P-mode (tube A); (b) S-mode
(tube B).

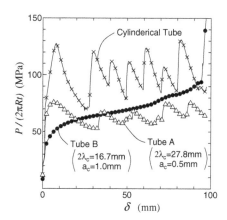

FIGURE 3.8
Crushing force-distance curve for tubes A and B shown in Fig. 3.7 [42].

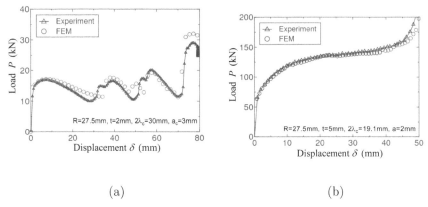

(a) (b)

FIGURE 3.9
Compressive load-displacement curves obtained from experiment [151]: (a) tube showing P-mode ($R = 27.5$ mm, $t = 2$ mm, $2\lambda_c = 30$ mm, $a_c = 3$ mm); (b) tube showing S-mode ($R = 27.5$ mm, $t = 5$ mm, $2\lambda_c = 19.1$ mm, $a_c = 2$ mm).

for corrugated circular tubes made of aluminum alloy A5052 with mechanical properties of Young's modulus $E = 70.6$GPa, initial yield stress $\sigma_s = 91.0$MPa and Poisson's ratio $\nu = 0.3$ [151]. The detailed stress-strain relation for this material is shown in Fig. 3.10. For the two corrugated tubes shown in Figs. 3.9(a) and (b), the axial crushing deformations are P-mode and S-mode, respectively. Both characteristics of P- and S-modes, such as the load fluctuation and the monotonically increasing of compressive stress accompanied with compression, are confirmed. In Figs. 3.9(a) and (b), the compressive load-displacement curves obtained by FEM analysis are also shown by circles. Good agreement is observed between the numerical and experimental load-displacement curves. The similar deformation characteristics during crushing are also observed from the experimental results of the deformation views as shown in Figs. 3.11(a) and (b), which show the deformation behavior corresponding to several process points in Figs. 3.9(a) and (b), respectively.

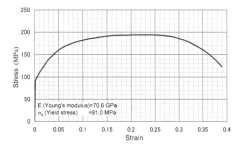

FIGURE 3.10
Nominal stress-strain relation for aluminum alloy A5052 used in the experiment [151].

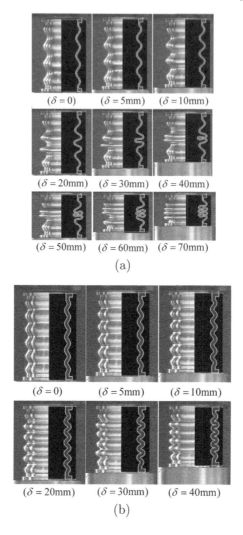

FIGURE 3.11
Deformation behavior observed in experiment for tubes used in Fig. 3.9[151]:
(a) tubes used in Fig. 3.9(a); (b) tubes used in Fig. 3.9(b).

Among the various parameters that control the deformation mode of a
corrugated tube, the most important are the corrugation wavelength $2\lambda_c$ and
amplitude a_c. In Figs. 3.12(a) and (b), the collapse deformations of the various
corrugated tubes with the wall thickness $t = 0.5$ mm and $t = 1.5$ mm are
summarized on a chart where the horizontal axis shows the wavelength $2\lambda_c$
and the vertical axis shows the amplitude a_c. Note here that the broken line in

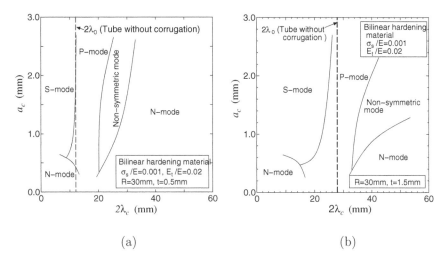

(a)   (b)

FIGURE 3.12
Mode classification chart of deformation mode for tubes with $E_t/E = 1/50$
[42]: (a) $t = 0.5$ mm; (b) $t = 1.5$ mm.

the figure shows the fold wavelength $2\lambda_0$ of axisymmetric deformation in axial
crushing of smooth circular tubes. Axisymmetric deformation tends to occur
if the corrugation wavelength is close to the fold wavelength $2\lambda_0$ for a smooth
tube, and another type of deformation non-corresponding to corrugations (N-
mode) tends to occur if the corrugation wavelength is far from $2\lambda_0$, especially
when amplitude a_c is small. In the corrugated tubes that show axisymmetric
deformation modes on the chart, whether the P- or S-mode occurs depends
mainly on the corrugation wavelength $2\lambda_c$; the S-mode preferentially occurs
when $2\lambda_c$ is small, and thereby the influence of amplitude a_c is small. In
addition, as shown by the comparison between (a) and (b) in the figure, the
boundary between deformation modes in the λ_c-a_c chart moves to the right
as wall thickness t increases. This is attributed to the fold wavelength $2\lambda_0$ in
a smooth tube increasing with wall thickness t.

Figs. 3.13 and 3.14 show the relationship between crushing stress $P/(2\pi Rt)$
and axial displacement δ with respect to corrugation wavelength $2\lambda_c$ and amplitude a_c. All results in the figure (a) are S-mode, where crushing stress monotonically increases as deformation proceeds; the upward slope of the crushing
stress increases as $2\lambda_c$ decreases as shown in Fig. 3.13(a), whereas crushing
stress increases as corrugation amplitude a_c decreases as shown in Fig. 3.14(a).
All results in the figure (b) show the P-mode, where crushing stress periodically fluctuates as deformation proceeds; the fluctuation of crushing stress
decreases as $2\lambda_c$ decreases as shown in Fig. 3.13(b), whereas crushing stress
and force fluctuation decrease as corrugation amplitude a_c increases as shown
in Fig. 3.14(b).

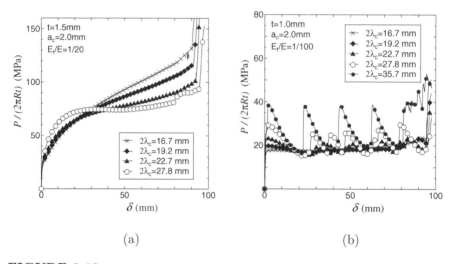

(a)   (b)

FIGURE 3.13
Force-displacement curve for various values of wavelength $2\lambda_c$ with fixed amplitude $a_c = 2$ mm (bilinear hardening material, $\sigma_s/E = 0.001, R = 30$ mm) [42]: (a) S-mode ($t = 1.5$ mm, $E_t/E = 1/20$); (b) P-mode ($t = 1.0$ mm, $E_t/E = 1/100$).

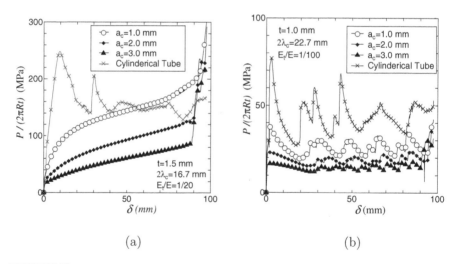

(a)   (b)

FIGURE 3.14
Force-displacement curve for various values of amplitude a_c with fixed $2\lambda_c$ (bilinear hardening material, $\sigma_s/E = 0.001, R = 30$ mm) [42]: (a) S-mode ($2\lambda_c = 16.7$ mm, $t = 1.5$ mm, $E_t/E = 1/20$), and (b) P-mode ($2\lambda_c = 22.7$ mm, $t = 1.0$ mm, $E_t/E = 1/100$).

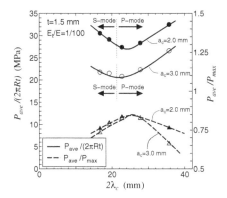

FIGURE 3.15
Change in average force P_{ave} and load efficiency P_{ave}/P_{max} with wavelength $2\lambda_c$ (bilinear hardening material, $\sigma_s/E = 0.001$, $R = 30$ mm) [42].

Fig. 3.15 shows the dependence of average crushing stress $P_{ave}/(2\pi Rt)$ (solid line) and **load efficiency** P_{ave}/P_{max} (dashed line) on corrugation wavelength $2\lambda_c$ assuming two values of amplitude ($a_c = 2.0$, 3.0 mm) in a corrugated circular tube. Load efficiency is an index of the change in compressive force during crushing; a large load efficiency means a small change of force, that is, as the load efficiency increases, the force fluctuation decreases in P-mode crushing, and the force curve becomes more horizontal in S-mode crushing.

In the figure, the left side of the vertical dotted line for the average crushing stress corresponds to the S-mode and the right side to the P-mode. The figure shows that the average crushing stress reaches the minimum when wavelength $2\lambda_c$ is close to the boundary between the S- and P-modes, and increases as the wavelength $2\lambda_c$ becomes farther from the boundary. In addition, as shown by the comparison between the two curves with amplitudes $a_c = 2.0$ mm and $a_c = 3.0$ mm, the average crushing stress decreases with increasing corrugation amplitude a_c for any wavelength $2\lambda_c$.

In addition, as the figure shows, load efficiency reaches the maximum when wavelength $2\lambda_c$ is close to the boundary between the S- and P-modes. Therefore, if a constant crushing stress is desirable in the relation between crushing stress and collapse displacement for an energy-absorbing member, the corrugation design must use a set of $2\lambda_c$ and a_c near the boundary between the P- and S-modes on the λ_c-a_c chart. Thereby, to increase average crushing force while maintaining load efficiency, it is necessary to find and use a lower value of a_c along the boundary between the P- and S-modes. In addition, if it is desirable that the crushing force gradually increases as deformation proceeds in the relation between crushing force and collapse displacement for an energy-absorbing member, the design should find and use a set of $2\lambda_c$ and a_c inside the S-mode region on the λ_c-a_c chart. Thereby, one can decrease corrugation amplitude a_c to increase average crushing force, and can decrease corrugation wavelength $2\lambda_c$ to increase both average crushing force and force curve slope.

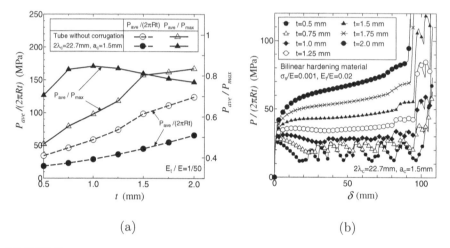

(a)   (b)

FIGURE 3.16
Effect of thickness t on the crushing force (bilinear hardening material, $\sigma_s/E = 0.001, R = 30$ mm): (a) change in average force P_{ave} and load efficiency P_{ave}/P_{max} (b) force-displacement curves [42].

Fig. 3.16(a) shows the variation of average crushing stress $P_{ave}/(2\pi Rt)$ and **load efficiency** P_{ave}/P_{max} for various values of wall thickness t in corrugated circular tubes. In the figure, for comparison, the average crushing stress and load efficiency of smooth tubes are also shown. As the figure shows, the average crushing stress of a corrugated tube increases with the wall thickness. This resembles the case of a smooth tube. In addition, it is shown that the load efficiency of a smooth tube monotonically increases with wall thickness, while the load efficiency of a corrugated tube first increases to a point and then decreases. This is understood from Fig. 3.16(b), which shows the relationship between crushing stress and displacement in the corrugated tube discussed in Fig. 3.16(a). As the wall thickness t increases, as shown in the figure, the deformation mode changes from the P-mode to the S-mode, and thereby the amplitude of force fluctuation decreases while in the P-mode and the upward slope of the force increases in the S-mode. Therefore, as wall thickness t increases, the load efficiency of a corrugated tube increases to a point and then decreases, as shown in Fig. 3.16(a). Also, it is found from Fig. 3.16(a) that the load efficiency improves at the sacrifice of the average crushing force by having corrugations in a tube; however, the load efficiency could be lower than that of a smooth tube.

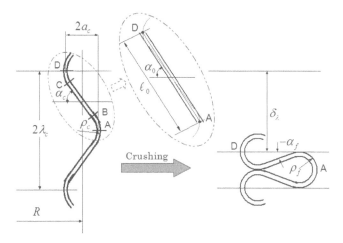

FIGURE 3.17
Analytic model of deformation behavior of tube wall for crushing of a corrugated circular tube.

The crushing force can be predicted using a theoretical model of axial crushing of a corrugated circular tube. The analysis is shown in detail in [43]. Also the average crushing force can be predicted by a simple approximation method. Fig. 3.17 shows an analytical model for deformation of the corrugated wall in crushing of a corrugated circular tube. The corrugations in a range of half-wavelength λ_c are approximated using arcs AB and CD with radius ρ_c and its tangential straight line BC having an inclination angle α_c. When a corrugated tube is subjected to an axial compressive load, it is assumed that the straight line BC rotates, reducing the angle α_c to α; simultaneously, arcs AB and CD are compressed, reducing their radius of curvature from ρ_c to ρ. Here, the minimum values of α and ρ are defined as $\alpha = -\alpha_f$ and $\rho = \rho_f$.

In deriving the approximation formula, it is assumed that the external work is consumed by the strain energy of bending at the fold crest and by the strain energy of expansion and shrinkage in the radial direction. In other words, for one folding wave assuming strain energy consumption E_b of bending and strain energy consumption E_m of expansion and shrinkage in the radial direction, the equilibrium equation between the external work and the internal energy consumption is expressed as

$$P_{ave} = \frac{E_b + E_m}{\delta_\lambda} \tag{3.1}$$

where δ_λ is the **effective crushing distance** in one folding wave as shown in Fig. 3.17.

In order to evaluate simply the strain energy consumption E_m of expansion and shrinkage in the radial direction, the curved wall ABCD is approximated by a virtual straight wall AD as shown in the inset of Fig. 3.17. According to such approximation, E_m can be given as

$$E_m = \pi \sigma_0^N t l_0^2 (1 - \cos \alpha_0) \tag{3.2}$$

where α_0 represents the inclination angle of AD, and can be obtained by

$$\alpha_0 = \tan^{-1} \left(\frac{\lambda_c}{2a_c} \right) \tag{3.3}$$

Also, the length of AD described by l_0 is given by

$$l_0 = \sqrt{4a_c^2 + \lambda_c^2} \tag{3.4}$$

In Eq. (3.2), σ_0^N is the **energy equivalent flow stress** for extensional deformation. Based on the study of Abramowicz and Jones [5], the energy equivalent flow stress σ_0^N is defined by

$$\sigma_0^N = \frac{\displaystyle\int_0^{\varepsilon_\theta|_{max}} f_\sigma(\varepsilon) d\varepsilon_\theta}{\varepsilon_\theta|_{max}}, \tag{3.5}$$

where $f_\sigma(\varepsilon)$ is the stress-strain relation of material, and $\varepsilon_\theta|_{max}$ is the maximum circumferential strain. Based on the study of Chen and Ozaki [41], $\varepsilon_\theta|_{max}$ is given as (see Eq. (3.21) in Section 3.2.3.1)

$$\varepsilon_\theta|_{max} = \log \left(1 + \frac{0.9\sqrt{\lambda_c^2/4 + a_c^2} - a_c}{R + a_c} \right) \tag{3.6}$$

The strain energy E_b needed to change the curvature radius at the fold crest to ρ_f is given by

$$E_b = 8\pi R M_0 (\alpha_c - \alpha_f) \cong 8\pi R M_0 \alpha_0 \tag{3.7}$$

Here, in order to calculate simply, it is assumed that $\alpha_c \cong \alpha_0$ and $\alpha_f \cong 0$.

In Eq. (3.7), M_0 is the full plastic bending moment per unit circumferential length, and is defined as

$$M_0 = \frac{\sigma_0^M t^2}{4} \tag{3.8}$$

Here, σ_0^M is the **energy equivalent flow stress** for bending and is given by

$$\sigma_0^M = \frac{2}{\left(\varepsilon_x|_{max} \right)^2} \int_0^{\varepsilon_x|_{max}} f_\sigma(\varepsilon) \varepsilon d\varepsilon \tag{3.9}$$

where $\varepsilon_x|_{max}$ is the maximum bending stress at the fold crest.

For the bending deformation of arc AB or CD,

$$\rho_c \cong \rho_f \frac{\pi}{\pi - 2\alpha_0} \tag{3.10}$$

holds. Therefore,

$$\varepsilon_x|_{max} = \frac{t}{2}\left(\frac{1}{\rho_c} - \frac{1}{\rho_f}\right) = \frac{t/2}{\rho_f} \times \left(\frac{2\alpha_0}{\pi}\right) \tag{3.11}$$

In order to calculate $\varepsilon_x|_{max}$, it is necessary to evaluate the **curvature radius** ρ_f at the crest of folds. Here, the evaluation of ρ_f is given approximately as follows.

According to the study of Abramowicz and Jones [5], the ratio of the **effective crushing distance** δ_{eff} to the length of a smooth circular tube L_{cir} in axially crushing can be evaluated by

$$\frac{\delta_{eff}}{L_{cir}} \cong 0.75 \tag{3.12}$$

If the corrugation of a corrugated tube can be assumed to approximately be the same form appearing in the middle process of axially crushing of a smooth circular tube, then the following equation is obtained from Eq. (3.12):

$$\frac{\delta_\lambda + 2(l_0 - \lambda_c)}{2l_0} \cong 0.75 \tag{3.13}$$

As shown in Fig. 3.17, δ_λ can be written as

$$\delta_\lambda \cong 2\lambda_c - 2\rho_f \tag{3.14}$$

Substituting Eq. (3.14) into Eq. (3.13) yields

$$\rho_f = 0.25l_0 \tag{3.15}$$

for the radius ρ_f and

$$\frac{\delta_\lambda}{2\lambda_c} = 1 - 0.25\frac{l_0}{\lambda_c} \tag{3.16}$$

for the ratio of effective crushing distance in crushing of a corrugated circular tube.

Substituting Eqs. (3.2), (3.7) and (3.8) into Eq. (3.1) yields

$$P_{ave} = \frac{\pi t\left[\sigma_0^N l_0^2 (1 - \cos\alpha_0) + 2R\sigma_0^M t\alpha_0\right]}{2\lambda_c} \times \left(\frac{\delta_\lambda}{2\lambda_c}\right)^{-1} \tag{3.17}$$

where the value of $\delta_\lambda/(2\lambda_c)$ is taken from Eq. (3.16).

Fig. 3.18 compares the average crushing load P_{ave} from FEM analysis with the approximation using Eq. (3.17). Furthermore, the average loads P_{ave} in the quasi-static crushing experiments for the corrugated tube shown in Fig. 3.9 are also calculated using Eq. (3.17) as shown in Table 3.2. It is seen from Fig. 3.18 and Table 3.2 that the values of P_{ave} using Eq. (3.17) are almost in agreement with the result of FEM analysis.

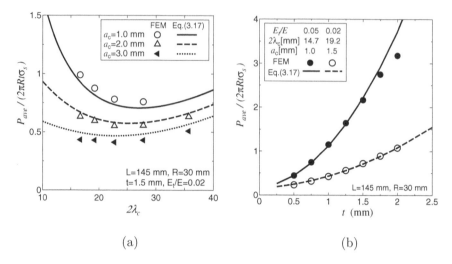

(a)   **(b)**

FIGURE 3.18
P_{ave} obtained from Eq. (3.17) and FEM: (a) P_{ave}-λ_c curve; (b) P_{ave}-t curve.

TABLE 3.2
Average loads P_{ave} in the quasi-static crushing experiments for corrugated
tube shown in Fig. 3.9.

	Fig. 3.9(a)	Fig. 3.9(b)
Experiment	14.9 kN	130 kN
Eq. (3.17)	15.9 kN	127 kN

3.2.2   Axially corrugated rectangular tubes

Axial crushing of a corrugated rectangular tube shown in Fig. 3.19 is now
discussed. Here, radius r_s at the corner satisfies $r_s/C_1 \leq 0.5$, $r_s/C_2 \leq 0.5$;
the cross section is a square if $C_1 = C_2 = C$, and the corrugated rectangular
tube becomes a corrugated circular tube if $r_s/C = 0.5$.

As in the deformation of corrugated circular tubes discussed in the preceding section, the deformation of corrugated rectangular tubes can be first classified into the two types of axial deformations: along and not along the corrugations. In the former, deformation can be further classified into two subtypes
according to circumferential deformation: parallel with corrugations as shown
in Fig. 3.20(a) and not parallel with corrugations as shown in Fig. 3.20(b).
From the viewpoint of the folding process, the deformations circumferentially
and axially in accordance with corrugations are classified into the **progressive crushing mode** (P-mode) where the folding occurs one by one and the
simultaneous crushing mode (S-mode) where all folds are crushed simultaneously. In the P-mode, as shown in Fig. 3.21, the crushing force fluctuates

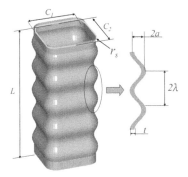

FIGURE 3.19
Geometry of rectangular tube with corrugated surface.

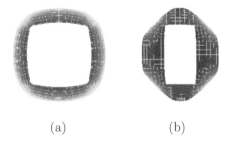

(a)   (b)

FIGURE 3.20
Classification of deformation type according to circumferential deformation
[47]: (a) parallel with corrugations; (b) not parallel with corrugations.

as deformation proceeds, but the amplitude of fluctuation is smaller than in
smooth tubes. In the S-mode, the force does not repeatedly fluctuate and
gradually increases as deformation proceeds.

Fig. 3.22 shows classification charts of collapse deformation mode for various corrugated square tubes with the wall thickness $t = 1.5$ mm and side
length $C = 50$ mm for the cases of $r_s/C = 0.5$ (corrugated circular tube)
and $r_s/C = 0.2$ (square tube). In the figure, the broken line shows the fold
wavelength $2\lambda_0$ of axisymmetric deformation in the axial crushing of smooth
circular tubes ($R = C/2$). As the figure shows, in a corrugated square tube,
similarly to the case of a corrugated circular tube, deformations tend to occur
both circumferentially and axially in accordance with corrugations if the corrugation wavelength is close to the fold wavelength $2\lambda_0$ in a smooth circular
tube, while deformations tend to be not in accordance with corrugations if the
corrugation wavelength $2\lambda_c$ is far from the fold wavelength $2\lambda_0$ of a smooth
circular tube, especially if the corrugation amplitude a_c is small. In addition,
a comparison between Fig. 3.22(a) and (b) shows that, in a square tube, the
boundary between the S- and P-modes in the λ_c-a_c chart is similar to that for

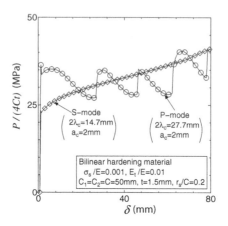

FIGURE 3.21
Force-displacement curves for tubes exhibiting S-mode and P-mode deformation [47].

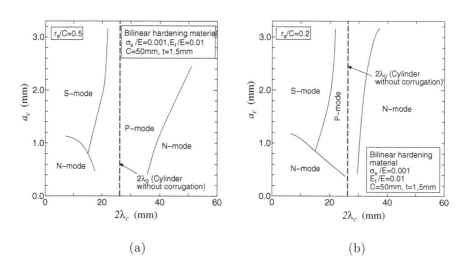

(a)   (b)

FIGURE 3.22
Classification chart of deformation mode for tubes with $t = 1.5$ mm [47]: (a) $r_s/C = 0.5$ (circular tube); (b) $r_s/C = 0.2$.

a corrugated circular tube, but there is a wider area in which the deformation proceeds regardless of corrugation.

In addition, the influence of the corrugation shape (wavelength $2\lambda_c$ and amplitude a_c) on the axial crushing force of a corrugated rectangular tube is similar to that for a corrugated circular tube; the details are discussed in [47].

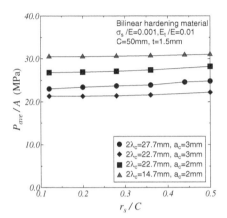

FIGURE 3.23
Effect of various values of radius r_s on average P_{ave}/A [47].

Fig. 3.23 shows the variation of average crushing stress $\sigma_{ave} = (P_{ave}/A)$ against the corner radius r_s of a corrugated square tube for several combinations of wavelength $2\lambda_c$ and amplitude a_c. Here, A is the mean cross-sectional area of a corrugated rectangular tube given by

$$A = \left[4(C - 2r_s) + 2\pi r_s \right] \times t \tag{3.18}$$

In Fig. 3.23, the case of '$2\lambda_c = 14.7$ mm, $a_c = 2$ mm' exhibits the S-mode and the other cases exhibit the P-mode. The average stress σ_{ave} increases only slightly with corner radius r_s/C, indicating that average stress is basically independent of corner radius.

Fig. 3.24 shows the relationship between crushing stress P/A and axial displacement δ for various ratios $C_1 : C_2$ between the two sides of a rectangle while keeping C_1+C_2 constant ($C_1+C_2 = 100$ mm). The figure shows that the crushing stress curves for the rectangles are almost identical. In other words, the corrugated rectangular tube (where the ratio between long and short sides is less than 2) and the corrugated square tube of the same $C_1 + C_2$ both have almost the same average crushing stress σ_{ave}.

Based on the data in Figs. 3.23 and 3.24, average crushing stress σ_{ave} for a corrugated rectangular tube, provided the aspect ratio between long and short sides is less than 2, can be approximately evaluated using a corrugated

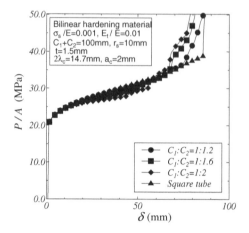

FIGURE 3.24
Crushing force for various corrugated rectangular tubes with same $C_1 + C_2$
[47].

circular tube with the same material, the same wall thickness and the same
corrugation dimensions, and with a radius of $R = (C_1 + C_2)/4$, as shown
in Fig. 3.25. Fig. 3.26 shows average crushing stresses for various corrugated
rectangular tubes, including a circular tube, with the same $C_1 + C_2$.

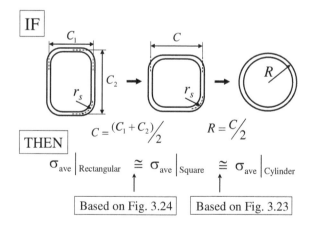

FIGURE 3.25
Evaluation of average crushing stress for corrugated rectangular tube from
corrugated circular tube [47].

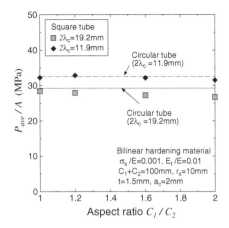

FIGURE 3.26
Average crushing stresses for various corrugated rectangular tubes, including circular tube, with same $C_1 + C_2$ [47].

3.2.3 Strain concentration in axial crushing of corrugated tubes

In axial crushing of a corrugated circular tube, there are cases where the results of energy absorption are not in line with expectations because the circumferential strain concentrates at the fold crest, causing axial cracking. Therefore, it is also important to evaluate the circumferential strain concentration at the fold crest.

3.2.3.1 Strain concentration in corrugated circular tubes

As discussed in Section 3.2, there are two axisymmetric deformation modes, P-mode and S-mode, for corrugated circular tubes. In Fig. 3.27(b) and (c), changes of the cross-sectional shape are shown for two tubes exhibiting P- and S-modes during axial crushing at six time points marked by N_k ($k = 1, \cdots, 6$) in Fig. 3.27(a). Fig. 3.28(a) and (b) shows the variation of circumferential strain ε_θ during crushing at the fold crests A, B and C (Fig. 3.27(b)) and D, E and F (Fig. 3.27(c)) for the tubes shown in Fig. 3.27(b) and (c). The maximum circumferential strain is approximately the same between the crests, though there are small fluctuations. In the P-mode deformation of axial crushing, since folds are formed and crushed one by one, as shown in Fig. 3.28(a), the value of ε_θ at the fold crest B increases until the next fold begins to buckle and becomes constant after that. Then, while circumferential strain ε_θ at fold crests A and C similarly increases as the respective folds are buckled, for both folds the strain is about the same as the maximum value $\varepsilon_\theta|_{max}$ at point B.

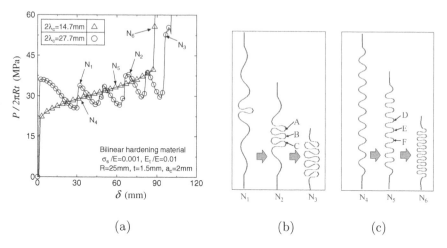

(a)   (b)   (c)

FIGURE 3.27
Variation of longitudinal cross-sectional shape during the crushing of two types
of corrugated circular tubes [41]: (a) force-displacement curve; (b) tube exhibiting P-mode deformation; (c) tube exhibiting S-mode deformation.

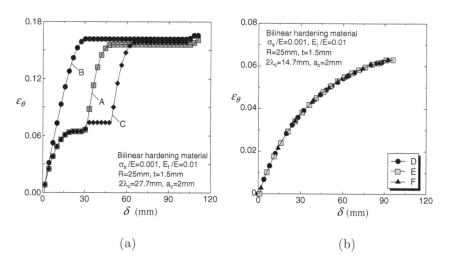

(a)   (b)

FIGURE 3.28
Variation of circumferential strain ε_θ at the vertex of outward corrugation [41]:
(a) for the tube shown in Fig. 3.27(b); (b) for the tube shown in Fig. 3.27(c).

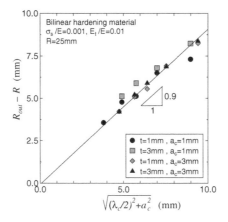

FIGURE 3.29
Relation between $R_{out} - R$ and $\sqrt{(\lambda_c/2)^2 + a_c^2}$ [41].

In contrast, in the S-mode deformation of axial crushing, all folds are folded simultaneously, as shown in Fig. 3.28(b), so the values of ε_θ at fold crests D, E and F are almost the same, increasing at a similar pace while deformation proceeds and reaching a maximum when all folds are completely buckled.

Assuming the radius R_{out} of a fold crest, the maximum circumferential strain $\varepsilon_\theta|_{max}$ is given by

$$\varepsilon_\theta|_{max} = \log\left(\frac{R_{out}}{R + a_c}\right) \tag{3.19}$$

On the other hand, as Fig. 3.29 shows, the value of $R_{out} - R$ is almost independent of tube wall thickness t and is approximately given by

$$R_{out} - R = 0.9\sqrt{\left(\frac{\lambda_c}{2}\right)^2 + a_c^2} \tag{3.20}$$

as a function of corrugation wavelength $2\lambda_c$ and amplitude a_c. By substituting Eq. (3.20) into Eq. (3.19), an approximation formula to evaluate $\varepsilon_\theta|_{max}$ is obtained as [41]

$$\varepsilon_\theta|_{max} = \log\left(1 + \frac{0.9\sqrt{\left(\frac{\lambda_c}{2}\right)^2 + a_c^2} - a_c}{R + a_c}\right) \tag{3.21}$$

Fig. 3.30 compares the values of $\varepsilon_\theta|_{max}$ obtained from FEM analysis and from Eq. (3.21). The figure shows a good agreement between them.

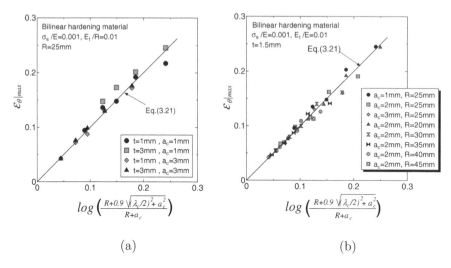

(a)   (b)

FIGURE 3.30
Comparison of $\varepsilon_\theta|_{max}$ obtained from FEM analysis and Eq. (3.21) [41]: (a) for tubes with R fixed at $R = 25$ mm; (b) for tubes with t fixed at $t = 1.5$ mm.

3.2.3.2   Strain concentration in corrugated square tubes

Fig. 3.31 shows contours of circumferential strain ε_θ (lighter color shows larger strain) in a corrugated square tube under an axial compressive force. In axial crushing of a square tube, strain concentrates at the outer corners of corrugations.

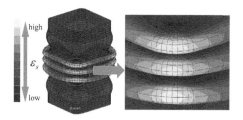

FIGURE 3.31
Concentration of strain at the corner of corrugated square tube [41].

Fig. 3.32(a) and (b) shows the variation of the maximum circumferential strain $\varepsilon_\theta|_{max}$ versus corner radius r_s/C for several wall thicknesses and materials (with different yield stress and plastic strain hardening coefficient). The maximum circumferential strain $\varepsilon_\theta|_{max}$ increases as r_s/C decreases. Further, in a corrugated circular tube, the maximum circumferential strain $\varepsilon_\theta|_{max}$ is

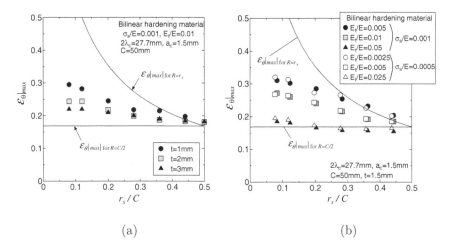

(a)   (b)

FIGURE 3.32
Relationship between strain concentration and corner radius r_s/C in corrugated square tubes [41]: (a) for different thickness; (b) for different yield stress and plastic strain hardening.

barely affected by wall thickness t, strain hardening coefficient or yield stress ($\varepsilon_\theta|_{max}$ is approximately the same for various wall thickness and materials when $r_s/C = 0.5$ as seen in the figure); in a corrugated square tube, however, wall thickness t, strain hardening coefficient and yield stress strongly affect maximum circumferential strain $\varepsilon_\theta|_{max}$. As Fig. 3.32(a) shows, $\varepsilon_\theta|_{max}$ increases as t decreases. In addition, considering that the strain hardening coefficients are $\sigma_s/E_t = 1/5, 1/10, 1/50$ corresponding to the two kinds of yield stress $\sigma_s/E = 0.001, 0.0005$ used here, and that the calculated values in the figure corresponding to the two kinds of yield stress roughly agree with each other, the strain hardening coefficient and yield stress affect $\varepsilon_\theta|_{max}$ through the parameter σ_s/E_t alone, and the maximum circumferential strain $\varepsilon_\theta|_{max}$ increases with σ_s/E_t [41].

Thus, it has been demonstrated that $\varepsilon_\theta|_{max}$ in square corrugated tubes is influenced by the yield strength, the hardening coefficient, the wall thickness and other parameters, and unlike in circular tubes, cannot be predicted solely on the basis of the corrugation wavelength, amplitude and tube size. Here, if one, using the approximate expression given in Eq. (3.21) to calculate the maximum stain $\varepsilon_\theta|_{max}|_{for\ R=C/2}$, which is corresponding to the case when the tube is changed to be circular with its cross section represented by a circle inscribed within the square tube with side length C, and the maximum stain $\varepsilon_\theta|_{max}|_{for\ R=r_s}$, which is corresponding to the case when the tube is changed to be circular but with its radius R equal the corner radius r_s of the square tube, $R = r_s$, and plot the values of $\varepsilon_\theta|_{max}|_{for\ R=C/2}$ and $\varepsilon_\theta|_{max}|_{for\ R=r_s}$ in

Fig. 3.32, then it is found from the figure that values of $\varepsilon_\theta|_{max}$ for square corrugated tubes with various r_s/C mainly lie between these two lines of $\varepsilon_\theta|_{max}|_{for\ R=C/2}$ and $\varepsilon_\theta|_{max}|_{for\ R=r_s}$. In other words, the following relationship holds:

$$\varepsilon_\theta|_{max}\ \big|_{for\ R=C/2} \leq \varepsilon_\theta|_{max} \leq \varepsilon_\theta|_{max}\ \big|_{for\ R=r_s} \tag{3.22}$$

3.2.4   Axially corrugated frusta

An actual impact energy absorber usually has a taper in the longer direction, assuming that the force is typically applied from an oblique direction. Compared with tubes, frusta are advantageous in that the initial peak force is suppressed by the axial taper; further, stable energy absorption is realized just as in axial crushing along the perpendicular direction even in the case of oblique collision (e.g., [139, 140]). Mamalis and Johnson [136] examined axial compressive deformation of frusta made of an aluminum alloy and pointed out that there are two types of deformation modes in axial crushing of frusta: in one mode a switch from axisymmetric to non-axisymmetric deformation occurs in the crushing process, and in the other mode the deformation is non-axisymmetric from the beginning. Singace et al. [186] pointed out that, compared with cylinders, frusta are more stable as a structure with higher Euler buckling force, and the modes of frustum collapse are invariably mixed, including the '**inversion deformation mode**", in which the inversion starts from the upper or lower part when the radial displacement at the ends is not constrained, and further that energy absorption capacity increases with the application of radial end constraints. It is found that the characteristic feature of axial crushing of a truncated cone lies in the unique deformation where the fold formed at the top part of a truncated cone sinks into its interior [206]. The deformation occurs more easily as the taper angle increases; thereby average force becomes small but maximum crushing distance increases, for example, the **effective crushing distance ratio** increases up to 0.8 while the upper limit is utmost 0.73 in the case of a circular tube. Paying attention to the folding of each fold, one can predict the average force during folding by evaluating the examined part of the frustum using average force of a circular tube having the average radius [206].

In this section, axial crushing of a thin-walled corrugated frustum shown in Fig. 3.33 shall be discussed.

A corrugated frustum has the features as follows in comparison with the corrugated circular tube [37].

(**1**) Because of the difference in radius between the top and bottom ends, deformation easily starts from the upper end, and P-mode deformation occurs more easily than in a corrugated circular tube.

(**2**) If the taper angle is small, a corrugated frustum shows almost the same crushing force as a corrugated circular tube, and the crushing force decreases as the taper angle increases. This is because, if the taper angle

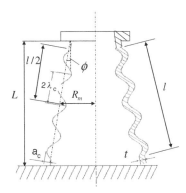

FIGURE 3.33
Thin-walled, axial corrugated frustum.

ϕ is large, the fold formed at the upper end bends inward and does not vertically pile onto the successive fold and sinks deeply into the frustum's interior.

(3) In a corrugated frustum, since the folds formed sink into the cone, the effective crushing distance tends to be large compared with that in a corrugated circular tube.

Fig. 3.34 plots the crushing force of corrugated frusta against taper angle ϕ with the average radius fixed at $R_m = 45$ mm. The corrugated frusta in the figures (a) and (b) show the S-mode and P-mode deformations, respectively, if their taper angle $\phi=0°$.

The inward inversion phenomenon that the fold sinks into the subsequent fold becomes evident as taper angle ϕ is increased. Fig. 3.35 shows the deformation behavior in axial crushing of corrugated frusta with $\phi = 5°$ and $15°$. When $\phi = 15°$, the created fold sinks into the lower part of the frustum, and the crushing force fluctuates. Here, this type of deformation is called M-mode.

Fig. 3.36 shows the classification chart of deformation mode, which shows the relationship between deformation mode and wavelength λ_c and amplitude a_c for various values of taper angle ϕ. The figure shows that the area of the M-mode region increases as the taper angle increases. The change from $\phi = 0°$ to $\phi = 5°$ has little effect on the deformation mode, except for a slight increase in the area of the P-mode. If $\phi = 10°$ is reached, the M-mode appears; the transition into the M-mode becomes easier as amplitude a_c decreases and becomes more difficult as wavelength λ_c decreases.

Fig. 3.37 shows the average crushing force and absorbed energy per unit volume versus taper angle ϕ, with wavelength λ_c and amplitude a_c as parameters. As Fig. 3.37(a) shows, for any set of λ_c and a_c, average force scarcely decreases from the value at $\phi = 0°$ up to $\phi = 15°$ but decreases at and above

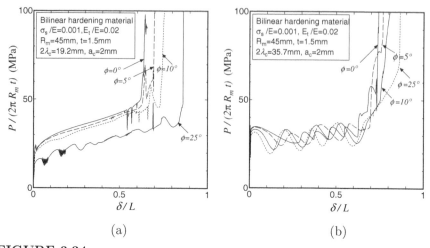

(a)   **(b)**

FIGURE 3.34
Crushing force of corrugated frusta for various ϕ [37]: (a) $2\lambda_c = 19.2$ mm; (b) $2\lambda_c = 35.7$ mm.

$\phi = 20°$. In addition, Fig. 3.37(b) shows that, for any set of λ_c and a_c, the absorbed energy per unit volume E_{Ab}/V increases up to $\phi = 20°$ but the value at $\phi = 25°$ clearly shows a downward slope. Examination of deformation mode around $\phi = 20°$ shows the S-mode up to $\phi = 15°$ and the M-mode at and above $\phi = 20°$. The increase of the absorbed energy up to $\phi = 15°$ is due to the increase of effective crushing distance ratio, and the decrease of the absorbed energy from $\phi = 20°$ is due to the deformation mode change from the S-mode to M-mode. where average force decreases.

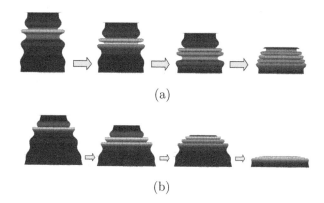

(a)

(b)

FIGURE 3.35
Deformation behavior in axial crushing of corrugated frusta: (a) $\phi = 5°$; (b) $\phi = 15°$.

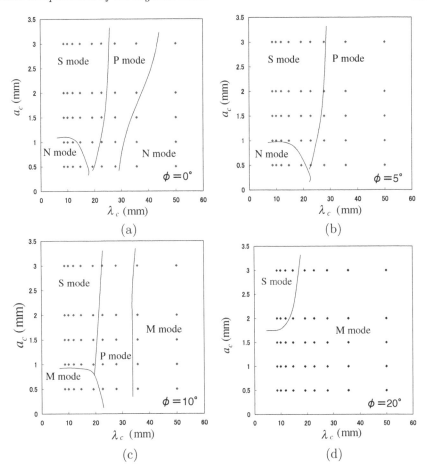

FIGURE 3.36
Classification chart of deformation mode for axial crushing of corrugated circular truncated cones (bilinear hardening material, $\sigma_s/E = 0.001$, $E_t/E = 0.02$, $R_m = 45$ mm, $t = 1.5$ mm): (a) $\phi = 0°$; (b) $\phi = 5°$; (c) $\phi = 10°$; (d) $\phi = 20°$.

3.2.5 Stepped circular tubes

In order to improve the compressive force-displacement response of shock absorbers, various ideas and devices have been proposed. One of them is shown in Fig. 3.38, proposed by Stangl and Meguid [192], and is called stepped circular tube, which consists of several cylinders with different radii. The important features of this device are that the compressive force goes up gradually and there is no initial peak force in the axial crushing process.

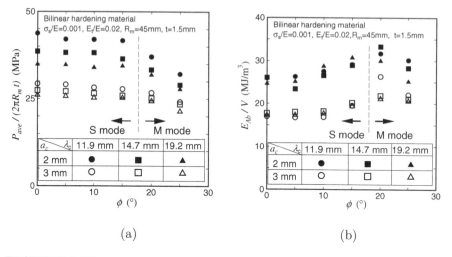

FIGURE 3.37
Influence of taper angle ϕ on average force and energy absorption efficiency.

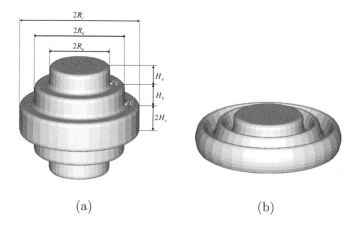

(a)   (b)

FIGURE 3.38
Stepped circular tube: (a) before crushing; (b) after crushing.

Fig. 3.39 shows the relationship between crushing force P and crushing distance δ for a three-stepped circular tube subjected to axial compression with $R_a = 15$ mm, $R_b = 30$ mm, $R_c = 45$ mm and $H_a = H_b = H_c = 15$ mm. It is seen from the figure that there is no peak force in the initial crushing and the compressive force goes up gradually during the crushing deformation. Also it is found from the deformed forms shown in Fig. 3.39 that the crushing of the

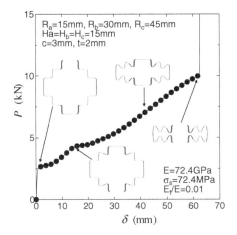

FIGURE 3.39
Crushing force for a three-stepped circular tube under axial compression [48].

tube consists of two telescopic deformations. One is, at first, an entering of the smallest cylinder with radius of R_a into the cylinder with radius of R_b, and the other is, at second, an entering of the cylinder with radius of R_b into the cylinder with radius of R_c. The deformation of a three-stepped circular tube is a sum of these two deformation processes. Each deformation process can be analyzed by a kinematically admissible plastic hinge model for a two-stepped circular tube.

Stangl and Meguid [192] proposed three kinds of theoretic models for the deformation mechanism. Here, two of the three theoretic models are shown in Fig. 3.40. In the analysis of the theoretical model, the material is assumed to be rigid perfectly plastic with an energy equivalent flow stress σ_0.

(1) Model with the vertical wall of the small cylinder expanding.

As shown in Fig. 3.40(a), in this model localized bending occurs at three hinges (A, B and C), and the plastic strain energy dissipation due to bending is

$$E_{b1} = M_{p0} 2\pi \left[R_1 \alpha + R'(\pi/2 - \beta) + R_2 \phi \right] \tag{3.23}$$

where $M_{p0} = \sigma_0 t^2/(2\sqrt{3})$ and

$$\begin{cases} \alpha &= \sin^{-1}\left[\ell_0(1 - \cos\phi)/H\right] \\ \beta &= \pi/2 + \alpha - \phi \\ R' &= R_1 + H\sin\alpha \\ \ell_0 &= R_2 - R_1 \end{cases} \tag{3.24}$$

Moreover, the plastic strain energy dissipation due to the stretching of the

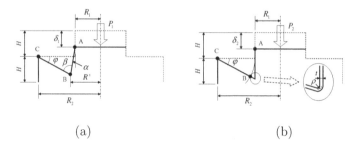

(a)   (b)

FIGURE 3.40
Plastic hinge model proposed by Stangl and Meguid [192]: (a) model with
the vertical wall of the small cylinder expanding; (b) model with a travelling
hinge.

elements between hinges A and B and between hinges B and C is given by

$$
E_{s1} = \int_0^H 2\pi\sigma_0 R_1 t \log\left(\frac{R_1 + x\sin\alpha}{R_1}\right) dx +
$$
$$
\int_0^{\ell_0} 2\pi\sigma_0 (R_2 - x)t \log\left(\frac{R_2 - x\cos\phi}{R_2 - x}\right) dx
\tag{3.25}
$$

Thus, the crushing force P_1 is given by

$$
P_1 = \frac{dE_{b1} + dE_{s1}}{d\delta_1}
\tag{3.26}
$$

where δ_1 is

$$
\delta_1 = H(1 - \cos\alpha) + \ell_0 \sin\phi
\tag{3.27}
$$

(2) Model with a travelling hinge.

As shown in Fig. 3.40(b), in this model the vertical wall of the small
cylinder is restricted to move in the axial direction only so that a travelling
hinge occurs near the point B. The plastic strain energy dissipation due
to this travelling hinge is given by

$$
E_{travel} = 2 \times 2\pi R_1 \ell_2 \frac{M_{p0}}{\rho}
\tag{3.28}
$$

where ρ is the radius of the travelling hinge, and ℓ_2 is the swept distance
by the travelling hinge

$$
\ell_2 = \ell_0 \left[\frac{1}{\cos\phi} - 1\right]
\tag{3.29}
$$

Moreover, the plastic strain energy dissipation due to bending at hinge C
is

$$
E_{b2} = M_{p0} 2\pi R_2 \phi
\tag{3.30}
$$

and the plastic strain energy dissipation due to the stretching of the elements between the two circular tubes is

$$E_{s2} = \int_0^{\ell_2} 2\pi\sigma_0 R_1 t \log\left(\frac{R_1 + x\cos\phi}{R_1}\right) dx +$$
$$\int_0^{\ell_0} 2\pi\sigma_0(R_2 - x)t \log\left(\frac{R_2 - x\cos\phi}{R_2 - x}\right) dx \qquad (3.31)$$

Thus, the crushing force P_2 is given by

$$P_2 = \frac{dE_{b2} + dE_{travel} + dE_{s2}}{d\delta_2} \qquad (3.32)$$

where δ_2 is

$$\delta_2 = \ell_2(1 + \sin\phi) + \ell_0 \sin\phi \qquad (3.33)$$

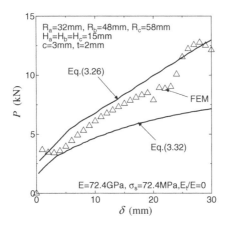

FIGURE 3.41
Comparison of $P - \delta$ curve obtained from FEM and theoretical analysis.

Fig. 3.41 shows load-displacement curves for a three-stepped circular tube made of an elastic perfectly plastic material obtained from FEM and theoretical analysis using Eqs. (3.26) and (3.32). The error of Eq. (3.32) compared with the numerical results of FEM seems large. This could be guessed from the fact that the travelling of the hinge, which is assumed in the theoretic model shown in Fig. 3.40(b), is not found in numerical analysis as shown in Fig. 3.39.

Moreover, the crushing force-distance response of the stepped circular tube is not sensitive to the inclination of compression direction [50], because its crushing deformation is an entering process of a small cylinder into a large cylinder, which is not sensitive to the inclination of the load direction. Fig. 3.42

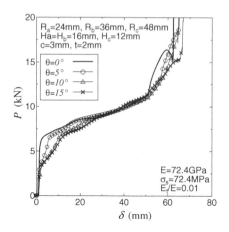

FIGURE 3.42
$P - \delta$ curves of a stepped circular tube for the load inclination angle $\theta = 0, 5°, 10°$ and $15°$.

shows the relationship between P and δ for a stepped circular tube subjected to a compressive load with inclination angle $\theta = 0, 5°, 10°$ and $15°$. It is seen from the figure that although initial load decreases by the increase in the inclination angle, these curves for various inclination angle θ are almost the same, except for the initial portion. Such a feature can be understood from its deformation behaviors under oblique compressive load, shown in Fig. 3.43.

3.3   Circumferentially corrugated tubes

Generally, in axial crushing of a thin-walled axially corrugated shell, the corrugation serves as a starting point for buckling and the formation of folds on the bellows leads to stable crushing deformation. However, although the folds are easily produced, the simultaneous pursuit of energy absorption efficiency is difficult. On the other hand, a circumferentially corrugated tube is today attracting attention for high-efficiency energy absorption. Abdewi et al. [1] performed a crushing test of impact energy absorbers consisting of a composite material and showed that the circumferentially corrugated structure could cope with the high crushing force of axial crushing more efficiently than a smooth tube.

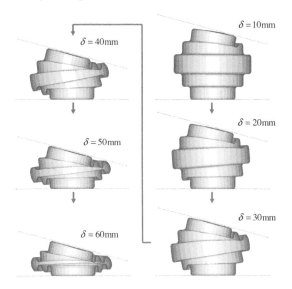

FIGURE 3.43
Deformed shapes of stepped circular tube under oblique compressive load with
$\theta = 15°$ at crushing distance of $\delta = 10, 20, 30, 40, 50$ and 60 mm.

3.3.1 Circumferentially corrugated circular tubes

In this section, based on the FEM numerical simulation [33], the effect of the
geometrical shape on the deformation mode and crushing force of a circumferentially corrugated tube is discussed assuming the two types of circumferentially corrugated tubes: a circumferentially corrugated circular tube with
sinusoidal profile (Fig. 3.44(b); RCT) and a circumferentially corrugated circular tube with corners (Fig. 3.44(c); RCTC). For RCT and RCTC, the shape
of the corrugation is defined as a sine wave and a rectangular wave, respectively, of n units and amplitude a_c. In addition, to investigate the effects of the
corrugation, the same numerical analysis for a basic non-corrugated circular
tube (Fig. 3.44(a); referred to as CT hereafter) is also performed. The materials used in the FEM analysis are assumed to obey the bilinear hardening rule
with the parameters of $\sigma_s/E = 0.001$ and $E_t/E = 0.01$.

Fig. 3.45(a) shows the relationship between crushing stress P/A and compressive displacement δ in RCTs with length $L = 400$ mm, assuming the
corrugation amplitude $a_c = 4$ and 10 mm. For comparison, Fig. 3.45(a) also
includes the results of CT of the same length L. In addition, Fig. 3.45(b) shows
the deformation behavior of each cylinder shown in Fig. 3.45(a) at compressive displacement $\delta/L = 30\%$ and 50%. As can be seen from these figures,
the half-wavelength of the buckling folds becomes long due to the manufactured corrugation in RCT. In the examples shown in Fig. 3.45, although the

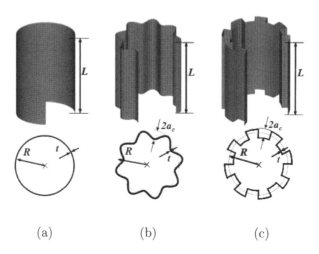

(a)   (b)   (c)

FIGURE 3.44
Geometry of circular tubes with circumferential corrugation [33]: (a) CT; (b) RCT; (c) RCTC.

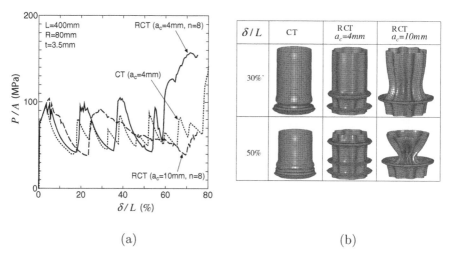

(a)   (b)

FIGURE 3.45
Axial crushing of a RCT ($L = 400$ mm) [33]: (a) force-displacement curves; (b) deformed shapes of a CT and RCT.

folding half-wavelength of a CT is approximately $\lambda = 70$ mm, the folding half-wavelength of an RCT with corrugation amplitude $a_c = 4$ mm is about $\lambda = 130$ mm, about 1.9 times that of a CT. In axial crushing, the folding wavelength of an RCT is greater because the corrugated profile adds to the equivalent bending rigidity of the tube wall.

In axial crushing of a CT, from Eq. (1.14) of Section 1.1.1, the folding half-wavelength λ is given by

$$\lambda_{CT} = \pi \sqrt{R\sqrt{\frac{D_{CT}}{Et}}} \tag{3.34}$$

where D_{CT} is the bending rigidity per unit width of CT tube given by

$$D_{CT} = \frac{Et^3}{12(1-\nu^2)} \tag{3.35}$$

Since the bending rigidity per unit width D_{RCT} in RCT is different due to the shape of the corrugation, the bending rigidity D_{RCT} is calculated by using

$$D_{RCT} \cong \frac{Et \int_0^{\pi R/n} \left(a_c \sin \frac{nx}{R}\right)^2 \sqrt{1 + \left(\frac{na_c}{R}\right)^2 \cos^2 \frac{nx}{R}} \, dx}{\pi R/n} \tag{3.36}$$

From Eqs. (3.34) and (3.36), the ratio between the half-wavelength λ_{RCT} of an RCT and the half-wavelength λ_{CT} of a CT is approximately given by

$$\frac{\lambda_{RCT}}{\lambda_{CT}} = \left(\frac{D_{RCT}}{D_{CT}}\right)^{1/4} \tag{3.37}$$

Table 3.3 compares the folding half-wavelength of an RCT obtained from FEM numerical analysis and from Eq. (3.37) for several values of amplitude a_c. The table shows good agreement between them.

TABLE 3.3
Ratio of λ_{RCT} to λ_{CT} for various values of amplitude a_c ($R = 80$ mm, $t = 3.5$ mm) [33].

a_c (mm)	2	3	4	5	10
$\frac{\lambda_{RCT}}{\lambda_{CT}}$ from FEM	1.1	1.5	1.9	2.0	2.8
$\frac{\lambda_{RCT}}{\lambda_{CT}}$ from Eq. (3.37)	1.2	1.5	1.7	1.9	2.7

In an RCT, the fold wavelength is rather long, but Fig. 3.45(b) shows that the folding deformation is concentrated in a partial region of a fold due to the local buckling in the side surface of the corrugations. As a result, the RCT has a longer fold half-wavelength λ than that of a CT, but the radial expansion due to folding is almost the same in an RCT and CT. Therefore, compared with a CT, the deformation of an RCT is characterized by (1) a longer fold half-wavelength and (2) the folding deformation occurring in only a part of a half-wavelength, with small expansion in the radial direction. Therefore, it is generally considered difficult to obtain a good RCT for use as a collision energy absorber.

However, if the tube is so short that the folding half-wavelength is limited by tube length L, an RCT behaves differently in axial crushing. In the following, by assuming $L = 150$ mm, a short RCT for which the length is comparable with the tube diameter is discussed.

Fig. 3.46(a) shows the relationship between crushing stress P/A and compressive displacement δ for RCTs with length $L = 150$ mm for various values of amplitude a_c $(= 0\text{--}35$ mm); Fig. 3.46(b) shows the crushing deformation behavior of the various types of tubes used.

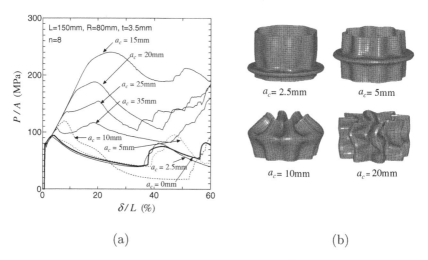

(a)   (b)

FIGURE 3.46
Axial crushing of an RCT with sinusoidal profile ($L = 150$ mm) [33]: (a) force-displacement curves; (b) deformed shapes.

As shown in Fig. 3.46(b), the crushing behavior of an RCT depends on the corrugation amplitude. If the amplitude is small ($a_c = 2.5$ mm), the deformation behavior of the tube is, in axial crushing, almost the same as that of a CT. If the amplitude is large, in the case of for example $a_c = 10$ mm, only a single axisymmetric fold is formed in the RCT. The middle portion of the axial length becomes the apex of the fold during compression and spreads in the radial direction; both the upper and lower edges fold down and inside the cylinder. This deformation mode is often called the "**mushroom mode.**"

When an RCT shows this type of deformation, the peak axial crushing stress is greater than that of a longer RCT having the same amplitude a_c. In addition, in the mushroom mode shown in Fig. 3.46, the crushing stress of an RCT with corrugation amplitude $a_c = 15$ mm is greater than that of one with $a_c = 10$ mm; this shows that, in this deformation mode of an RCT, the crushing stress of axial crushing increases with corrugation amplitude.

As the corrugation amplitude further increases, for example, when $a_c = 20$ mm, the crushing stress is smaller than that when $a_c = 15$ mm. This type of deformation is called the "**side collapse mode**" where, unlike in the mushroom mode, the sidewalls of each corrugation are twisted as they are crushed.

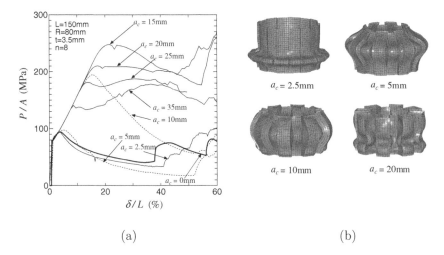

(a) (b)

FIGURE 3.47
Axial crushing of an RCTC with corners ($L = 150$ mm) [33]: (a) force-displacement curves; (b) deformed shapes.

Fig. 3.47 shows the behavior of crushing force and crushing deformation of an RCTC with length $L = 150$ mm shown in Fig. 3.44(c). In axial crushing, the deformation behavior of an RCTC is basically similar to that of an RCT. In other words, the deformation mode depends on the corrugation amplitude; if the amplitude a_c is small, the effect of corrugations is small and the crushing behavior is roughly similar to that of a CT; if the amplitude a_c is large, the mushroom mode deformation occurs, which is not found in a CT, and the crushing force increases with a_c. In addition, if the corrugation amplitude further increases, side collapse mode appears; conversely, the crushing force decreases as the amplitude a_c increases. However, the crushing force of an RCTC is greater than that of an RCT with a similar magnitude of corrugation amplitude. In particular, in the side collapse mode of an RCTC, the sidewall is not twisted, remaining in the standing position during the crushing, and so the crushing force does not decrease much or even slightly increases as compression proceeds (see the case of $a_c = 35$ mm). For this reason, the average crushing force P_{ave} and **load efficiency** P_{ave}/P_{max} are larger in the side collapse mode of an RCTC than those of an RCT.

3.3.2 Circumferentially corrugated rectangular tubes

Rectangular tubes with circumferential corrugation are attracting attention as promising shock energy absorbers to support high crushing forces because these tubes have an in-plane stress-holding capability in the axial direction due to the circumferential corrugations and the corners.

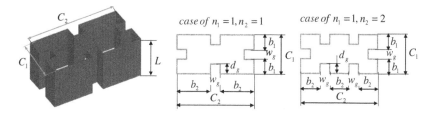

FIGURE 3.48
Geometry of square tubes with circumferential corrugation.

In this section, on the basis of FEM numerical analysis [53], the axial crushing behavior of a thin-walled, circumferentially corrugated rectangular tube, as shown in Fig. 3.48, is discussed. The corrugated rectangular tube has geometry of length L, sides C_1 and C_2 and thickness t; and each side of length C_1 and C_2 has n_1 and n_2 corrugations, respectively, with depth d_g and width w_g. In addition, the materials used for FEM analysis are assumed to obey the bilinear hardening rule with the parameters $\sigma_s/E = 0.001$ and $E_t/E = 0.01$.

Fig. 3.49(a) shows the relationship between crushing stress P/A and dimensionless compressive displacement δ/L in the circumferentially corrugated rectangular tubes (length $L = 150, 200, 250, 300$ mm, $d_g = 40$ mm and $n_1 = n_2 = 1$). In the figure, STC means the circumferentially corrugated square tube and ST means the square tube without corrugation. In addition, the results for RCTCs (with $n = 4$ and amplitude $a_c = 20$ mm) are also shown in Fig. 3.49(a). The crushing stress of the circumferentially corrugated rectangular tube is greater than that of a smooth tube or an RCTC when all else is equal.

In addition, in a long circumferentially corrugated circular tube, the crushing stress decreases sharply after peaking, whereas in a circumferentially corrugated rectangular tube, the crushing stress remains high even when the tube is long. This can be explained by comparing the deformation behavior of a circumferentially corrugated rectangular tube and a circumferentially corrugated

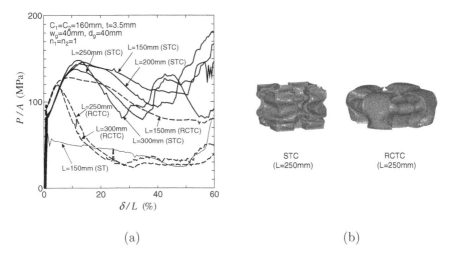

(a) (b)

FIGURE 3.49
Axial crushing of circumferentially corrugated rectangular tubes (d_g = 40 mm) [34]: (a) crushing force; (b) deformation behavior ($\delta/L = 50\%$).

circular tube both with length $L = 250$ mm (Fig. 3.49(b)), and their crushing displacement $\delta/L = 50\%$. In a circumferentially corrugated rectangular tube, the deformation mode does not change up to a certain length and while axial compression continues the folds are continuously buckled on the sidewalls of corrugations and the tube does not fall over; thus, the average crushing stress and force efficiency do not appreciably decrease. On the other hand, in the deformation of a circumferentially corrugated circular tube, the mushroom mode appears where the fold crest expands outward, with both ends folding inward; if the tube is long, average crushing stress and force efficiency decrease as the force decreases after the peak.

Fig. 3.50 shows the crushing force of a circumferentially corrugated rectangular tube for various values of corrugation depth d_g and shows the deformation behavior for axial crushing distance $\delta/L = 30\%$. In the circumferentially corrugated rectangular tube, if the corrugation is deep, the folds are formed on the sidewalls of corrugations without the tube falling over as axial compression progresses (see the case of $d_g = 30$ mm in Fig. 3.50(b)), and the decrease of crushing force after its peak is small. The deformation mode with only a small decrease in force after its peak is similar to the side collapse mode of the RCTC discussed above.

The average crushing force P_{ave} of a circumferentially corrugated rectangular tube can be analytically estimated as follows. First, for similarly shaped corrugations (with common values of d_g and w_g) the average crushing stress $\sigma_{ave}(= P_{ave}/A)$ of a circumferentially corrugated rectangular tube is a function of only corrugation interval b_k/t $(k = 1, 2)$, as shown in Fig. 3.51.

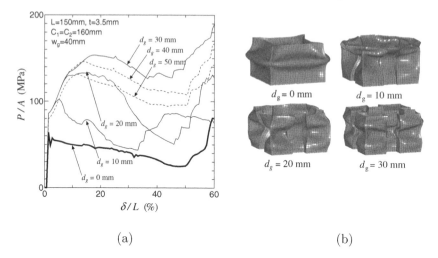

(a)   (b)

FIGURE 3.50
Axial crushing of circumferentially corrugated rectangular tubes (L = 150 mm) [34]: (a) crushing force, and (b) deformation behavior ($\delta/L = 30\%$).

Fig. 3.51 shows average crushing stress versus corrugation interval in the range $b_1 = b_2 = 50$–150 mm for a corrugated square tube ($C_1 = C_2$). Average crushing stress σ_{ave} depends on only the corrugation interval, independent of the number of corrugations. Therefore, the average crushing force of a circumferentially corrugated tube can be evaluated by a sum of the crushing force on each side:

$$P_{ave} = 2\left[P_{ave}|_1 + P_{ave}|_2\right] \tag{3.38}$$

Here, $P_{ave}|_1$ and $P_{ave}|_2$ applied to sides 1 and 2 can be obtained from the master curve of Fig. 3.51 for similarly shaped corrugations.

In addition, it is known that there are the inextensible mode and the extensible mode in axial crushing of a rectangular tube and the average crushing stress of the former is smaller than that of the latter. In order to consider this, Eq. (3.38) can be rewritten as follows:

$$P_{ave} = 2\left[P_{ave}|_1 + P_{ave}|_2\right] - \Delta P_{ave} \tag{3.39}$$

where ΔP_{ave} is the difference between the average crushing forces of the two modes and is approximately evaluated using formulas of the average crushing force for the extensible and inextensible modes, respectively, in a rectangular tube discussed in Section 2.4.3.

For a circumferentially corrugated rectangular tube, Fig. 3.52 plots the average crushing force obtained by FEM analysis versus C_2, and a comparison is made with the approximation formulas, Eqs. (3.38) and (3.39). For a corrugated rectangular tube with $C_1 = 200$ and 280 mm, both the prediction

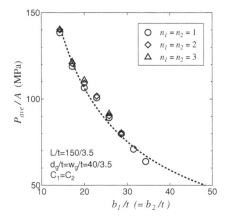

FIGURE 3.51
Relationship between average crushing stress and corrugation interval.

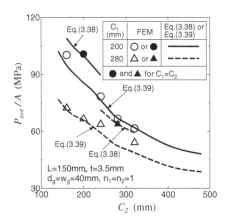

FIGURE 3.52
Comparison between the predicted average crushing force of a circumferentially corrugated rectangular tube and the FEM calculation.

using the approximation formula Eq. (3.38) for the vicinity of $C_2 \cong C_1$ and the prediction using the approximation formula Eq. (3.39) for the other areas of C_2 roughly agree with the FEM calculation.

4

Bending of tubes

In general, not only an axial compressive load but also a bending moment is applied to all members of an automobile when a collision occurs. It is therefore important to understand the collapse behavior of members during the bending deformation to evaluate strength and energy absorption characteristics. In this chapter, the bending deformation of circular or square tubes is discussed.

4.1 Bending deformation of a cylinder

Many researchers have studied the bending deformation of cylinders. The most important feature of a cylinder subjected to bending is the cross section flattening as the deformation proceeds. As the cross section flattens, the bending stiffness of the cylinder decreases, the relationship between bending moment and curvature becomes nonlinear, and the bending moment reaches a maximum at a certain curvature. Therefore, the factors leading to the **bending collapse** of a cylinder should include not only buckling of the bottom surface due to the maximum compressive stress imposed by the bending, but also the maximum bending moment due to flattening.

4.1.1 Flattening in bending deformation of a cylinder

Brazier [26] first pointed out the **flattening** associated with the bending deformation of a cylinder. This is thus called the **Brazier effect**, and has been studied experimentally and theoretically by many researchers, including Ades [8], Stephens et al. [193], Tugoce and Schroeder [204], Gellin [78], Fabian [70], Bushnell [29], and Kyriakides and Shaw [117]. The thin line in Fig. 4.1 shows the FEM simulation of the cross section of a cylinder with radius $R = 25$ mm and thickness $t = 2$ mm, to which a pure bending moment is applied to yield a bending deformation with curvature (of the cylinder's central axis) of $\kappa_g = 1.49 \mathrm{m}^{-1}$. The flattened cross section is usually approximated as an ellipse. Fig. 4.1 compares the actual, flattened cross section with the theoretical cross section (denoted by $\square$) of the ellipse $(x/a_{el})^2 + (y/b_{el})^2 = 1$ with lateral radius a_{el} and vertical radius b_{el} ($a_{el} = 26.67$ mm, $b_{el} = 23.23$ mm in the figure).

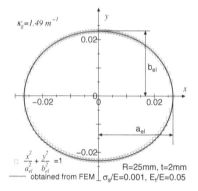

FIGURE 4.1
Cross-sectional shape of a cylinder under bending deformation.

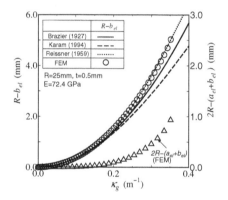

FIGURE 4.2
Change in lateral and vertical radii a_{el} and b_{el} in the cross section of a cylinder during elastic bending (note that $R - b_{el} > a_{el} - R$ as κ_g increases).

Fig. 4.2 shows the relationship between the reduction of vertical radius $R - b_{el}$ in the cross section and the bending curvature κ_g in elastic bending deformation of a cylinder, as evaluated by FEM simulation. The figure also shows the difference $(R - b_{el}) - (a_{el} - R) = 2R - (a_{el} + b_{el})$ between vertical shrinkage and lateral stretch. As the figure shows, $R - b_{el} \cong a_{el} - R$ if the bending curvature κ_g is small, but $R - b_{el} > a_{el} - R$ as κ_g increases.

A nondimensional value $1 - b_{el}/R$, which is obtained by dividing $R - b_{el}$ by R, is normally used as an index of flatness. The nondimensional parameter $1 - b_{el}/R$ is called the "**flattening ratio**" and is denoted by the symbol μ.

$$\mu = 1 - b_{el}/R$$

In the flattened cross section, if the flattening ratio or the vertical radius b_{el} is known, the lateral radius a_{el} of the ellipse is determined from the following equation:

$$\int_0^{a_{el}/2} \sqrt{\frac{a_{el}^4 - (a_{el}^2 - b_{el}^2)x^2}{a_{el}^2(a_{el}^2 - x^2)}}\, dx = \frac{\pi R}{2} \tag{4.1}$$

assuming that the cross-sectional perimeter does not change with deformation.

The cross-sectional flattening that occurs in the bending deformation of a cylinder can also be understood based on the mechanical model for flattening shown in Fig. 4.3. As shown in Fig. 4.3(a), assuming a bending curvature κ_g in the axial direction of the cylinder, bending stress σ_x appears at both end surfaces of a ring of width $dx = d\phi/\kappa_g$ (length in the axial direction at the center of the cylinder), as given by

$$\sigma_x = E\left(\kappa_g R \sin\theta\right) \tag{4.2}$$

if the bending is elastic. A downward force occurs due to the stress σ_x perpendicularly acting on both end surfaces of the ring of width dx. The force density q is given by

$$q = \sigma_x t \kappa_g \tag{4.3}$$

Therefore, flattening of a cylinder can be treated as a deformation problem of a curved beam subjected to the distributed force density q shown in Fig. 4.3(b).

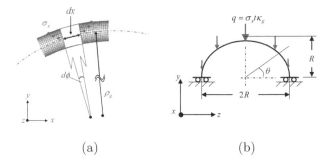

(a)   (b)

FIGURE 4.3
Mechanical model for flattening: (a) bending stress σ_x; (b) curved beam subjected to distributed force.

Brazier [26], using a moving system of displacements u, v, w, as shown in Fig. 4.4, sought a theoretical solution for flattening during the elastic bending of a cylinder. He discussed the strain energy of a cylinder subjected to bending deformation based on two factors, the one associated with bending in the axial direction of the cylinder, and the other associated with flattening of the cross section, and used as a criterion that the sum of the strain energy is minimized to derive the displacement components w in the radius direction and v in the

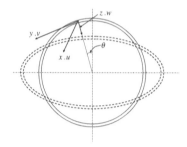

FIGURE 4.4
Coordinates and shape of the cross section for Brazier's analysis of a flattened cylinder.

circumferential direction of the cross section as functions of cylinder curvature κ_g, as follows:

$$
\begin{cases}
v &= -(1-\nu^2)\dfrac{R^5}{2t^2}\kappa_g^2 \sin 2\theta \\[2mm]
w &= -(1-\nu^2)\dfrac{R^5}{t^2}\kappa_g^2 \cos 2\theta + \nu R^2 \kappa_g \sin \theta
\end{cases}
\tag{4.4}
$$

In Eq. (4.4), the second term $(\nu R^2 \kappa_g \sin \theta)$ for w is St. Venant's displacement with respect to the Poisson's ratio, and is usually omitted because it is smaller than the first term.

Brazier's analysis shows that the shape of the flattened cross section is given by

$$
\begin{cases}
x &= \left[R+(1-\nu^2)\dfrac{R^5}{t^2}\kappa_g^2\right]\cos\theta - (1-\nu^2)\dfrac{R^5}{2t^2}\kappa_g^2 \sin 2\theta \sin \theta \\[2mm]
y &= \left[R-(1-\nu^2)\dfrac{R^5}{t^2}\kappa_g^2\right]\sin\theta + (1-\nu^2)\dfrac{R^5}{2t^2}\kappa_g^2 \sin 2\theta \cos \theta
\end{cases}
\tag{4.5}
$$

where x, y are the coordinates shown in Fig. 4.1. In Eq. (4.5), the shape of the flattened cross section is mainly approximated by the first term. Note that here the shape given by the first term is an ellipse with the length of the semi-major axis $a_{el} = R + (1 - \nu^2)R^5\kappa_g^2/t^2$ and the length of the semi-minor axis $b_{el} = R - (1 - \nu^2)R^5\kappa_g^2/t^2$. Therefore, in elastic bending, the **flattening ratio** is approximately given by

$$
\mu = (1-\nu^2)\left(\frac{\kappa_g R^2}{t}\right)^2
\tag{4.6}
$$

as a function of the parameter $\kappa_g R^2/t$, which is composed of the curvature κ_g, and the wall thickness t and radius R of the cylinder.

For comparison, Fig. 4.2 also shows Brazier's theoretical solution for the elastic bending of a cylinder, in other words $R - b_{el}$ in Eq. (4.6). As the figure shows, Brazier's solution agrees well with that of the FEM numerical analysis if the curvature κ_g is small, but the discrepancy grows as κ_g increases. In Brazier's theoretical analysis, assuming that the displacement v, w corresponding to the cylindrical flattening is sufficiently small, the terms consisting of their products are omitted. Note, however, that the analytical accuracy does not clearly improve even if it adopts such terms. For example, Karam [110] showed that, by considering the higher order terms of strain energy, the flattening ratio is given by

$$\mu\big|_{Karam} = \mu \times \frac{1}{1 + \frac{5}{6}\mu} \tag{4.7}$$

where μ is the flattening ratio obtained using Eq. (4.6). Fig. 4.2 also shows the flattening ratio obtained by Eq. (4.7). As the figure shows, the discrepancy in the flattening ratio from Eq. (4.7) is great compared with the analytical result from Eq. (4.6).

Reissner [170] and Fabian [69] conducted more exact theoretical analyses of flattening in relation to bending of a cylinder. The flattening ratio obtained by Reissner's method is also shown in Fig. 4.2. Through comparison with the results of FEM numerical analysis, it is known that the analytical result obtained using Reissner's method is clearly more accurate than Brazier's. In comparison with Reissner's analysis, a smaller flattening ratio is found in Brazier's analysis, for example, by about 12% at $\kappa_g = 0.35\text{m}^{-1}$.

Fig. 4.5 shows the relationship between flattening ratio μ and nondimensional curvature $\kappa_g \times R^2/t$ obtained by FEM numerical analysis of cylinders, assuming a material that obeys the bilinear hardening rule: in figure (a) it analyzed about four thicknesses $t = 0.25, 0.5, 1,$ and 2 mm, assuming a yield stress of $\sigma_s/E = 0.001$, and a strain hardening coefficient of $E_t/E = 0.01$; in figure (b) it analyzed about three strain hardening coefficients $E_t/E = 0.005, 0.01$ and 0.02, assuming a yield stress of $\sigma_s/E = 0.001$ and a thickness $t = 2$ mm, and to three yield stresses $\sigma_s/E = 0.001, 0.003$ and 0.005, assuming a strain hardening coefficient of $E_t/E = 0.01$ and a thickness $t = 2$ mm. For comparison, Fig. 4.5 also shows the flattening curves for elastic bending of cylinders of thicknesses $t = 0.5$ and 2 mm (assuming a perfectly elastic body). For the same curvature, the flattening ratio is smaller in plastic bending than in elastic bending. Furthermore, in elastic bending, the flattening ratio μ can be expressed by a sole function of nondimensional curvature $\kappa_g \times R^2/t$, independent of cylinder thickness; in plastic bending, as shown in Fig. 4.5(a), the relation between flattening ratio μ and $\kappa_g \times R^2/t$ depends on the thickness t; the flattening ratio μ decreases against the thickness t for the same $\kappa_g \times R^2/t$. Also it is seen from Fig. 4.5(b) that the effect of strain hardening on the flattening ratio is small.

Fig. 4.6 re-plots the flattening curves on double logarithmic axes for elastic and plastic bending for $t = 0.5$ and 2 mm, assuming $\sigma_s/E = 0.001$ and

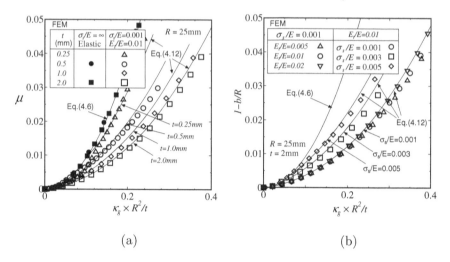

FIGURE 4.5
Change of cross-sectional flattening ratio in a cylinder due to plastic bending
deformation: (a) for different thicknesses t; (b) for different yield stresses σ_s
and different strain hardening coefficients E_t.

$E_t/E = 0.01$. As shown in Fig. 4.6, in elastic bending, the flattening ratio μ is
proportional to the square of nondimensional curvature $\kappa_g \times R^2/t$ where the
proportionality coefficient is about 1. On the other hand, in plastic bending,
the slope of the double logarithmic plot of a flattening curve is 2 in the initial
phase, then temporarily decreases due to plastic yielding occurring at the
upper and lower parts of the cylinder, and after that gradually increases up
to 2 due to the widening of plastic deformation at both the left and the right
sides of the cylinder. This is easily understood from the mechanical model
of flattening shown in Fig. 4.3. In the elastic bending, the beam bends in
proportion to the load, and the load density q is proportional to the square of
the curvature κ_g as shown by the formula

$$q = ERt \sin \theta \kappa_g^2 \qquad (4.8)$$

which is obtained by substituting Eq. (4.2) into Eq. (4.3), and in this way, the
degree of flattening is proportional to the square of curvature κ_g. In plastic
bending, however, once plastic yielding occurs due to bending near the upper
and lower parts of the cylinder, the increasing rate dq/dk_g of the load density
q becomes smaller than that in elastic bending. Therefore, the slope of a
flattening curve in the double logarithmic plot of flattening ratio μ versus
$\kappa_g R^2/t$ becomes smaller than 2.

Based on Fig. 4.6, flattening associated with plastic bending of a cylinder
is divided into three phases—elastic deformation with the slope of the double
logarithmic plot being 2, initial plastic deformation with the slope increasing

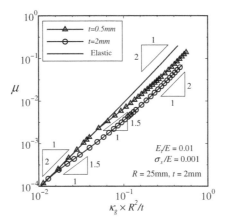

FIGURE 4.6
Flattening ratio in a double logarithmic plot.

from a value smaller than 2 to 2, and plastic deformation with the slope approximately being 2—and the slop can be approximately expressed as [36]:

$$\frac{d(\log \mu)}{d(\log \chi)} = \begin{cases} 2 & \chi \le \chi_1 \\ d_1\chi + d_2 & \chi_1 < \chi < \chi_2 \\ 2 & \chi_2 \ge \chi \end{cases} \tag{4.9}$$

where the nondimensional bending curvature is

$$\chi = \kappa_g \times R^2/t \tag{4.10}$$

and χ_1 is the value of χ at the start of plastic yield, which is determined from

$$\chi_1 = \frac{\sigma_s R}{Et} \tag{4.11}$$

From Eq. (4.9) the **flattening ratio** can be approximately obtained as

$$\mu = \begin{cases} c_1\chi^2 & \chi \le \chi_1 \\ c_2 e^{d_1\chi}\chi^{d_2} & \chi_1 < \chi < \chi_2 \\ c_3\chi^2 & \chi_2 \ge \chi \end{cases} \tag{4.12}$$

where

$$c_1 \cong 1 \tag{4.13}$$

The other unknown parameters in Eq. (4.12) χ_2, c_2, c_3, d_1 and d_2 are determined by solving the following simultaneous equations:

$$d_1 \chi_1 + d_2 \cong 1.5 \tag{4.14a}$$

$$\frac{\mu_2(3 - \mu_2)}{R(1 - \mu_2)^2} = 6\frac{\sigma_s}{Et} \tag{4.14b}$$

$$\chi_1^2 = c_2 e^{d_1 \chi_1} \chi_1^{d_2} \tag{4.14c}$$

$$c_3 \chi_2^2 = c_2 \chi_2^{d_2} e^{d_1 \chi_2} \tag{4.14d}$$

$$d_1 \chi_2 + d_2 \cong 2 \tag{4.14e}$$

where μ_2 is the flattening ratio at $\chi = \chi_2$, $\mu_2 = c_3 \chi_2^2$.

In deriving Eq. (4.14), Chen et al. [36] made the following two assumptions based on a lot of FEM numerical results about plastic bending deformations of cylinders:

(1) The slope of the double logarithmic plot $d(\log \mu)/d(\log \chi)$ at the starting point of the 2nd phase is about 1.5 as shown in Fig. 4.6, namely,

$$d_1 \chi_1 + d_2 \cong 1.5$$

(2) The 2nd-phase end is dependent on the widening of plastic deformation at both the left and the right sides of the cylinder, which is caused by bending in the circumferential direction due to flattening. As an index of the widening of plastic deformation at both the sides the ratio of the curvature in the circumferential direction at the sides of the cylinder $\kappa_l = a_{el}/b_{el}^2 - 1/R$ to a curvature corresponding to an initial plastic yielding $\kappa_{ls} = 2\sigma_s/Et$ is used, and it is assumed that the 2nd phase will be completed if the ratio of κ_l/κ_{ls} amounts to about 3, namely,

$$\frac{\mu_2(3 - \mu_2)}{R(1 - \mu_2)^2} = 6\frac{\sigma_s}{Et}$$

Therefore, it is seen that Eqs. (4.14)(a) and (b) correspond to the above-mentioned assumptions (1) and (2), respectively, and Eqs. (4.14)(c) and (d) correspond to the continuation conditions of the flattening ratio μ at $\chi = \chi_1$ and $\chi = \chi_2$, respectively, and Eq. (4.14)(e) corresponds to the continuation conditions of the slope of the double logarithmic plot at $\chi = \chi_2$. For the details of this analysis, see [36].

Fig. 4.7 compares the FEM analysis and the approximation formula Eq. (4.12) for flattening ratio μ in pure bending deformation of cylinders, assuming a material that obeys the following Ramberg-Osgood equation with $N = 5$:

$$\varepsilon = \frac{\sigma}{E}\left[1 + \frac{3}{7}\left(\frac{\sigma}{\sigma_s}\right)^{N-1}\right] \tag{4.15}$$

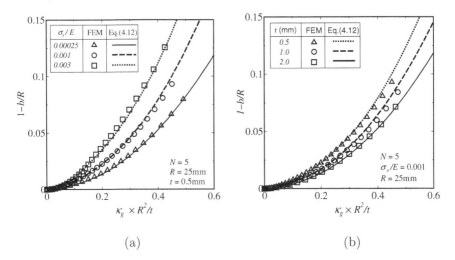

FIGURE 4.7
Comparison of flattening ratio μ obtained from Eq. (4.12) and FEM analysis for a material obeying Ramberg-Osgood equation with $N = 5$ [36]: (a) for different yield stresses; (b) for different thicknesses.

In Fig. 4.7(a) it analyzed about three yield stresses $\sigma_s/E = 0.00025, 0.001$ and 0.003, assuming a thickness $t = 0.5$ mm; and in Fig. 4.7(b) it analyzed about three thicknesses $t = 0.5, 1$ and 2 mm, assuming a yield stress of $\sigma_s/E = 0.001$. In Figs. 4.5(a) and (b) the flattening ratios derived using the approximation formula Eq. (4.12) are also shown for comparison with the FEM results. As these figures show, the approximation formula Eq. (4.12) gives an available prediction for flattening ratio in the plastic bending of cylinders.

No reasonable analytic solution as good as Eq. (4.6) has been proposed for flattening in plastic bending of a cylinder, because the phenomenon is more complicated than that of elastic bending. In the analytical expressions, the displacements v and w in the flattening of a cylinder cannot be expressed as simple functions like Eq. (4.4), and are normally expanded in a series:

$$v = R \sum_{n=1}^{N} b_n \sin n\theta, \quad w = R \sum_{n=0}^{N} a_n \cos n\theta \qquad (4.16)$$

Consequently, a frequently used method is to determine the unknown parameters $b_1, b_2, \cdots, b_N Ca_0, a_1, \cdots, a_N$ in Eq. (4.16) by the principle of virtual work under an assumption that the cylinder circumference is inextensible. For the detail of the analysis, see [78, 70, 104].

Furthermore, flattening due to bending of a cylinder also depends on the length of the cylinder because end parts of the cylinder cannot be flattened.

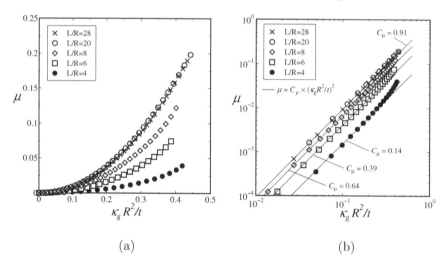

(a)   (b)

FIGURE 4.8
Relationship between flattening ratio and length in elastic bending of a cylinder: (a) flattening ratio; (b) flattening ratio in a double logarithmic plot.

Then, when the cylinder is short, the cross-sectional flattening is also blocked in the middle part. Fig. 4.8(a) shows the relationship between cross-sectional flattening in the middle part and length in elastic bending of a cylinder. Here, Fig. 4.8(b) shows the flattening curves in (a) on a double logarithmic scale. Even when the cylinder is short, the flattening ratio $\mu|_{L\neq\infty}$ is, as in the case of a long cylinder, proportional to the square of nondimensional curvature $\kappa_g R^2/t$ and is approximately given by

$$\mu|_{L\neq\infty} = C_\mu \times \left(\frac{\kappa_g R^2}{t}\right)^2 \tag{4.17}$$

where C_μ is called the coefficient of flattening ratio and is smaller for a shorter cylinder, as shown in the figure.

Fig. 4.9 shows the relationship between cross-sectional flattening in the middle part and length in plastic bending of a cylinder; the flattening ratio becomes small as the cylinder becomes short, just as in elastic bending. Further, the ratio between the flattening ratios of short and long cylinders is almost the same as that in elastic bending, so the flattening ratio $\mu|_{L\neq\infty}$ in plastic bending of a cylinder of finite length can be approximated by

$$\mu|_{L\neq\infty} = C_\mu \times \Big(\mu \text{ given by Eq. (4.12)}\Big) \tag{4.18}$$

assuming the same coefficient C_μ as in Eq. (4.17). Note that the thin lines in Fig. 4.9 show the flattening ratio derived using the approximation formula, Eq. (4.18).

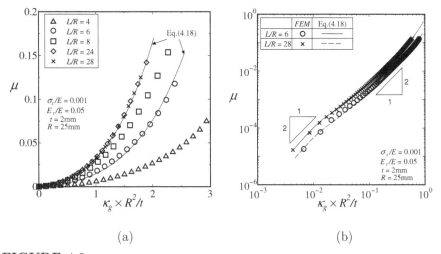

(a)   (b)

FIGURE 4.9
Relationship between flattening ratio and length in plastic bending of a cylinder: (a) flattening ratio; (b) flattening ratio in the double logarithmic plot.

4.1.2 Maximum bending moment due to cross-sectional flattening

As a result of cross-sectional flattening due to the bending deformation of a cylinder, the bending stiffness of the cylinder decreases, and the relationship between bending moment and bending curvature of the cylinder becomes nonlinear. Thus, if the decreasing rate of bending stiffness exceeds the increasing rate of bending moment due to the increment of bending curvature, the moment of the cylinder should begin decreasing. As a result, the maximum bending moment M_{lim}[1] appears in the curve relating bending moment and bending curvature of the cylinder. After the maximum moment is reached, the bending deformation is concentrated in a specific area of the cylinder, and overall bending collapse occurs.

Assuming the flattened cross section is an ellipse as shown in Fig. 4.1, the bending moment M can be evaluated by

$$M = 4t \int_0^{\pi/2} f_\sigma \left(k_g b_{el} \sin \theta \right) b_{el} \sin \theta \sqrt{a_{el}^2 \sin^2 \theta + b_{el}^2 \cos^2 \theta} d\theta \qquad (4.19)$$

where $f_\sigma(\varepsilon)$ is the stress-strain formula of the material.

---

[1]In this book, the maximum moment M_{max} is also called the collapse moment M_{col}. In discussing the bending collapse of a tube, it is notable that there exist two type of collapse: collapse caused by the critical stress (e.g., buckling at the bottom surface) and collapse caused by flattening of the cross section, as shown later in Eq. (4.41). Therefore, in order to emphasize the cause of collapse, here the maximum moment is expressed as M_{cri} or M_{lim} depending on the cause that generates it.

Considering a_{el} and b_{el} are functions of the flattening ratio μ, Eq. (4.19) can be written in the following form:

$$M = f_M(k_g, \mu) \qquad (4.20)$$

Therefore, the maximum bending moment M_{lim} appears when

$$\frac{dM}{dk_g} = 0 \qquad (4.21)$$

In elastic deformation, the **flattening ratio** of a cylinder is evaluated by the theoretical solution Eq. (4.6) [26]. Fig. 4.10 shows the relationship between bending moment and bending curvature (solid line) and between compressive stress at the bottom surface and bending curvature (broken line), based on the cross section derived from Eq. (4.6), assuming the flattened cross section as elliptical. As the figure shows, the bending moment has the maximum at a certain curvature. The theoretical analysis of Brazier [26] shows that if the bending curvature k_{lim}^e satisfies

$$k_{lim}^e = \frac{0.8165t}{\sqrt{3(1-\nu^2)}R^2} \qquad (4.22)$$

the bending moment is maximized as

$$M_{lim}^e = 0.5443 \frac{E\pi Rt^2}{\sqrt{3(1-\nu^2)}} \qquad (4.23)$$

Here, the compressive stress at the bottom surface of the cylinder is given by

$$\sigma_{lim}^e = 0.635 \frac{Et}{\sqrt{3(1-\nu^2)}R} \qquad (4.24)$$

and the flattening ratio is given by

$$\mu_{lim} = 1 - \left.\frac{b_{el}}{R}\right|_{lim} = \frac{2}{9} \qquad (4.25)$$

Fig. 4.10 also shows the relationship between bending moment and bending curvature (○) and between compressive stress at the bottom surface and bending curvature (△) obtained from FEM numerical analysis. Until the maximum moment is reached, the result of FEM numerical analysis roughly agrees with the result of theoretical analysis using Eq. (4.6). In particular, the maximum moment and the corresponding compressive stress at the bottom part closely agree with theoretical Eqs. (4.23) and (4.24). However, the bending curvature at the maximum moment is about $k_g/[t/\sqrt{3(1-\nu^2)}R^2] = 0.726$ by FEM numerical analysis, which is about 12% smaller than the theoretical value of 0.8165 by Eq. (4.22). This is likely due to the tendency of Brazier's theoretical solution for the flattening ratio to be low, as shown in Fig. 4.2.

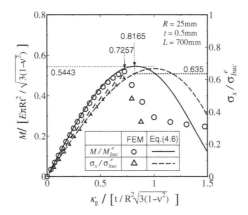

FIGURE 4.10
Relationship between moment and curvature in elastic bending.

Furthermore, after the maximum moment is reached, the moment and stress obtained by the FEM numerical analysis greatly decrease, showing a feature very different from that of the theoretical analysis. In particular, the stress at the bottom surface increases even after the maximum moment is reached in the theoretical analysis, but rapidly decreases after the maximum moment is reached in the FEM analysis. This suggests bifurcation of the deformation near the bottom surface of the cylinder [193, 69]. This is discussed in the following section.

Even in plastic bending of a cylinder, if the cylinder is long the maximum bending moment equals the limit value M_{lim} due to flattening, similar to elastic bending. Fig. 4.11 shows the changes in bending moment and compressive stress at the bottom surface for long cylinders $(L/R = 28)$ with thin $(t/R = 0.02)$ and thick $(t/R = 0.08)$ walls subjected to pure bending. From the fact that the compressive stress increases due to bending even after the maximum moment is reached (point A in the figure), clearly the maximum bending moment is the limit value due to flattening in both cases of $t/R = 0.02$ and $t/R = 0.08$. In a thin-walled cylinder $(t/R = 0.02)$, on the one hand, since the bending compressive stress at the bottom surface rapidly decreases at point B in the figure after the maximum moment is reached, it is quite natural to suppose that the bottom surface of the cylinder undergoes accidental buckling by compressive stress at this time. In a thick-walled cylinder $(t/R = 0.08)$, on the other hand, at point B in the figure, the compressive stress at the bottom surface begins to decrease, but since the decrease is slow, it is attributable to flattening rather than buckling.

In plastic deformation, if the flattening ratio is evaluated as a function of bending curvature κ_g of the cylinder, the maximum moment M_{lim} due to flattening of the cylinder can also be derived from the condition for an

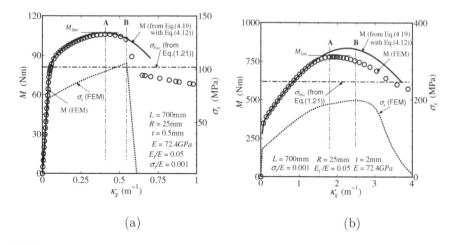

(a)   (b)

FIGURE 4.11

Change of moment and bending stress at the bottom surface in plastic bending of a long cylinder: (a) $t/R = 0.02$; (b) $t/R = 0.08$.

extremum, $dM/d\kappa_g = 0$, as shown in Eq. (4.21). Fig. 4.11 also shows the bending moment M derived from Eq. (4.19) using the approximation formula Eq. (4.12) for the flattening ratio μ. Although a good result is obtained for the cylinder with $t/R = 0.02$ as shown in Fig. 4.11(a), the prediction accuracy of the moment M using Eq. (4.12) is not so good for the cylinder with $t/R = 0.08$ as shown in Fig. 4.11(b). Since M_{lim} is dependent on the conditions of $dM/d\kappa_g = 0$, in order to obtain an accurate M_{lim}, for the approximate formula of flattening ratio, high accuracy is required not only to the value itself, but also to the rate of change of the value. As discussed above, it is difficult to derive an accurate analytical formula of the flattening ratio for the plastic bending of a cylinder. Therefore, many numerical analysis techniques have been proposed to derive M_{lim} (for example [78, 70, 104, 197]).

Regarding these methods, many reports investigate bending curvature κ_{lim} and maximum compressive strain ε_{lim} in the bottom part of a cylinder when the limit moment M_{lim} is reached [83, 152, 205, 230]. Fig. 4.12(a) shows the maximum bending moment M_{lim} for cylinders with different thicknesses obtained by FEM simulation, for three types of yield stress ($E_t/E = 0.05C\sigma_s/E = 0.001, 0.002, 0.003$) in the case of materials obeying the bilinear hardening rule, and for four types of yield stress ($N = 3C\sigma_s/E = 0.00025, 0.001, 0.003, 0.01$) in the case of materials whose stress-strain relation obeys the following Ramberg-Osgood equation shown by Eq. (4.15). In the figure, the horizontal axis shows $(\sigma_s/E) \times (R/t)$. The nondimensional maximum moment M_{lim}/M_{buc}^e, which depends on yield stress and cylinder wall thickness, can be expressed as a function of $(\sigma_s/E) \times (R/t)$ for a material obeying the bilinear hardening rule and also for a material obeying Eq. (4.15)

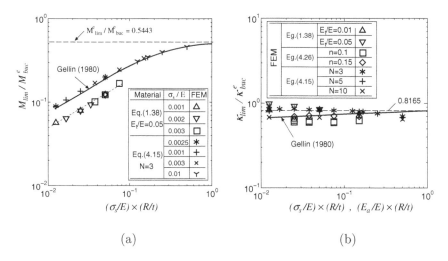

(a)   (b)

FIGURE 4.12
Maximum bending moment M_{lim} and curvature κ_{lim} of a cylinder at $M = M_{lim}$ as functions of $(\sigma_s/E) \times (R/t)$: (a) M_{lim}; (b) κ_{lim}.

with $N = 3$. The ratio M_{lim}/M_{buc}^e increases toward the value for an elastic material as R/t or σ_s/E increases. This was also confirmed by Gellin [78]. The solid line in Fig. 4.12 shows the maximum bending moment M_{lim} of a cylinder obtained by Gellin [78] for a material obeying Eq. (4.15).

Fig. 4.12(b) shows bending curvature κ_{lim} of a cylinder at $M = M_{lim}$ as a function of $(\sigma_s/E) \times (R/t)$ obtained from the FEM simulation for materials obeying the bilinear hardening rule, the hardening rule given by Eq. (4.15), or the nth-power hardening rule given by

$$\sigma = E_a(\varepsilon^p)^n \tag{4.26}$$

The curvature κ_{lim} at $M = M_{lim}$ is almost the same among the materials.

$$\kappa_{lim} \cong 0.6\kappa_{buc}^e - 1.0\kappa_{buc}^e \tag{4.27}$$

The dashed lines in the figure show the curvature κ_{lim} in elastic bending, and the solid line shows Gellin's work [78] for a material obeying Eq. (4.15).

Fig. 4.13 shows the relationship between the maximum compressive strain ε_{lim} at the bottom part of a cylinder and the cylinder wall thickness when the bending moment reaches its maximum value M_{lim} as analyzed by FEM numerical simulation of the pure bending deformation of a cylinder consisting of materials obeying the bilinear hardening rule and the n-power hardening rule. In Fig. 4.13 are also shown experimental results obtained by other researchers [25, 219, 167]. In any material, as the figure shows, the maximum compressive strain ε_{lim} is roughly proportional to t/R, and the proportionality coefficient

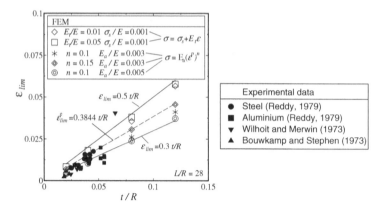

FIGURE 4.13
Maximum strain ε_{lim} at the bottom surface when the bending moment reaches
the maximum value M_{lim}.

is roughly 0.3–0.5.

$$\varepsilon_{lim} \cong 0.3\frac{t}{R} - 0.5\frac{t}{R} \qquad (4.28)$$

The dashed line in the figure shows the relationship between the maximum
compressive strain ε_{lim} at the bottom of the cylinder and the cylinder wall
thickness in elastic bending.[2]

Approximation methods have also been proposed to predict the limit value
M_{lim} of the moment in bending of a cylinder. For example, approximation
methods were proposed [83, 205] to predict the limit value of the moment
by focusing attention on the relationship between the maximum compressive
strain at the bottom surface and wall thickness by assuming

$$\varepsilon_{lim} \cong t/R \qquad (4.29)$$

In extracting the limit value of the moment M_{lim}, instead of using the
relational formula between flattening ratio and curvature, Gerber [83] introduced an unknown constant α, which is used to derive M_{lim} by assuming that
the reduction of the moment due to flattening can be evaluated through the
function $(1 - \alpha\theta)$; that is, assuming

$$M = M|_{circular} \times (1 - \alpha\theta) \qquad (4.30)$$

where θ is the rotation angle, and $M|_{circular}$ is the bending moment of the
cylinder under the assumption that there is no flattening and it is proportional

---

[2]Many researchers assume that the maximum compressive strain ε_{lim} is approximately
equal to $0.6t/R$, $\varepsilon_{lim} \cong 0.6t/R$, but this applies only to collapse due to buckling.

to θ^n for a material obeying the nth-power hardening rule given by Eq. (4.26). The unknown constant α is determined by assuming that the moment takes an extremum when the strain ε_x at the bottom surface of the cylinder is equal to the value $\varepsilon_x = \varepsilon_{lim} \cong t/R$ in Eq. (4.29). Thus, to derive the limit moment M_{lim} in the pure bending of a cylinder with an external diameter R_o and an internal diameter R_i, Gerber [83] proposed the formula

$$M_{lim} = 4E_a t R_o^2 \left(\frac{t}{R_o}\right)^{n-1} \frac{I_n}{(n+3)(n+1)}\left[1 - \left(\frac{R_i}{R_o}\right)^{n+3}\right] \tag{4.31}$$

for the material obeying the nth-power hardening rule given by Eq. (4.26). In Eq. (4.31),

$$I_n = \int_0^{\pi/2} \sin^{n+1}\theta\, d\theta \cong 1 - 0.291n + 0.076n^2 \tag{4.32}$$

Further, in using the approximation formula $\mu \cong C_0\varepsilon_x^2 = C_0(\kappa_g R)^2$ for the relationship between flattening ratio μ and curvature, Ueda [205] proposed a method for determining the unknown coefficient C_0 in the formula, which is needed to derive μ, adopting the same assumption as used by Gerber, namely, that the moment takes an extremum when the strain ε_x reaches the value $\varepsilon_x = \varepsilon_{lim} \cong t/R$ in Eq. (4.29).

Therefore, for the material obeying the nth-power hardening rule given by Eq. (4.26) M_{lim} is given by

$$
\begin{aligned}
M_{lim} = {} & 4E_a t R_o^2 \left(\frac{t}{R_o}\right)^{n-1} \frac{I_n}{(n+3)}\left[1 - \left(\frac{R_i}{R_o}\right)^{n+3}\right]\left[1 - C_0\left(\frac{t}{R_o}\right)^2\right]^2 \\
& \times \left[1 + C_0\left(\frac{t}{R_o}\right)^2\right]
\end{aligned}
\tag{4.33}
$$

where I_n is given by Eq. (4.32), and

$$C_0 = \frac{-1 + \sqrt{1 + 6n + n^2}}{n+6}\left(\frac{R_o}{t}\right)^2 \tag{4.34}$$

On the other hand, to evaluate the limit value of the moment M_{lim} using the formula

$$M = 4tR \int_0^{\pi/2} f_\sigma(\varepsilon)b_{el}\sin\theta d\theta \tag{4.35}$$

in the pure bending of a cylinder, one can assume the maximum compressive strain ε_{lim} in the bottom part of the cylinder at $M = M_{lim}$ to be

$$\varepsilon_{lim} \cong 0.5t/R \tag{4.36}$$

from Fig. 4.13 and the curvature κ_{lim} to be

$$\kappa_{lim} \cong \kappa_{buc}^e \tag{4.37}$$

from Fig. 4.12(b). Note that $f_\sigma(\varepsilon)$ in Eq. (4.35) is the stress-strain formula of the material.

From the assumption of Eq. (4.36), the strain ε at $M = M_{lim}$ is given by

$$\varepsilon = \varepsilon_{lim} \sin\theta \cong 0.5\frac{t}{R}\sin\theta \tag{4.38}$$

and the length of the semi-minor axis b_{el} of the flattened ellipse is given by

$$b_{el}/R \cong 0.5\sqrt{3(1-\nu^2)} \tag{4.39}$$

from the relation $\varepsilon_{lim} = b_{el}\kappa_{lim}$ and the assumptions of Eq. (4.37). Therefore, the limit moment M_{lim} is evaluated as

$$M_{lim} = 4tR \int_0^{\pi/2} 0.5R\sqrt{3(1-\nu^2)}\sin\theta f_\sigma(\varepsilon)d\theta \tag{4.40}$$

where ε is given by Eq. (4.38).

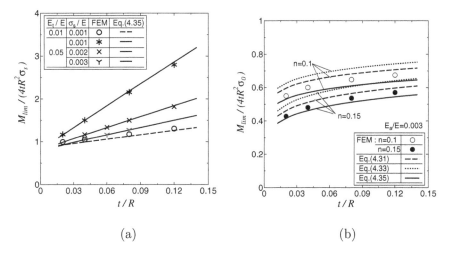

(a)   (b)

FIGURE 4.14
Comparison between FEM simulation and approximation formulas for the maximum bending moment M_{lim}: (a) materials obeying the bilinear hardening rule; (b) materials obeying the nth-power hardening rule.

Fig. 4.14(a) and (b) compare the FEM simulation and the approximation formulas discussed above for the limit moment M_{lim} in the pure bending deformation of a cylinder assuming materials obeying the bilinear hardening rule and the nth-power hardening rule. As the figure shows, the simple approximation formula Eq. (4.35) gives good prediction for materials obeying both the bilinear hardening rule and the nth-power hardening rule.

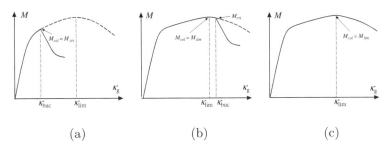

FIGURE 4.15
Schematic relationship between bending moment and bending curvature: (a) $2R/t > 50$; (b) $2R/t = 25$–45; (c) $2R/t < 20$.

4.1.3 Buckling at the bottom wall in bending of a cylinder

In bending deformation of a cylinder, compressive stress occurs due to the bending moment at the cylinder's bottom surface, and bending collapse may occur due to buckling if the compressive stress reaches the critical stress for buckling. The bending moment will begin to decrease once the bottom surface of the cylinder buckles. Therefore, in discussing the bending collapse of a cylinder, it is first necessary to determine whether the maximum bending moment is due to flattening or buckling at the bottom surface.

The relationship between bending moment and curvature in a cylinder is divided into three categories as schematically shown in Fig. 4.15 [118]:

(1) If the wall is very thin (e.g., $2R/t > 50$), then as figure (a) shows, buckling occurs at the bottom surface before the maximum moment due to flattening is reached.

(2) If the wall is thin (e.g., $2R/t \cong 25$–45), then as figure (b) shows, buckling occurs at the bottom surface after the maximum moment due to flattening is reached.

(3) If the wall is thick (e.g., $2R/t < 20$), then as figure (c) shows, buckling does not occur at the bottom surface.

As Fig. 4.15 shows, **the maximum moment** M_{max}, namely the collapse moment M_{col}, in the bending deformation of a cylinder equals one of either **the critical moment** M_{cri} caused by the buckling at the bottom surface due to the compressive stress of bending or **the limit bending moment** M_{lim} due to flattening, depending on which bending curvature is small.

$$M_{max} = M_{col} = f(\kappa_{col}), \quad \kappa_{col} = \min(\kappa_{buc}, \kappa_{lim}) \quad (4.41)$$

With respect to the elastic bending deformation of a cylinder, Gellin [78] pointed out that if cross-sectional flattening does not occur (e.g., if flattening is blocked by constraints at both ends as in the case of a short cylinder), the maximum compressive stress $\sigma^e_{buc}|_{bent}$ that causes buckling at the bottom surface of the cylinder[3] is a slightly greater than or about the same as the buckling stress $\sigma^e_{buc}|_{comp}$ in the axial collapse of the cylinder,[4] and can be evaluated by

$$\sigma^e_{buc}|_{bent} \cong \sigma^e_{buc}|_{comp} = \frac{Et}{\sqrt{3(1-\nu^2)}R} \qquad (4.42)$$

Therefore, in bending collapse due to buckling at the bottom surface of a cylinder subjected to bending compressive stress, the moment at collapse will correspond to the buckling stress of Eq. (4.42), and is given by

$$M^e_{buc} = \frac{E\pi Rt^2}{\sqrt{3(1-\nu^2)}} \qquad (4.43)$$

The bending curvature of the cylinder κ^e_{buc} is thus given by

$$\kappa^e_{buc} = \frac{t}{\sqrt{3(1-\nu^2)}R^2} \qquad (4.44)$$

Comparing Eq. (4.44) with Eq. (4.22) and Eq. (4.43) with Eq. (4.23), one can know

$$\kappa^e_{lim} / \kappa^e_{buc} = 0.8165 < 1 \qquad (4.45)$$

and

$$M^e_{lim} / M^e_{buc} = 0.5443 \qquad (4.46)$$

Therefore, in elastic bending of a long cylinder, the maximum bending moment usually equals the limit value M^e_{lim} due to flattening. However, if the cylinder is short, the cross-sectional flattening is blocked and the limit bending moment due to flattening becomes large. Fig. 4.16 shows the relationship, as evaluated by FEM simulation, between moment and curvature in elastic bending of cylinders of different lengths. The figure also shows the change of the bending moment due to flattening evaluated by Eq. (4.17) for cylinders of length $L/R = 28$ and $L/R = 6$. In the long cylinder of length $L/R = 28$, the limit value M_{lim} of the bending moment obtained by Eq. (4.17) agrees well with Brazier's theory [26] (where $M^e_{lim} / \left(E\pi Rt^2/\sqrt{3(1-\nu^2)}\right) = 0.544$), as well as the maximum moment evaluated by FEM numerical analysis. In the cylinder of $L/R = 6$, the maximum moment obtained by FEM numerical

---

[3]Here, the bending buckling stress is expressed as $\sigma_{buc}|_{bent}$ to distinguish from the buckling stress appearing in the axial compression of the cylinder, but can be simply expressed as σ_{buc} if there is no confusion.

[4]Timoshenko and Gere [203] assumed that, in bending of a cylinder, the buckling stress $\sigma^e_{buc}|_{bent}$ at the bottom surface is 1.3 times as great as the buckling stress $\sigma^e_{buc}|_{comp}$ of axial collapse $\sigma^e_{buc}|_{bent}/\sigma^e_{buc}|_{comp} \cong 1.3$. However this conclusion was based on the research results of Flugge [73] alone, who used a cylinder of $m\pi R/L = 1$, and the conclusion is not universally valid. Seide and Weingarten [181] showed that the ratio $\sigma^e_{buc}|_{bent}/\sigma^e_{buc}|_{comp}$ depends on the ratio $m\pi R/L$ of the semiperimeter of a cylinder and the half-wavelength, and the minimum value is about 1.

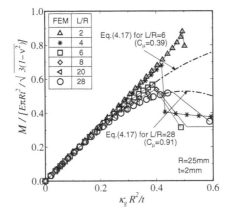

FIGURE 4.16

Relationship between elastic bending moment and curvature for cylinders of different lengths.

analysis is greater than that for a longer cylinder, but smaller than the limit value M_{lim} for the same length of $L/R = 6$ obtained by Eq. (4.17). It is clear that, at the bottom surface, buckling occurs due to the compressive stress generated by bending before the limit bending moment is reached due to flattening if the cylinder is short.

Fig. 4.17 shows the change in bending moment and compressive stress at the bottom surface during plastic bending of long ($L/R = 24$) and short ($L/R = 6$) cylinders, as obtained from the FEM simulation. In a long cylinder, on the one hand, because the compressive stress caused by bending increases even after the maximum moment is reached (point A in the figure), it is clear that the maximum bending moment equals the limit value due to flattening. In a short cylinder, on the other hand, because the time points for the maximum moment and maximum stress are roughly the same and thereafter the bending compressive stress at the bottom surface rapidly decreases, it is reasonable to assume that the bottom surface of the cylinder buckles due to the bending compressive stress at this time point, and that the maximum bending moment is due to buckling at the bottom surface.

In Eq. (4.42), it is assumed that the buckling stress in elastic bending of a cylinder equals that in axial collapse of the cylinder. However, since flattening occurs during bending of the cylinder, the circumferential curvature radius $\rho_b \cong a_{el}^2/b_{el}$ at the bottom surface increases with bending and is greater than the cylinder radius R. Therefore, buckling stress is smaller in bending of a cylinder than in axial collapse of a similarly shaped cylinder [197, 87]. In Fig. 4.18, the symbol ▲ shows buckling stress at the bottom surface in the bending of cylinders, and the symbol × shows, for comparison, the buckling

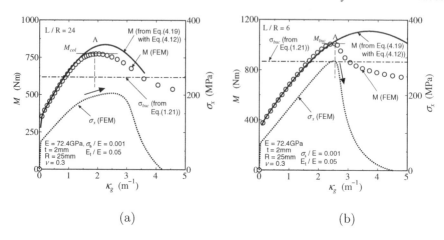

FIGURE 4.17

Moment and bending stress at the bottom surface in plastic bending of a cylinder: (a) $L/R = 24$; (b) $L/R = 6$.

stress in axial collapse of the same cylinders obtained by FEM numerical analysis. The buckling stress in bending of a cylinder decreases as the cylinder becomes longer. This seems attributable to the increase of the circumferential curvature radius ρ_b of the bottom surface in association with the flattening of the cylinder due to bending. To help understand this, the figure also shows the predicted values of the buckling stress given by the theoretical formula with ρ_b, which is obtained from FEM numerical analysis (O) and from approximate value of flattening ratio μ evaluated by Eq. (4.12) (solid line), substituted for R, in order to consider the change of curvature radius ρ_b of the bottom surface. The formula for buckling stress uses Eq. (4.42) for the elastic bending of Fig. 4.18(a) and Eq. (1.21), which is based on the J_2 deformation theory of plasticity of Batterman [20], for the plastic bending of Fig. 4.18(b). As the figure shows, by considering the change in curvature radius ρ_b of the bottom surface, the buckling stress in bending of a cylinder can be estimated from Eq. (4.42) in the case of elastic bending and from Eq. (1.21) in the case of plastic bending. Note that there are also many reports on the analysis of plastic buckling due to bending of a cylinder (for example [168, 78, 154]).

Based on the above discussions, Fig. 4.19 shows a flow chart for determining the critical moment M_{cri} caused by the buckling at the bottom surface due to the compressive stress of bending and the limit bending moment M_{lim} due to flattening.

Fig. 4.20 shows the relationship between the maximum moment M_{col} and cylinder length in the bending of cylinders. The symbols O and ●, indicating collapse due to buckling and flattening, respectively, are obtained from the FEM simulation. The symbols △ and ▲, corresponding to the collapse by buckling and flattening, respectively, show the analytical results based on the

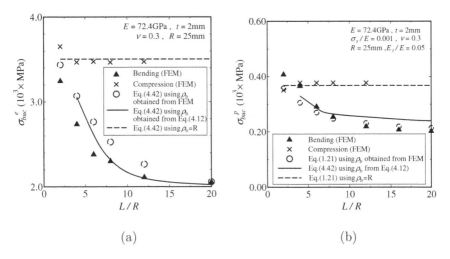

(a)   (b)

FIGURE 4.18
Buckling stress at the bottom surface in bending of a cylinder (including cases
of $\kappa_{buc} > \kappa_{lim}$): (a) elastic bending; (b) plastic bending.

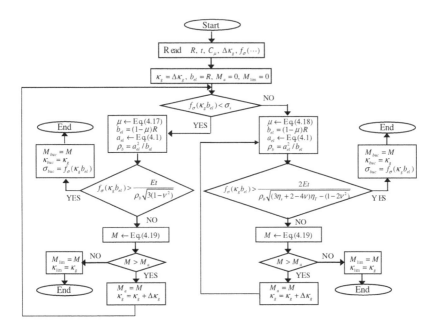

FIGURE 4.19
Flow chart of the calculation method for the critical moment and the limit
bending moment.

flow chart in Fig. 4.19 using theoretical formulae for flattening curves and buckling stress. Furthermore, Fig. 4.20(a) also shows the theoretical analysis by Tatting et al. [197]. Tatting et al. [197] investigated the Brazier effect for cylinders of finite length under bending using semi-membrane constitutive theory [9, 125].

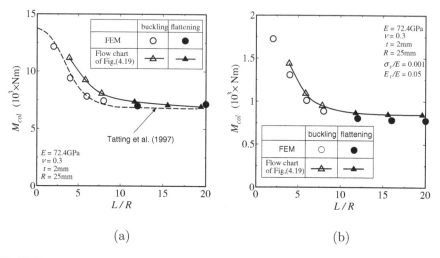

FIGURE 4.20
Relationship between cylinder length and maximum moment M_{col} in bending of a cylinder: (a) elastic bending; (b) plastic bending.

If the tube wall is thick ($2R/t < 20$) and the material is a perfectly elasto-plastic material, since the limit value of stress is the yield stress of the material σ_s, the maximum bending moment can be evaluated by

$$M_{cri} = \sigma_s Z_p \tag{4.47}$$

where Z_p is the plastic section modulus as given by

$$Z_p = 4tR^2 \tag{4.48}$$

Fig. 4.21 shows the change of moment M and compressive stress σ_x at the bottom surface in the bending deformation of a cylinder made of a perfectly elasto-plastic material. As the figure shows, although the lowering of bending moment occurs due to the flattening of the cylinder, the maximum bending moment is equal to $M_{lim} = 4tR^2\sigma_s$, as obtained from Eq. (4.47).

4.1.4   Bending collapse of thin-walled cylinders

The relationship between bending moment and curvature in the thin-walled cylinder shown in Fig. 4.15(a) and (b) is such that, since the deformation

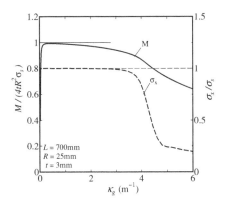

FIGURE 4.21
Change of the bending moment of a cylinder made of a perfectly elasto-plastic material.

is concentrated in one area after the maximum moment M_{col} is reached in the bending deformation, the bending deformation switches from a stable cross-sectional flattening deformation mode to a **plastic collapse** deformation mode, where the deformation is concentrated in the plastic hinge, and as a result, the moment applied to the cylinder rapidly decreases (see Fig. 4.15(a) and (b)). Fig. 4.22 shows the deformation behavior of a cylinder and also a schematic representation of the deformation mode.

In relation to this deformation mode in plastic collapse, Mamalis et al. [137] proposed a mechanism of bending collapse in a thin-walled cylinder. In this mechanism, the bending deformation of a cylinder is, after the maximum moment is reached, concentrated in an area of length $2l$, where only plastic bending occurs without causing in-plane expansion and contraction. Based on the assumption of inextensibility the length $2l$ is shown to be

$$2l \cong 2R \tag{4.49}$$

Therefore as shown in Fig. 4.23, the two cross-sectional planes FG and EH, which are separated by $2l$ from each other, rotate by an angle $\theta_2 = \theta/2$ keeping the shapes as they are; as a result of the rotation, a horizontal hinge AC appears on the top cylinder surface, and thereby triangular planes ACF and ACE appear on both sides of hinge AC. As the bending collapse of the cylinder proceeds, the horizontal hinge AC slowly moves into the inside of the cylinder. The parameters δ_y, ϕ_0 and ϕ_0' (see Fig. 4.23(c)), which are related to the position of hinge AC, are given as follows as functions of rotation angle θ:

$$\delta_y = 2R - y_B \tag{4.50}$$

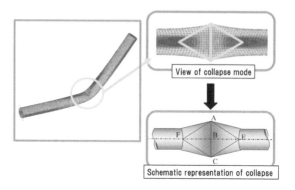

FIGURE 4.22
Bending collapse behavior of a cylinder and the schematic representation.

where y_B is the y coordinate of point B,

$$y_B = 2R\cos\theta_2 - 2\sqrt{\{R\sin\theta_2(l - R\sin\theta_2)\}} \qquad (4.51)$$

and

$$\phi_0 = \arccos\left(\frac{R - \delta_y}{R}\right) \qquad (4.52)$$

$$\sin\phi_0' = \frac{\phi_0(\pi - \phi_0')}{\pi - \phi_0} \qquad (4.53)$$

In this type of bending collapse, the strain energy, which is needed for the deformation to develop, is made up of the following four components.

(1) Strain energy U_1 for deformation from the original cylinder surface to the triangular planes ACF and ACE:

$$U_1 = 4M_0R\phi_0^2 \qquad (4.54)$$

where M_0 is the plastic moment per unit length of plate of thickness t.

(2) Strain energy U_2 associated with the flattening of the circular region:

$$U_2 = 4M_0R(\pi - \phi_0)(\phi_0' - \phi_0) \qquad (4.55)$$

(3) Strain energy U_3 for rotation about hinge AC:

$$U_3 = 2\phi_0 R M_0(\pi - 2\alpha) \qquad (4.56)$$

where

$$\alpha = \arcsin\left(1 - \frac{2R}{l}\sin\theta_2\right) \qquad (4.57)$$

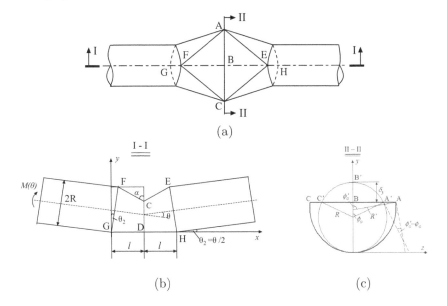

(a)

I - I   II – II

(b)   (c)

FIGURE 4.23
Bending collapse mechanism of a thin-walled cylinder: (a) a plain view of
deformation behavior of tube under bending; (b) a cross section I-I of (a); (c)
a cross section II-II of (a).

(4) Strain energy U_4 for rotation about hinges AE, AF, CE and CF is given
by

$$U_4 = 2M_0 \frac{l_h^2}{l} \phi_0'$$ (4.58)

where l_h is the length of hinges AE, AF, CE and CF as given by

$$l_h = \sqrt{l^2 + (\phi_0 R)^2}$$ (4.59)

Thus, the energy $U(\theta)$ needed for bending deformation with an angle θ is
given by

$$U(\theta) = U_1 + U_2 + U_3 + U_4$$ (4.60)

and the bending moment $M(\theta)$ is given by

$$M(\theta) = \frac{dU(\theta)}{d\theta}$$ (4.61)

Fig. 4.24 compares the moment derived using Mamalis et al.'s Eq. (4.61)
(dashed lines in the figure) and the one obtained from FEM simulation for the
bending deformation of thin-walled cylinders ($t/R = 0.02$ and 0.04) made of a
perfectly elasto-plastic material. As the figure shows, there is a large discrepancy between the analytic theory of Mamalis et al. and the FEM numerical
analysis.

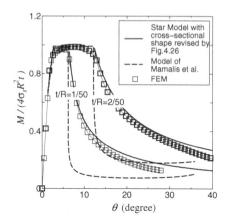

FIGURE 4.24
Relationship between moment and rotational angle in bending of a cylinder.

To improve the model of Mamalis et al., Elchalakani et al. [66] proposed, based on experimentally observed behavior in cylinder collapse, a star model where the collapsed area of the cylinder is star-shaped and proposed a formula for the bending moment. Fig. 4.25 shows the mechanism of the bending collapse of a cylinder proposed by Elchalakani et al. [66]. Their theoretical model differs from Mamalis et al.'s in two ways: (1) the deformed area in bending collapse of a cylinder extends over its entire length and (2) star-shaped, flattened regions ACE_1SE_2 and ACF_1UF_2 appear on both sides of hinge AC as it develops at the center of the cylinder.

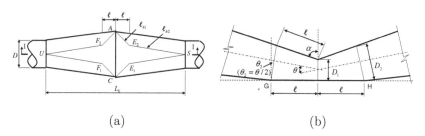

(a)   (b)

FIGURE 4.25
Elchalakani et al.'s deformation model for the bending collapse of a cylinder: (a) a plain view; (b) a cross section I-I of (a).

On the basis of this model, in a similar way to Mamalis et al., Elchalakani et al. [66] derived strain energies U_1–U_4 needed to create plastic hinges and to change the cross-sectional shape as follows:

$$\begin{cases} U_1 &= 2BM_0R\phi_0^2 \\ U_2 &= 2BM_0R(\pi - \phi_0)(\phi_0' - \phi_0) \\ U_3 &= 2\phi_0 RM_0(\pi - 2\alpha) \\ U_4 &= M_0\phi_0'\left(\dfrac{l_{h1}^2}{l} + 2l_{h1} + \dfrac{l_{h2}^2}{(L_0/2 - l)}\right) \end{cases} \tag{4.62}$$

Here,

$$\begin{cases} B = 3 + \dfrac{L_0/2 - R}{R} \\ \alpha = \arcsin\left(1 - \dfrac{D_2}{l}\sin\theta_2\right) \\ l_{h1} = \sqrt{l^2 + (AC/4)^2} \\ l_{h2} = \sqrt{(L_0/2 - l)^2 + (AC/4)^2} \end{cases} \tag{4.63}$$

In the analysis, Elchalakani et al. [66] used an analysis method different from that of Mamalis et al. for the relationship between the parameters ϕ_0 and ϕ_0' for the deformation behavior and rotational angle θ of the cylinder. In the analysis of Mamalis et al., based on Fig. 4.23(c), Eq. (4.51) is substituted into Eq. (4.50), and the angles ϕ_0 and ϕ_0' are derived from Eqs. (4.52) and (4.53). In the analysis of Elchalakani et al., instead of using Fig. 4.23(c), the following approximation formula was proposed:

$$\begin{cases} \phi_0 = a + b\dfrac{\theta}{\theta_s} \\ \phi_0' = \arccos\left(\dfrac{A_1^2 - 1}{A_1^2 + 1}\right) \end{cases} \tag{4.64}$$

where

$$\begin{cases} \theta_s = \dfrac{2\sigma_s L_0}{E(2R + t)}, \\ A_1 = 2\left(\dfrac{c + k\theta/\theta_s}{a + b\theta/\theta_s}\right) \end{cases} \tag{4.65}$$

Further, for the parameters a, b, c and k used in the approximation formula, they proposed the empirical values of $a = 0.6541$, $b = 0.0244$, $c = 1.259$ and

$$k = \min(\,0.373\alpha_1 - 5.356,\ 0.00314\alpha_1 - 0.174\,) \tag{4.66}$$

where

$$\alpha_1 = \dfrac{Et}{2\sigma_s R}$$

However, it is found that the applicable range of the approximation formula, Eq. (4.64), is limited and that the cross section passing through hinge AC shown in Fig. 4.23(c) is not accurate and should be corrected by Fig. 4.26. Based on Figs. 4.26 and 4.25(b), the relationship between angle ϕ_0' and rotational angle θ of the cylinder can be obtained as the root of the following equation:

$$D_2^2 - 2D_2(l\sin\theta_2 + D_1\cos\theta_2) + D_1^2 = 0 \tag{4.67}$$

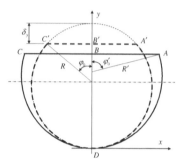

FIGURE 4.26
Cross-sectional shape in bending collapse of a cylinder (with the cross section passing through hinge AC).

Once ϕ_0' is derived, ϕ_0 is given by

$$\phi_0 = \frac{\pi \sin \phi_0'}{\pi - \phi_0' + \sin \phi_0'} \tag{4.68}$$

Fig. 4.24 also shows the results of the theoretical analysis using the star model in Fig. 4.25 proposed by Elchalakani et al., but with the cross-sectional deformation model revised by Fig. 4.26.

Furthermore, various analysis methods have been proposed for the bending collapse behavior of thick-walled cylinders (for example [163, 228].

4.2   Bending deformation of rectangular tubes

Many researchers have studied bending deformation of rectangular tubes (e.g., [111, 218, 112]). Such bending deformation is similar to that of a cylinder: cross-sectional flattening is observed in the bending of rectangular tubes as well. Therefore, the contributing factors to bending collapse of a rectangular tube will include not only the critical compressive stress applied to the bottom wall of the tube but also the limit bending moment due to flattening.

The critical values of compressive stress applied to the bottom wall of a rectangular tube are mainly based on compression buckling at the bottom wall and plastic yielding of the full cross section. As discussed in Chapter 2, however, in axial compressive deformation of a rectangular tube, a compressive force is continuously applied to the corners of the tube even after buckling occurs at the wall plate. This also occurs in bending deformation of a rectangular tube.

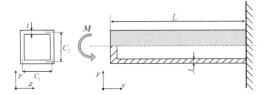

FIGURE 4.27
Rectangular tube to which a pure bending moment is applied.

In this section, bending deformation of a rectangular tube without partition walls is first discussed, followed by a tube with partition walls for use in bumpers, for example. Here, a tube with a cross-sectional area $C_1 \times C_2$, wall thickness t, and length L as shown in Fig. 4.27 is considered, and particularly, unless otherwise noted, a square tube ($C_1 = C_2 = C$ in the figure) is considered.

4.2.1 Flattening in bending deformation of a rectangular tube

Timoshenko was the first to point out the flattening of a rectangular tube [200].

Fig. 4.28 shows the cross section of a square tube deformed by bending as analyzed by FEM simulation. The cross section shrinks vertically and stretches laterally. The flattened cross section is almost vertically symmetric, and the vertical distance is the smallest at the center. This distance C_{2f} is shown in the figure, and the vertical shrinkage is evaluated by the **flattening ratio** as defined by

$$\mu = 1 - \frac{C_{2f}}{C_2} \tag{4.69}$$

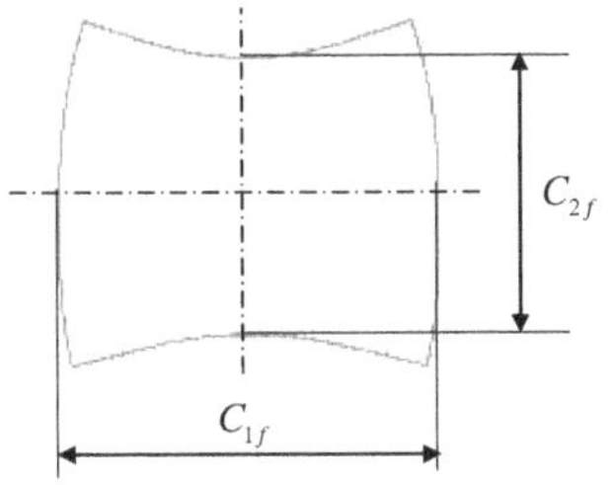

FIGURE 4.28
Cross-sectional shape of a square tube in bending deformation obtained from FEM ($C = 50$ mm, $t = 3$ mm, $\kappa_g = 3.24$ m^{-1}).

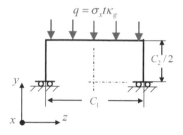

FIGURE 4.29
Mechanical model for the flattening of a rectangular tube.

The cross-sectional flattening in bending deformation of a rectangular tube can be treated as a deformation problem of a Π-shaped frame under the distributing force density q as shown in Fig. 4.29.

In elastic bending, the density q of the distributing force is evaluated by

$$q = E\kappa_g^2 t C_2/2 \tag{4.70}$$

where κ_g is the bending curvature of the tube's central axis. Therefore, by analyzing the deformation of the frame in Fig. 4.29 as a beam problem, the flattening ratio μ is given by

$$\mu = (1 - \nu^2)\frac{(C_1 + 5C_2)}{32(C_1 + C_2)}\left(\frac{C_1}{C_2}\right)^4\left(\frac{\kappa_g C_2^2}{t}\right)^2 \tag{4.71}$$

Furthermore, the ratio of vertical shrinkage and lateral stretch can be evaluated by

$$\frac{1 - \dfrac{C_{2f}}{C_2}}{\dfrac{C_{1f}}{C_1} - 1} = \frac{C_1^2(C_1 + 5C_2)}{4C_2^3} \tag{4.72}$$

For a square tube with $C_1 = C_2 = C$, Eq. (4.71) is transformed into

$$\mu = (1 - \nu^2)\frac{3}{32}\left(\frac{\kappa_g C^2}{t}\right)^2 \tag{4.73}$$

This shows that, in a square tube, the flattening ratio μ is a sole function of nondimensional curvature $\kappa_g \times (C/2)^2/t$. Furthermore, for $C_1 = C_2 = C$ Eq. (4.72) becomes

$$\frac{1 - \dfrac{C_{2f}}{C}}{\dfrac{C_{1f}}{C} - 1} = 1.5 \tag{4.74}$$

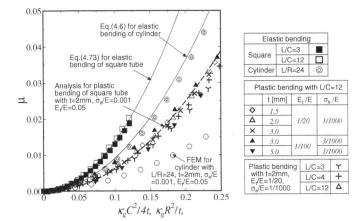

FIGURE 4.30
Change of flattening ratio μ in bending deformation of a square tube.

This shows that the amount of vertical shrinkage is 1.5 times the lateral stretch of a square tube.

Many reports have covered cross-sectional flattening due to plastic bending of a square tube (e.g., [90, 59]). Fig. 4.30 shows the relationship between flattening ratio μ and nondimensional curvature $\kappa_g \times C^2/4t$ in the bending deformation of a square tube obtained from the FEM simulation.

The symbol $\square$ in Fig. 4.30 shows the flattening ratio in elastic bending deformation of a square tube ($L/C = 12$), which agrees well with the theoretical formula, Eq. (4.73), as shown in the figure.

Also, Fig. 4.30 shows the flattening ratio μ in plastic bending deformation of a square tube, as indicated by the symbols $\triangle$, $\blacktriangle$, $\blacktriangledown$, etc. It is assumed that the material obeys the bilinear hardening rule. Assuming the same curvature, the flattening ratio is smaller in plastic bending than in elastic bending. Furthermore, in elastic bending, the flattening ratio μ can be expressed by a sole function of nondimensional curvature $\kappa_g C^2/4t$ independently of tube thickness; in plastic bending, it seems that the relation between flattening ratio μ and $\kappa_g C^2/4t$ is influenced by the tube thickness and the material parameters of yield stress and strain hardening. However, these effects are very small. Paulsen and Welo [162] also showed that for the flattening deformation the importance of the yield stress as well as the hardening of the material are surprisingly small.

For plastic bending deformation of a square tube, the cross-sectional flattening ratio can also be analyzed by using the mechanical model of a Π-shaped frame as shown in Fig. 4.29, similarly to the elastic tube. The broken line in Fig. 4.30 shows the flattening ratio derived through this analysis, which roughly agrees with the result of FEM numerical analysis.

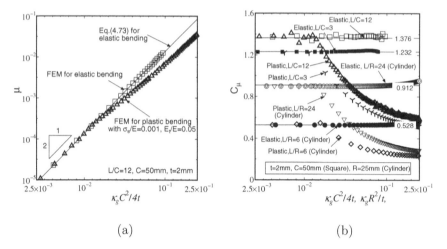

(a)   (b)

FIGURE 4.31
Bending flattening of a square tube: (a) flattening ratio in a double logarithmic plot; (b) coefficient of flattening ratio $C_\mu = \mu \, / \, [\kappa_g C^2/4t]^2$.

On double logarithmic axes, Fig. 4.31(a) re-plots the flattening curve of a square tube for the material with $\sigma_s/E = 0.001$, $E_t/E = 0.05$ and thickness $t = 2$ mm in Fig. 4.30 and also shows the flattening curve for elastic bending. In plastic bending, the slope of the flattening curve in double logarithmic plot is initially equal to 2 before plastic yielding, drops afterward due to the plastic yielding caused by bending stress at the upper and lower parts of the tube, and then slowly increases back to 2 as the plastic deformation proceeds.

Fig. 4.30 also shows the flattening ratio of a square tube of different lengths. To investigate the influence of length, first, the **coefficient of flattening ratio** C_μ is defined as the ratio between flattening ratio μ and the square of $\kappa_g \times C^2/4t$:

$$C_\mu = \frac{\mu}{\left(\kappa_g \times C^2/4t\right)^2} \tag{4.75}$$

Fig. 4.31(b) shows the change of the coefficient of flattening ratio C_μ in bending deformation. In elastic bending, if the square tube is long the coefficient C_μ is 1.376, which agrees with Eq. (4.73). On the other hand, when the tube is short, the flattening is blocked at the end parts and also at the central part; as a result, the flattening ratio becomes small. The coefficient C_μ was 1.232 in the tube of length $L/C = 3$ in Fig. 4.31(b).

For comparison, Figs. 4.30 and 4.31(b) also show the flattening in the bending of cylinders. Corresponding to C_μ of a square tube, the coefficient of flattening ratio C_μ of a cylinder is given by $C_\mu = \mu/\left(\kappa_g \times R^2/t\right)^2$. The coefficient C_μ of a cylinder is smaller than that of a square tube. For example, the coefficients C_μ of square tubes with $L/C = 12$ and $L/C = 3$ are about 1.376 and 1.232, respectively; the coefficients C_μ of cylinders with lengths $L/2R = 12$ and $L/2R = 3$ are about 0.912 and 0.528, respectively.

4.2.2 Maximum bending moment due to cross-sectional flattening

In bending deformation of a square tube, because of cross-sectional flattening, bending stiffness of the tube decreases and thus the relationship between bending moment and bending curvature becomes nonlinear. If the rate at which the bending stiffness lowers becomes greater than the rate at which the bending moment increases with increasing strain, the moment applied to the tube begins to decrease. As a result, the maximum bending moment M_{lim} appears in the curve relating bending moment and bending curvature for a square tube [146].

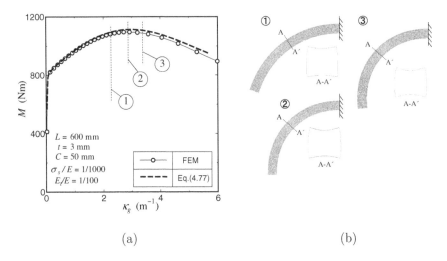

(a)   (b)

FIGURE 4.32
Collapse of a square tube due to flattening: (a) relations of M and κ_g; (b) deformation behavior.

To demonstrate the maximum bending moment due to the cross-sectional flattening, assuming a square tube of $\sigma_s/E = 0.001$, $E_t/E = 0.01$, $C = 50$ mm and thickness $t = 3$ mm, Fig. 4.32 shows the relationship between bending moment M and curvature κ_g in pure bending deformation and also the bending deformational behavior. Even after the maximum bending moment is reached, buckling ripples are not created at the bottom surface on the compression side of the square tube, and the cross section A-A' in (b) is dented. Therefore, it is

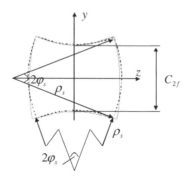

FIGURE 4.33
Approximate cross-sectional shape of a square tube deformed by bending.

thought that the maximum bending moment is attributed to cross-sectional flattening of the tube.

Analysis of flattening of a square tube also shows that the change of the moment is due to cross-sectional flattening, as shown in Fig. 4.32. First, it is necessary to approximate the flattened cross section. As a simple approximation shown in Fig. 4.33, here, the top and bottom walls and both side walls in the deformed cross section are all approximated by a circular arc of radius ρ_s and central angle $2\phi_s$. The radius ρ_s and angle ϕ_s are determined by solving the following simultaneous equations:

$$\begin{cases} C &=& 2\phi_s \rho_s, \\ C_{2f} &=& 2\rho_s(\sin\phi_s + \cos\phi_s - 1) \end{cases} \tag{4.76}$$

so that the length of the arc equals C and the vertical distance at the center equals C_{2f}. Under these approximations, the cross-sectional shape of a square tube can be expressed using a parameter C_{2f}, that is, flattening ratio μ.

Thus, if the flattening ratio of a square tube is evaluated using the mechanical model in Fig. 4.29, the cross-sectional shape of the flattened tube can be determined using Eq. (4.76), and therefore, in bending deformation, bending moment $M(\kappa_g)$ can be given by

$$M(\kappa_g) = \oint_\ell f_\sigma(\kappa_g y) y t \, ds \tag{4.77}$$

as a function of bending curvature κ_g. Here, ℓ is a closed curve of the center line of the flattened cross-sectional plate, y is the distance between the point on the closed curve and the central axis, and $f_\sigma(\varepsilon)$ is the stress-strain relation of the material.

The bending moment as obtained above is shown by the broken line in Fig. 4.32, agreeing well with the results of the FEM numerical analysis.

4.2.3 Maximum bending moment due to the critical stress at wall plates

The maximum bending moment M_{max} due to the critical stress at the tube wall, namely the **critical bending moment** M_{cri}, will be controlled by the buckling at the tube wall or the maximum stress in the tube material (e.g., the plastic yield stress in a perfectly elasto-plastic material). Kecman [111] was the first to predict the bending collapse load of a rectangular tube due to the critical stress. In the method of Kecman, the collapse moment in pure bending deformation of a rectangular tube is predicted from the maximum buckling moment at the bottom wall when the wall is thin, or by assuming that the material is a rigid perfectly plastic solid and the stress at the bottom wall equals the plastic yield stress when the wall is thick. The method is discussed in more detail in the following.

4.2.3.1 Kecman's method for predicting the maximum bending moment of rectangular tubes

Kecman [111] focused on buckling and plastic yielding at the compression bottom wall and proposed a formula to predict the collapse load or the maximum moment M_{max}. Depending on the value of buckling stress σ_{buc-1} of the compression bottom wall:

$$\sigma_{buc-1} = \frac{k_1 \pi^2 E}{12 \left(1 - \nu^2\right)} \left(\frac{t}{C_1}\right)^2 \tag{4.78}$$

The following three cases can be distinguished as shown in Fig. 4.34. In Eq. (4.78), k_1 is the buckling coefficient, which Kecman assumed to be

$$k_1 = 5.23 + 0.16 \frac{C_1}{C_2} \tag{4.79}$$

(1) Case 1 ($\sigma_{buc-1} \le \sigma_s$)

Elastic buckling occurs at the bottom wall in the compression side, and after that the stress increases up to the yield stress at both edges. To consider this, Kecman adopted an effective width a_e, defined by

$$a_e = C_1 \left(0.7 \frac{\sigma_{buc-1}}{\sigma_s} + 0.3\right) \tag{4.80}$$

and assumed the stress distribution on the cross section as shown in Fig. 4.34(a). Here, y_1 is the distance from the bottom wall in the compression side to the neutral axis obtained assuming that the resultant force is zero and is given by

$$\frac{y_1}{C_2} = \frac{C_1 + C_2}{a_e + C_1 + 2C_2} \tag{4.81}$$

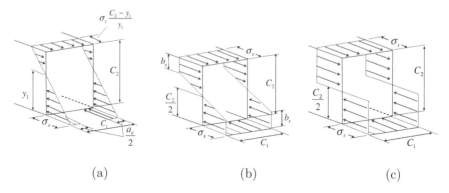

FIGURE 4.34

Schematic representation of axial stress distribution used in Kecman's method: (a) Case 1: $\sigma_{buc-1} < \sigma_s$; (b) Case 2: $\sigma_s \leq \sigma_{buc-1} < 2\sigma_s$; (c) Case 3: $\sigma_{buc-1} \geq 2\sigma_s$.

Thus, the maximum moment M_{max} for the rectangular tube is given by

$$\text{for Case 1:} \qquad M_{\max} = \sigma_s t C_2^2 \frac{2C_1 + C_2 + a_e \left(3\dfrac{C_1}{C_2} + 2 \right)}{3(C_1 + C_2)} \qquad (4.82)$$

(2) Case 2 ($\sigma_s < \sigma_{buc-1} < 2\sigma_s$)

Kecman supposed that the tube would approximate the cross-sectional fully plastic yielding if the elastic buckling stress of compression of the bottom wall σ_{buc-1} obtained from Eq. (4.78) reaches twice of material yield stress, and that when $\sigma_s < \sigma_{buc-1} < 2\sigma_s$ the maximum moment of the tube can be evaluated by linear interpolation of the maximum elastic moment M_{el} and the fully plastic bending moment M_{pl}.

$$\text{for Case 2:} \qquad M_{\max} = M_{el} + (M_{pl} - M_{el}) \frac{\sigma_{buc-1} - \sigma_s}{\sigma_s} \qquad (4.83)$$

Here, M_{el} is the maximum elastic moment, which is the moment realized when the yield stress σ_s is just reached at the bottom wall of the tube and is given by

$$M_{el} = \sigma_s t C_2 \left(C_1 + \frac{C_2}{3} \right) \qquad (4.84)$$

M_{pl} is the cross-sectional fully plastic bending moment of the tube and is given by

$$M_{pl} = \sigma_s t C_2 \left(C_1 + \frac{C_2}{2} \right) \qquad (4.85)$$

The cross-sectional stress distribution is shown in Fig. 4.34(b), in which b_s

shows the range of plastic yielding region in the sidewall. The maximum moment corresponding to Fig. 4.34(b) can also be evaluated using the value of b_s as follows:

$$\text{for Case 2:} \quad M_{max} = \sigma_s t \left[\frac{1}{6} \left(2C_2^2 + 2C_2 b_s - b_s^2 \right) + C_1 C_2 \right] \quad (4.86)$$

Substituting Eqs. (4.84), (4.85) and (4.86) into Eq. (4.83), b_s is obtained as

$$\frac{b_s}{C_2} = \begin{cases} 1 - \sqrt{2 - \left(\dfrac{t}{t_{ea}} \right)^2} & \text{(for } t_{ea} < t < \sqrt{2} t_{ea}) \\ 1 & \text{(for } t \geq \sqrt{2} t_{ea}) \end{cases} \quad (4.87)$$

where t_{ea} is the bottom wall thickness for which elastic buckling stress σ_{buc-1} obtained from Eq. (4.78) is equal to the yielding stress σ_s and is given by

$$t_{ea} = C_1 \sqrt{\frac{12(1 - \nu^2)}{k_1 \pi^2}} \sqrt{\frac{\sigma_s}{E}} \quad (4.88)$$

(3) Case 3 ($\sigma_{buc-1} \geq 2\sigma_s$)

The cross-sectional stress distribution is shown in Fig. 4.34(c). In other words, the maximum moment M_{max} of such a tube is assumed equal to the cross-sectional fully plastic bending moment of the tube M_{pl}.

$$\text{for Case 3:} \quad M_{\max} = M_{pl} \quad (4.89)$$

Using Eq. (4.88), the condition of $\sigma_{buc-1} \geq 2\sigma_s$ for Case 3 is equivalent to the following condition:

$$t \geq \sqrt{2} t_{ea} \quad (4.90)$$

Fig. 4.35 shows the results of FEM numerical analyses for tubes with aspect ratios $C_2/C_1 = 1$, 2 and 3, and the corresponding values obtained from Eq. (4.82), (4.83) and (4.89). As shown in the figures, the prediction of Kecman's method is well in agreement with the results of FEM when the relative thickness t/C_1 is not very small and the aspect ratio of sidewall to bottom wall C_2/C_1 is not large, for example, when the tube relative thickness is about $t/C_1 \geq 0.008$ for $C_2/C_1 = 1$ and is about $t/C_1 \geq 0.016$ for $C_2/C_1 = 2$. However, for large aspect ratios, there is a large discrepancy between the values of maximum moment obtained from Kecman's method and the FEM numerical results. It is thought that the errors originate in the compression buckling in the sidewall of the tube. In bending deformation of a rectangular tube, buckling can also occur in the sidewall earlier than at the bottom wall, and in particular, the maximum moment of a rectangular tube is determined by buckling in the sidewall in a tube having a cross section with a large aspect ratio C_2/C_1 [35]. The method to predict the maximum moment considering sidewall buckling in bending deformation of a rectangular tube is now discussed.

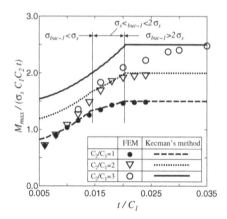

FIGURE 4.35
Comparison of Kecman's method and the FEM analysis.

4.2.3.2   Prediction of maximum moment considering sidewall buckling

Bending stress is applied to the sidewall of a tube. The problem of sidewall buckling is expressed by Fig. 4.36. In Fig. 4.36(a), displacement in the out-of-plane direction (displacement in the z direction) is fixed at both longitudinal edges (BC and DA) in plate ABCD of width b and thickness t; the bending and compression, which are linearly distributed on both edges (AB and CD), are applied through displacement control. For the ultimate loading, the distribution of compressive stress σ_x along the width direction is characterized by two effective widths, b_{e1} and b_{e2}, as shown in Fig. 4.36(b). In the figure, compressive stress is denoted by a positive value.

There have been many reports on the elastic buckling of a plate to which bending stress is applied and on the stress distribution at maximum load [173, 156, 172, 13, 227, 131]. For example, the effective widths b_{e1} and b_{e2} for a plate under stress gradient shown in Fig. 4.36 are given in AS/NZS 4600 standard (2005) [58] and NAS (2001) [11] as follows:

$$
\begin{cases}
b_{e1} = \dfrac{b_e}{3 - \psi} & \\[2mm]
b_{e2} = \begin{cases} b_e/2 & \text{when } \psi \leq -0.236 \\ b_e - b_{e1} & \text{when } \psi > -0.236 \end{cases}
\end{cases}
\tag{4.91}
$$

In addition, $b_{e1} + b_{e2}$ shall not exceed the compression portion of the sidewall.

Here, ψ is the ratio of f_1^* and f_2^*. f_1^* and f_2^* are sidewall stresses shown in Fig. 4.36(b):

$$
\psi = \frac{f_2^*}{f_1^*}
\tag{4.92}
$$

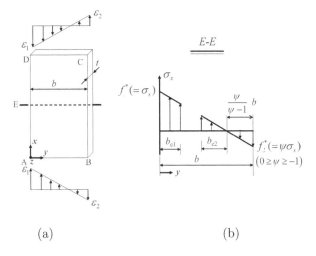

FIGURE 4.36
Plate subjected to compression and bending: (a) analyzed model; (b) axial compressive stress σ_x distribution on E-E cross section in (a).

λ is defined by

$$\lambda = \sqrt{\frac{\sigma_s}{\sigma_{buc-2}}} \tag{4.93}$$

The elastic buckling stress of sidewall σ_{buc-2} is calculated as follows:

$$\sigma_{buc-2} = \frac{k_2 \pi^2 E}{12 (1 - \nu^2)} \left(\frac{t}{C_2}\right)^2 \tag{4.94}$$

where the buckling coefficient k_2 is given by

$$k_2 = 4 + 2(1 - \dot{\psi})^3 + 2(1 - \psi) \tag{4.95}$$

b_e is given by

$$b_e = \rho \, C_2 \tag{4.96}$$

ρ is called the reduction factor and is given by

$$\rho = \frac{1}{\lambda} \tag{4.97}$$

which is proposed by von Karman et al. [211].

A different solution from Eq. (4.91) also exists. Masuda and Chen [145] used another solution published in the paper of Rusch and Lindner [177] to evaluate the effective width of sidewall after buckling. The solution in [177] is given for the same plate shown in Fig. 4.36(a) but with one of the two longitudinal edges BC being free. Although the free boundary condition at the longitudinal edge BC is different from the actual situation of sidewall constituting the tube, the effect of the boundary condition at the edge BC is assumed to be small, because the edge BC is under tension stress.

According to the solution published in [177] the effective widths b_{e1} and b_{e2} are given by

$$\begin{cases} b_{e1} & = & b_e - b_{e2} \\ \dfrac{b_{e2}}{C_2} & = & \dfrac{0.226}{\lambda^2} \end{cases} \tag{4.98}$$

where

$$\frac{b_e}{C_2} = \frac{\rho}{1 - \psi} \tag{4.99}$$

Here, λ and ρ are calculated by Eqs. (4.93) and (4.97), respectively, the buckling stress σ_{buc-2} is determined by Eq. (4.94) with k_2 determined as follows:

$$k_2 = 1.7 - 5\psi + 17.1\psi^2 \tag{4.100}$$

For tubes with large aspect ratio, as an effect of sidewall slenderness on the tube collapse, it is necessary to consider the possible buckling of sidewall, corresponding to which there are two possible critical stress distributions: Case 4 and Case 5, as shown in Fig. 4.37.

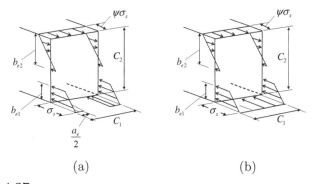

(a)   (b)

FIGURE 4.37
Schematic representation of axial stress distribution considering the buckling at sidewall when the maximum moment occurs: (a) Case 4: $\sigma_{buc-1} < \sigma_s$; (b) Case 5: $\sigma_{buc-1} > \sigma_s$.

For tubes with large aspect ratio, it is also necessary to consider that the condition for reaching the cross-sectional fully plastic yielding is also related to the sidewall slenderness. Therefore, the condition of $\sigma_{buc-1} \geq 2\sigma_s$ for Case 3 or for $M_{max} = M_{pl}$ in Kecman's method is replaced now by the following condition:

$$\begin{aligned} \sigma_{buc-1} &\geq 2\sigma_s \\ \sigma_{buc-2} &\geq 2\sigma_s \end{aligned} \tag{4.101}$$

Here, σ_{buc-2} is determined by assuming $\psi = -1$.

When Eq. (4.101) is not satisfied but $\sigma_{buc-1} > \sigma_s$ and $\sigma_{buc-2} > \sigma_s$, the stress on the cross section is expressed by Case 2 shown in Fig. 4.34(b), and $M_{\max}$ is determined as follows:

$$M_{\max} = \min(M_{\max 1}, M_{\max 2}) \tag{4.102}$$

where

$$M_{\max 1} = M_{el} + (M_{pl} - M_{el}) \frac{\sigma_{buc-1} - \sigma_s}{\sigma_s} \tag{4.103a}$$

$$M_{\max 2} = M_{el} + (M_{pl} - M_{el}) \frac{\sigma_{buc-2} - \sigma_s}{\sigma_s} \tag{4.103b}$$

Using Eq. (4.86), $M_{\max}$ for Case 2 can also be determined by the value of b_s. Based on Eq. (4.103a) b_s is calculated using Eq. (4.87), and based on Eq. (4.103b), b_s is calculated as

$$\frac{b_s}{C_2} = \begin{cases} 1 - \sqrt{2 - \left(\dfrac{t}{t_{eb}}\right)^2} & \text{(for } t_{eb} < t < \sqrt{2}t_{eb}) \\ 1 & \text{(for } t \geq \sqrt{2}t_{eb}) \end{cases} \tag{4.104}$$

where t_{eb} is the sidewall thickness for which the elastic buckling stress is equal to the yielding stress and is given by

$$t_{eb} = C_2 \sqrt{\frac{12(1 - \nu^2)}{k_2 \pi^2}} \sqrt{\frac{\sigma_s}{E}} \tag{4.105}$$

Based on Eq. (4.102), M_{max} is determined from the smaller one from both $M_{\max 1}$ and $M_{\max 2}$. Therefore, when using Eq. (4.86) to determine M_{max} for Case 2, the value of b_s is calculated using Eq. (4.87) if

$$t_{ea} > t_{eb} \tag{4.106}$$

and is calculated using Eq. (4.104) if

$$t_{ea} < t_{eb} \tag{4.107}$$

Using t_{ea} and t_{eb} the condition of Eq. (4.101) can be rewritten as

$$t \geq \sqrt{2}t_{ea} \quad \text{and} \quad t \geq \sqrt{2}t_{eb} \tag{4.108}$$

The above-stated method for predicting the maximum moment of tubes under pure bending is summarized and is shown by a flow chart in Fig. 4.38, from which it is seen that both the possible buckling at sidewall and the effect of sidewall slenderness on the cross-sectional fully plastic yielding are taken into account in the method. In the flow chart $\sigma_{buc-2,1}$ and $\sigma_{buc-2,2}$ are the buckling stress of sidewall assuming the stress ratio ψ to be

$$\psi = -\frac{C_2 - y_1}{y_1} = -\frac{a_e + C_2}{C_1 + C_2} \tag{4.109}$$

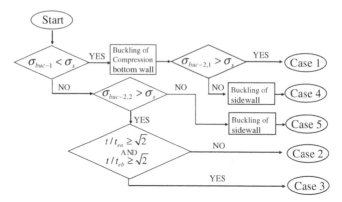

FIGURE 4.38
Flow chart for predicting the maximum bending moment considering sidewall
buckling.

and $\psi = -1$, respectively. Moreover, it is notable that in calculating the maximum bending moment for Case 4 and Case 5 the stress ratio ψ is also unknown, and shall be determined from the conditions of pure bending through trial and error. Using the determined value of ψ, the maximum moment for Case 4 and Case 5 is calculated as follows:

$$\text{for Case 4:} \quad \frac{M_{\max}}{\sigma_s t} = \frac{\psi}{2}[C_2^2 + d_1^2 - d_2^2] + \frac{1-\psi}{3C_2}[C_2^3 + d_1^3 - d_2^3] + 2a_e C_2 \quad (4.110)$$

$$\text{for Case 5:} \quad \frac{M_{\max}}{\sigma_s t} = \frac{\psi}{2}[C_2^2 + d_1^2 - d_2^2] + \frac{1-\psi}{3C_2}[C_2^3 + d_1^3 - d_2^3] + 2C_1 C_2 \quad (4.111)$$

In Eqs. (4.110), (4.111),

$$d_1 = b_{e2} + \frac{\psi}{\psi - 1}C_2 , \quad d_2 = C_2 - b_{e1} \quad (4.112)$$

In Fig. 4.39, the maximum moment predicted by the method shown in Fig. 4.38 with sidewall slenderness considered is compared with that of the FEM numerical analysis. As results of the prediction, both the analyses using Eq. (4.91) and using Eq. (4.98) for calculating the effective width are shown, respectively; the former is denoted as "Method-1" and the latter is denoted as "Method-2."

The case number of the collapse corresponding to each thickness is also shown in the figures. In Case 2 there are two possible subcases: (1) $t_{ea} > t_{eb}$ as shown in Figs. 4.39(a) and (b) and (2) $t_{ea} < t_{eb}$ as shown in Fig. 4.39(c); the maximum moment is determining by Eq. (4.103a) for the former and by Eq. (4.103b) for the latter. As shown in these figures, Eq. (4.103a) and Eq. (4.103b) give good prediction to the subcase corresponding, respectively.

For Case 4 and Case 5, although each result obtained from Method-1 and Method-2 is approximately in agreement with the analysis of FEM, it is found that there is a gap in the results between Method-1 and Method-2. When the buckling stress of the sidewall σ_{buc-2} is close to the yielding stress σ_s, the Method-1 gives a too large prediction as compared with FEM analysis. However, for small t/C_1 when σ_{buc-2} is very much less than σ_s, Method-1 is more accurate compared with the Method-2.

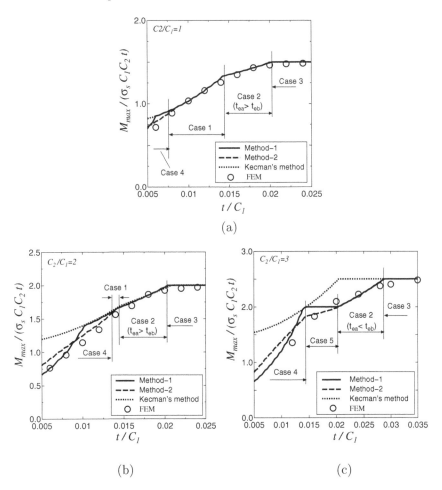

FIGURE 4.39

Prediction of the maximum bending moment M_{max} for rectangular tubes: (a) $C_2/C_1 = 1$; (b) $C_2/C_1 = 2$; (c) $C_2/C_1 = 3$.

4.2.4 Bending collapse of a thin-walled rectangular tube

In bending deformation of a rectangular tube, the analysis of **bending collapse** (bending deformation after the bending moment reaches the maximum) is generally evaluated assuming a rigid perfectly plastic material. Fig. 4.40 shows the collapsed shape due to bending deformation. In the collapse deformation of a thin-walled rectangular tube, plates protrude in both sides and plastic hinges appear in the sidewall and in the bottom surface to which bending compressive stress has been applied, and the tube bends as a result of the rotation about the hinges.

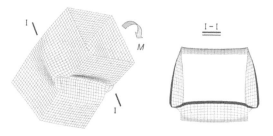

FIGURE 4.40
Collapse shape of a thin-walled square tube subjected to bending deformation as obtained from FEM analysis.

Based on these deformation characteristics, Kecman [111] proposed a collapse mechanism for a rectangular tube shown in Fig. 4.41 and analyzed the relationship between moment and rotational angle in the bending collapse of a thin-walled tube.

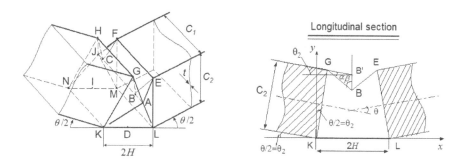

FIGURE 4.41
Theoretical model of bending collapse of a thin-walled rectangular tube.

In the mechanism of Fig. 4.41, the bending deformation of a rectangular tube is concentrated in the area of

$$2H = \min(C_1, C_2)$$

and the two planes $KGHN$ and $LEFM$, which are separated by distance $2H$ from each other, rotate toward the center by the angle $\theta/2$. Therefore, both end points A and J of hinge AJ move along the side wall, and wall plates bend along each hinge, indicated by the straight lines in the figure. Here, it is assumed that the wall plates are incompressible and inextensible.

In the analysis, rotational angle θ is used as a parameter of the deformation process, and $\theta_2 = \theta/2$.

First, from the geometrical relationship of deformation, the position of point A is derived as a function of θ.

By assuming the coordinate system x, y, z as shown in Fig. 4.41, the coordinates of point B are given by

$$x_B = H, \quad y_B = C_2 \cos\theta_2 - \sqrt{C_2 \sin\theta_2(2H - C_2 \sin\theta_2)}, \quad z_B = 0 \quad (4.113)$$

Therefore, the x, y coordinates of point A are given by

$$x_A = x_B, \quad y_A = y_B \quad (4.114)$$

and the z coordinate z_A is given by

$$z_A = C_2 \sin^2\theta_2 - H \sin\theta_2 + \cos\theta_2 \sqrt{C_2 \sin\theta_2(2H - C_2 \sin\theta_2)} \quad (4.115)$$

The strain energy is obtained against rotational angle θ from the sum of the strain energy of rotation about each plastic hinge as discussed below.

(1) Rotational energy U_1 around hinges GH, EF:

Since the rotational angle around hinges GH, EF is α, which is given by

$$\alpha = \frac{\pi}{2} - \beta - \theta_2 = \frac{\pi}{2} - \theta_2 - \arcsin\left(1 - \frac{C_2}{H}\sin\theta_2\right) \quad (4.116)$$

U_1 is given by

$$U_1 = 2M_0 C_1 \alpha \quad (4.117)$$

where M_0 is an average fully plastic bending moment of a unit width of the wall and is given by

$$M_0 = \frac{1}{4}\sigma_0 t^2 \quad (4.118)$$

(2) Rotational energy U_2 around hinge AJ:

Since the length of hinge AJ is $C_1 + 2z_A$ and the rotational angle around the hinge is $\pi - 2\beta = 2(\alpha + \theta_2)$, U_2 is given by

$$U_2 = 2M_0(\alpha + \theta_2)(C_1 + 2z_A) \quad (4.119)$$

(3) Rotational energy U_3 around hinges BG, BE, CH, CF:

Since the length of the hinge is H, and the rotational angle around the hinge is $\pi/2$, U_3 is given by

$$U_3 = 4M_0 H \frac{\pi}{2} = 2M_0 H \pi \qquad (4.120)$$

(4) Rotational energy U_4 around hinges GK, EL, HN, FM:

Denoting rotational angle around the hinges by ϕ, ϕ is given by

$$\phi = \arctan\left(\frac{z_A}{d}\right) \qquad (4.121)$$

where d is the distance between point B and line segment KG, and

$$d = x_B \cos\theta_2 - y_B \sin\theta_2$$

should hold. Therefore, by considering that the length of every hinge is C_2,

$$U_4 = 4M_0 C_2 \phi \qquad (4.122)$$

is obtained.

(5) Energy U_5 for movement of hinges GA, AE, CH, CF:

During deformation, these hinges travel as point A moves downward. For rotational angle θ, the area swept by the travelling of a hinge is given by

$$\Delta S = \frac{H z_A}{2}$$

Therefore, U_5 is given by

$$U_5 = 4\frac{2M_0 \Delta S}{r} = 4M_0 \frac{H}{r} z_A \qquad (4.123)$$

where r is the curvature radius around the hinge, which is assumed to be a function of θ

$$r = r(\theta) = \left(0.07 - \frac{\theta}{70}\right) H \qquad (4.124)$$

based on the experimental results [111].

(6) Energy U_6 for travelling of the hinges KA, LA, NJ, MJ:

During deformation, these hinges also travel. The curvature $1/r$ around the travelling hinges GA, AE, CH, CF can be assumed constant in the length direction of the hinges as Eq. (4.124) shows; on the other hand, for the travelling hinges KA, LA, NJ, MJ, the curvature along the length direction of each hinge (e.g., curvature $1/r_{KA}$ along hinge KA) is given by

$$\frac{1}{r_{KA}} = \frac{l_k}{KA}\left(\frac{1}{r}\right) \qquad (4.125)$$

so that it is proportional to the distance from point K, being equal to $1/r$ at point A. Here, l_k is the distance from point K on hinge KA. It is also assumed that the rolled length l_r is proportional to the distance from point K, which is equal to z_A at point A.

$$l_r = \frac{l_k}{KA} z_A \tag{4.126}$$

Using Eqs. (4.125) and (4.126), the rotational energy U_{KA} around hinge KA is given by

$$U_{KA} = \int_0^{KA} 2M_0 \frac{l_r}{r_{KA}} dl_k = 2M_0 \int_0^{KA} \frac{l_k}{KA} z_A \frac{l_k}{KA \cdot r} dl_k = \frac{2M_0 z_A . KA}{3r} \tag{4.127}$$

As a result, U_6 is given by

$$U_6 = \frac{8}{3} M_0 \frac{z_A}{r} \sqrt{H^2 + y_A^2 + z_A^2} \tag{4.128}$$

(7) Rotational energy U_7 around hinges KN, LM, KL, MN:

The length of hinges KN, LM is C_1, and the rotational angle is θ_2. Further, the length of hinges KL, MN is $2H$, and the rotational angle is $\arctan(z_A/y_A)$. Therefore, rotational energy U_7 around the hinges is given by

$$U_7 = 2M_0 \left[C_1 \theta_2 + 2H \arctan \left(\frac{z_A}{y_A} \right) \right] \tag{4.129}$$

Thus, for a rotational angle of θ, the sum of absorbed energy $U(\theta)$ around the hinges is given by

$$U(\theta) = \sum_{i=1}^7 U_i(\theta) \tag{4.130}$$

and the moment $M(\theta)$ is given by

$$M(\theta) = \frac{dU(\theta)}{d\theta} \tag{4.131}$$

Since these equations are based on deformation after the plastic hinges are formed, they cannot be used to predict the initial phase of collapse and the maximum moment M_{col}. Therefore, Kecman derived M_{col} using the method described in Section 4.2.3 and used the tangential line against the $M - \theta$ curve, which is obtained using the mechanism discussed above, drawn from point $(0, M_{col})$ in $M-\theta$ coordinates, as a relational curve of $M-\theta$ for the initial phase of collapse. In other words, denoting the tangent point as $(\theta_T, M(\theta_T))$, for the initial phase of $0 < \theta \le \theta_T$, $M(\theta)$ is given by

$$M(\theta) = M_{col} + (M(\theta_T) - M_{col}) \frac{\theta}{\theta_T} \tag{4.132}$$

Note here that the $M - \theta$ curve derived assuming the above mechanism can be applied to the deformation process until the two surfaces AEFJ and AGHJ in Fig. 4.41 contact each other. The corresponding rotational angle of jamming θ_j is

$$\theta_j = 2 \arcsin \left(\frac{H - 0.5t}{C_2} \right) \tag{4.133}$$

After jamming, namely, in the deformation phase $\theta > \theta_j$, from experimental results, the following formula has been proposed for the relationship between moment and rotational angle:

$$M(\theta) = M(\theta_j) + 1.4[M_{col} - M(\theta_j)](\theta - \theta_j) \tag{4.134}$$

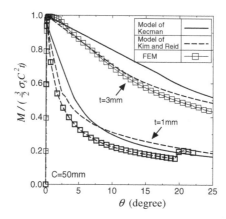

FIGURE 4.42
Relationship between moment and rotational angle in bending deformation of a square tube.

Fig. 4.42 shows the relationship between moment and bending curvature in bending deformation of square tubes with thicknesses $t/C = 1/50$ and $t/C = 3/50$, assuming a perfectly elasto-plastic material. The figure also shows the theoretical curve obtained by Kecman's method discussed above. The figure shows that Kecman's result roughly agrees with the FEM numerical analysis, but the moment is overestimated in Kecman's analysis, with the discrepancy increasing with wall thickness.

Kim and Reid [112] proposed a model to improve the analysis of the bending collapse of a thin-walled square tube. In the model, the neighborhood of the corner A in Fig. 4.41 is replaced by a toroidal area considering the movement of hinges GA and KA, similarly to the quasi-inextensional model in Fig. 2.32(b), which is applicable to the corner area crushing of inextensional folding mechanism as shown in Section 2.4.1. Moreover, they assumed that the half of folding length H and the curvature radius r of the travelling hinge are both unknown; H and r are determined under the criterion of minimizing average bending moment. For details, see [112]; here, their theoretical results are shown in Fig. 4.42 by dashed lines. The accuracy of the analytical method proposed by Kim and Reid [112] is better than that of Kecman's model.

Also, Chen et al. [56] proposed an approximate expression for the moment-rotation relation in bending collapse of thin-walled square tube as follows:

$$M(\theta) = \begin{cases} 4.65\sigma_0 C^{5/3} t^{4/3} & 0 < \theta < \theta_c \\ 5.52\sigma_0 C^{4/3} t^{5/3} \left(0.567 + \dfrac{1}{2\sqrt{\theta}}\right) & \theta \geq \theta_c \end{cases} \qquad (4.135)$$

where θ_c is the critical bending rotation for local sectional collapse and is given by

$$\theta_c = \frac{1}{4}\left[\frac{1}{0.8\left(\dfrac{C}{t}\right)^{1/3} - 0.576}\right]^2 \qquad (4.136)$$

4.2.5 Bending deformation of a square tube with partition walls

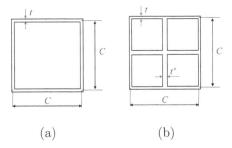

(a)   (b)

FIGURE 4.43
Cross section of a square tube: (a) without partition walls; (b) with partition walls.

Partition walls are often adopted in the design of square tubes to improve the energy absorption characteristics of thin-walled tubes [226, 222, 68]. In this section, the role of partition walls in a square tube under a bending load is discussed on the basis of FEM numerical analysis; the cross section with partition walls is such that the lateral and vertical partition walls divide each side in half, and thereby external wall thickness is t and partition wall thickness is t', as shown in Fig. 4.43(b).

Fig. 4.44 shows the relationship between bending moment M and curvature κ_g for square tubes with partition walls ($t = 1.5$ mm and $t' = 1.0$ mm)

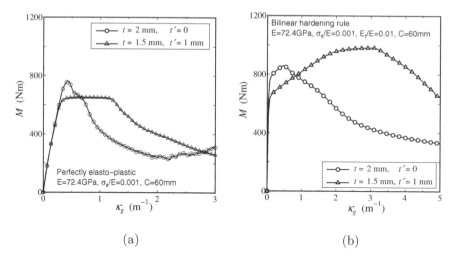

(a)   (b)

FIGURE 4.44
Comparison of bending moment of square tubes with and without partition
walls for the same cross-sectional area: (a) $E_t/E = 0$; (b) $E_t/E = 0.01$ [144].

and without partition walls ($t = 2.0$ mm). With length $L = 220$ mm and side
length $C = 60$ mm, both types of tubes have the same cross-sectional area of
$A = 480$ mm^2. In Fig. 4.44(a), the material is perfectly elasto-plastic, and the
maximum moment M_{col} of the tubes equals the full plastic moment, regardless of the presence of partition walls. Therefore, since both have the same
cross-sectional area, the maximum moment M_{col} of the tube with partition
walls is smaller than that of the tube without partition walls. However, in the
presence of partition walls, the curvature $\kappa_g |_{M=M_{col}}$ when the moment begins
to decrease is large. As a result, energy absorption is more efficient. On the
other hand, in Fig. 4.44(b), the material is strain-hardened, and the maximum
moment M_{col} and curvature $\kappa_g |_{M=M_{col}}$ at the maximum moment of the tube
with partition walls is greater than that of the tube without partition walls
and also the energy absorption is so large that the tube with partition walls
is expected to be far superior as a collision-energy-absorbing member.

In a square tube with partition walls, the partition walls suppress flattening
of the tube, and collapse occurs after both partition and external walls reach
the critical stress. In this way, two types of collapse behavior appear depending
on the thickness ratio between the partition wall and the external wall.

For a square tube with partition walls where the thickness of partition
walls and external walls is $t' = t = 1$ mm, Fig. 4.45(a) shows the relationship
between moment M and curvature κ_g during bending deformation, and also
shows the compressive stress σ_x in the axial direction at point P on the bottom wall and point Q on the partition wall (both are at the center of the walls
on the compression side, referring to the cross section A-A in Fig. 4.45(b)).

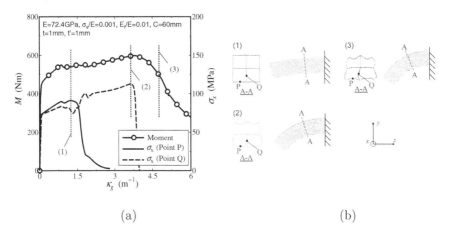

(a)   (b)

FIGURE 4.45

Typical bending collapse induced by buckling for a square tube with partition walls [144]: (a) relationship between bending moment M and curvature κ_g; (b) bending deformation behavior.

Fig. 4.45(b) shows the deformation behavior in the bending process at (1), (2) and (3) in Fig. 4.45(a). As the figure shows, even when the stress σ_x at point P becomes the largest (time point (1) in the figure), stress σ_x at point Q and moment M continue increasing. Furthermore, when stress σ_x at point Q becomes the largest (time point (2) in the figure), moment M achieves the highest value. In short, as Fig. 4.45(b) also shows, first, the bottom wall buckles due to stress σ_x caused by the bending load (time point (1) in the figure). Then, since flattening deformation on the bottom wall is blocked by the partition wall, the moment does not decrease, owing to the increase of stress σ_x at the corners (corners on the bottom wall and corners of the intersection between partition walls and the bottom wall), and buckling ripples are created all over the bottom wall, and moreover the partition walls buckle due to stress σ_x, and then collapse occurs (time point (2) in the figure). After collapse, the deformation is concentrated at one place, and even the partition walls show dent deformation in association with buckling caused by stress σ_x (time point (3) in the figure).

As an example of collapse due to cross-sectional flattening without the creation of ripples on the bottom surface of a square tube, Fig. 4.46(a) shows the relationship between moment M and curvature κ_g in bending deformation; the figure also shows compressive stress σ_x in the axial direction and compressive stress σ_y in the height direction in the cross section at point Q in the middle of the partition wall on the compression side. Fig. 4.46(b) shows the deformation behavior in the bending process at (1), (2) and (3) in Fig. 4.46(a). Moment M reaches a maximum value when compressive stress σ_y shows a peak (time point (2) in the figure), not when compressive stress σ_x at point Q does so.

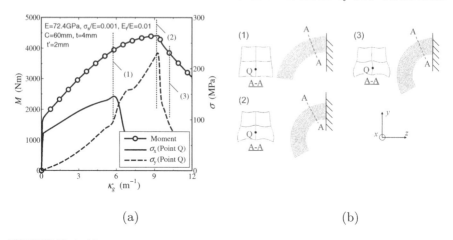

(a)   (b)

FIGURE 4.46
Typical bending collapse induced by the flattening of a square tube with partition walls [144]: (a) relationship between bending moment M and curvature κ_g; (b) bending deformation behavior.

In other words, using Fig. 4.46(b) to explain the collapse mode, under stress σ_x due to the bending load, buckling does not occur in the bottom wall but flattening occurs in the cross section. As a result, compressive stress σ_y occurs in the height direction in the cross section of a partition wall and the partition walls buckle; then, collapse occurs (time point (2) in the figure).

As discussed above, there are two **collapse modes** for both types of tubes irrespective of whether partition walls are present. That is, without partition walls, **buckling collapse** occurs due to buckling at the bottom wall and **flattening collapse** occurs due to the cross-sectional flattening; with partition walls, buckling collapse occurs in which the partition walls buckle because of stress σ_x after buckling of the bottom wall and flattening collapse occurs in which the partition walls buckle because of compressive stress σ_y in the height direction due to the cross-sectional flattening. However, the range of the partition thickness where the above-mentioned bending collapse mechanism can be applied is roughly $0.2 < t'/t < 2$.

5

Thin-walled structures with an open cross section

Thin-walled members with an open cross section are commonly used as structural components. In the axial crushing of a thin-walled structure with a closed cross section, such as a rectangular tube, after the collapse load has been reached, the compressive load periodically fluctuates when local buckles are formed continuously one after another. In contrast, in the axial crushing of a thin-walled structure with an open cross section, the load decreases monotonically after the collapse load is reached. Fig. 5.1 schematically shows the relationship between compressive load and displacement in a thin-walled member with an open cross section under axial loading. The deformation is divided into three stages: (1) OA, where the whole structure is uniformly deformed; (2) AB, where deformation is concentrated in a local region due to phenomena such as buckling; and (3) BC, where the whole structure collapses as a result of large plastic deformation concentrated in the local region. Furthermore, in a thin-walled member with an open cross section, similar behavior is also seen under bending.

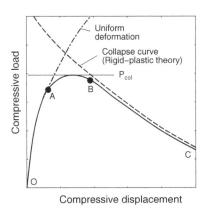

FIGURE 5.1
Relationship between compressive load and displacement in the axial crushing of thin-walled members with an open cross section.

In this section, the axial crushing and bending collapse of thin-walled members with an open cross section—in other words, the deformation behavior at the AB and BC stages in Fig. 5.1—is investigated. Such deformation is analyzed as a rotation about the plastic hinge formed in the thin-walled member. In this way, a rigid perfectly plastic material is assumed.

5.1   Basic mechanisms of plastic collapse of plates

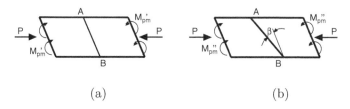

(a)   (b)

FIGURE 5.2
Moment about the plastic hinge of a plate: (a) vertical hinge; (b) inclined hinge.

The collapse behavior of a thin-walled structure with an open cross section is analyzed as a rotation about the plastic hinge produced in the thin-walled member. This basic approach was proposed by Murray and Khoo [155].

To find the axial collapse load of a thin-walled member, the basic approach is to calculate the **plastic moment** about plastic hinge AB produced when a compressive force P is applied in the axial direction to the plate of width b and thickness t (Fig. 5.2). In a rigid perfectly plastic material, in the absence of force P in the axial direction, the plastic moment M_{pm} about plastic hinge AB is given by

$$M_{pm} = \sigma_s \frac{bt^2}{4} \tag{5.1}$$

When the force P is applied in the axial direction (Fig. 5.2(a)), the plastic moment M'_{pm} about the hinge AB is given by

$$M'_{pm} = M_{pm} \left[1 - \left(\frac{P}{\sigma_s bt} \right)^2 \right] \tag{5.2}$$

In addition, according to Murray and Khoo [155], if plastic hinge AB is tilted, the plastic moment M''_{pm} is

$$M''_{pm} = M'_{pm} \sec^2 \beta \qquad (5.3)$$

where β is the inclined angle of the hinge as shown in Fig. 5.2(b).[1]

Murray and Khoo [155] proposed eight basic deformation mechanisms for analyzing the collapse characteristics during the axial crushing of a thin-walled member with an open cross section based on Eq. (5.2) and Eq. (5.3). These deformation mechanisms can be organized into three basic categories: (1) straight line hinge, (2) inclined line hinge and (3) curved hinge.

5.1.1 Straight line hinge

For the **straight line hinge**, the two cases shown in Fig. 5.3 [155] are discussed. For a single plastic hinge as shown in Fig. 5.3(a), based on the equi-

---

[1]The effect of the moment due to force P about a vertical hinge is given by Eq. (5.2), but it is not strictly accurate to say that the assumption also holds for an inclined hinge. Therefore, Zhao and Hancock [232] modified Eq. (5.3) considering the yield condition of the material. As a result, in the case that Tresca's yield condition is used, they proposed the following formula:

$$\frac{M''_{pt}}{M_{pt} \sec \beta} = (1 - k_n^2) \left(\frac{\Omega_1 + \Omega_2}{2} \right) \sec \beta \qquad (5.4)$$

where

$$k_n = \frac{2\alpha_{n0} \cos^2 \beta - (\Omega_1 + \Omega_2)}{\Omega_1 + \Omega_2}, \quad M_{pt} = \sigma_s \frac{bt^2}{4}$$

$$\Omega_1 = \frac{1 - 1/k_t^2 \alpha_{n0}^2 \cos^2 \beta \sin^2 \beta - \alpha_{n0} \sin^2 \beta}{1 - \alpha_{n0} \sin^2 \beta}$$

$$\Omega_2 = -\alpha_{n0} \sin^2 \beta + \sqrt{1 - \frac{1}{k_t^2} \alpha_{n0}^2 \sin^2 2\beta}$$

Furthermore, in the case that the von Mises's yield condition is used, Zhao and Hancock [232] proposed

$$\frac{M''_{pt}}{M_{pt} \sec \beta} = Q(1 - k_n^2) \sec \beta \qquad (5.5)$$

where

$$k_n = \frac{\alpha_{n0}}{Q} \left(\cos^2 \beta - \frac{1}{2} \sin^2 \beta \right)$$

$$Q = \sqrt{1 - \frac{3}{4} \alpha_{n0}^2 \sin^4 \beta - 3 \left(\frac{\alpha_{n0} \sin \beta \cos \beta}{k_t} \right)^2}$$

And in the above equations,

$$\alpha_{n0} = \frac{P}{\sigma_s bt}$$

$$k_t = \sqrt{1 + \left(\frac{C}{4} \right)^2} - \frac{C}{4}$$

$$C = \frac{M''_{pt}}{M_{pt} \sec \beta} \times \frac{1}{\alpha_{n0} \cos^2 \beta}$$

Note here that the analysis is complicated by the fact that C is a function of M''_{pt} and therefore deriving M''_{pt} using Eq. (5.4) or Eq. (5.5) requires iterative computation.

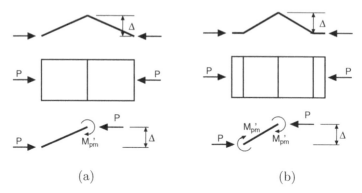

FIGURE 5.3
Straight line hinge: (a) single hinge; (b) three hinges.

librium for moment and Eq. (5.2),

$$PA = M'_{pm} = M_{pm}\left[1 - \left(\frac{P}{\sigma_s tb}\right)^2\right]$$ (5.6)

is obtained, where Δ is the out-of-plane displacement of the hinge. From Eq. (5.6), load P is obtained as

$$P = \sigma_s tb\left[\sqrt{\left\{\left(\frac{2\Delta}{t}\right)^2 + 1\right\}} - \frac{2\Delta}{t}\right]$$ (5.7)

Similarly, for the three plastic hinges in Fig. 5.3(b),

$$PA = M'_{pm} + M'_{pm} = 2M_{pm}\left[1 - \left(\frac{P}{\sigma_s tb}\right)^2\right]$$ (5.8)

is obtained. Therefore, load P is obtained as

$$P = \sigma_s tb\left[\sqrt{\left\{\left(\frac{\Delta}{t}\right)^2 + 1\right\}} - \frac{\Delta}{t}\right]$$ (5.9)

5.1.2   Inclined line hinge

For the **inclined line hinge**, the two cases shown in Fig. 5.4 [155] are discussed. In Fig. 5.4(a), first, consider strip AB of the plate. Taking the width dx of strip AB and the out-of-plane displacement Δ_c of the central hinge, the moment due to a load dP is given by

$$dP\Delta_c = M''_{pm} + M'_{pm} = k_{1m}M'_{pm} = \frac{k_{1m}\sigma_s t^2 dx}{4}\left[1 - \left(\frac{dP}{\sigma_s tdx}\right)^2\right]$$ (5.10)

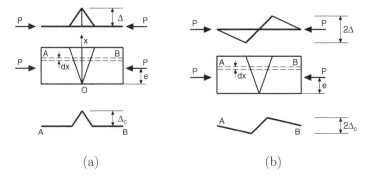

FIGURE 5.4
Inclined line hinge: (a) three hinges; (b) two hinges.

where
$$k_{1m} = 1 + \sec^2 \beta$$

Solving Eq. (5.10) for the unknown parameter dP yields

$$dP = \sigma_s t \left[\sqrt{\left\{ \left(\frac{2\Delta_c}{k_{1m}t} \right)^2 + 1 \right\}} - \frac{2\Delta_c}{k_{1m}t} \right] dx \qquad (5.11)$$

Starting from the point O, Δ_c increases linearly with x:

$$\Delta_c = \Delta \frac{x}{b}$$

Substituting this into Eq. (5.11) and integrating the resulting expression with respect to x, the load P is given by

$$
\begin{aligned}
P = \frac{\sigma_s t b}{2} &\left[\sqrt{\left\{ \left(\frac{2\Delta}{k_{1m}t} \right)^2 + 1 \right\}} - \frac{2\Delta}{k_{1m}t} \right.\\
&\left. + \frac{k_{1m}t}{2\Delta} \log \left(\sqrt{\left\{ \left(\frac{2\Delta}{k_{1m}t} \right)^2 + 1 \right\}} + \frac{2\Delta}{k_{1m}t} \right) \right]
\end{aligned}
\qquad (5.12)
$$

Further, in consideration of the equilibrium for moments about point O, the position e at which load P is concentrated is obtained from

$$Pe = \frac{\sigma_s t^3 b^2 k_{1m}^2}{12\Delta^2} \left[\left\{ \left(\frac{2\Delta}{k_{1m}t} \right)^2 + 1 \right\}^{3/2} - 1 - \left(\frac{2\Delta}{k_{1m}t} \right)^3 \right] \qquad (5.13)$$

Also, for Fig. 5.4(b), in consideration of strip AB of the plate, the following equation holds:

$$dP\Delta_c = M_{pm}'' = k_{2m} M_{pm}'$$

where

$$k_{2m} = \sec^2 \beta$$

Therefore, the load P and its position e where it is concentrated are obtained by substituting k_{2m} for k_{1m} in Eqs. (5.12) and (5.13).

5.1.3   Curved hinge

The collapse of a plate due to a **curved hinge** (Fig. 5.5) is called the **flip-disc mechanism** [155]. Boundary of the flip-disc is approximated using two parabolas:

$$z = \pm a \left(1 - \frac{4x^2}{b^2} \right) \tag{5.14}$$

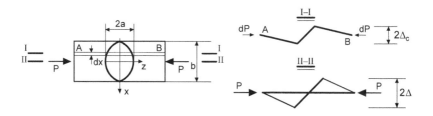

FIGURE 5.5
Curved hinge.

In strip AB of the plate, out-of-plane displacement $2\Delta_c$ of the hinge is

$$\Delta_c = \Delta \frac{z}{a} = \Delta \left(1 - \frac{4x^2}{b^2} \right)$$

The moment due to load dP is then given by

$$dP\Delta_c = M''_{pm} = M'_{pm} \sec^2 \beta$$

where

$$\sec^2 \beta = 1 + \left(\frac{dz}{dx} \right)^2 = 1 + \frac{64a^2x^2}{b^4}$$

Therefore,

$$dP = \sigma_s t \left[\sqrt{\left\{ \left(\frac{2\Delta b^2 (b^2 - 4x^2)}{t(b^4 + 64a^2x^2)} \right)^2 + 1 \right\}} - \frac{2\Delta b^2 (b^2 - 4x^2)}{t(b^4 + 64a^2x^2)} \right] dx \tag{5.15}$$

is obtained. Then, by using Simpson's rule for numerical integration and using the values of the function at $x = 0$, $b/4$, $b/2$, load P is obtained as

$$
P = \frac{\sigma_s t b}{6} \left[1 - \frac{2\Delta}{t} - \frac{6\Delta}{t(1 + 4a^2/b^2)} + \sqrt{\left\{ \left(\frac{2\Delta}{t}\right)^2 + 1 \right\}} \right. \\
\left. + 4\sqrt{\left\{ \left(\frac{3\Delta}{2t(1 + 4a^2/b^2)}\right)^2 + 1 \right\}} \right]
$$

(5.16)

5.2 Plastic collapse during axial crushing

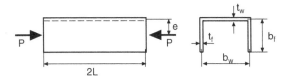

FIGURE 5.6
U-shaped thin-walled member.

The collapse behavior of a thin-walled structure with an open cross section can be analyzed using the basic deformation mechanisms discussed in the preceding section. In the following, the plastic collapse of a U-shaped thin-walled structure under axial compressive load will be analyzed. Fig. 5.6 shows the dimensions of the U-shaped thin-walled structure—web thickness t_w and web width b_w, and flange thickness t_f and flange width b_f—and the compressive load P applied at a distance e from the web center. In the plastic collapse of a U-shaped thin-walled structure under axial load, there are two possible cases, namely, compressive buckling at the end of the flange and compressive buckling at the web, depending on the position e at which the load is applied.

5.2.1 Collapse with compressive buckling of the flange

In this case, as shown in Fig. 5.7, three types of collapse modes are possible depending on the behavior of the plastic hinge [155]. Here, the collapse mode CF_1 is discussed as an example.

In collapse mode CF_1, as shown in Fig. 5.8, inclined plastic hinges are formed on the flange (Fig. 5.4(a)); correspondingly, the force and the position at which the force is applied are denoted by P_f and e_f, respectively. In addition, both plastic tensile (width d_1) and compressive (width $t_w - d_1$) yield zones are formed on the web by the bending. The total tensile force P_w and

CF1   CF2   CF3

FIGURE 5.7
Collapse mode with compressive buckling of the flange.

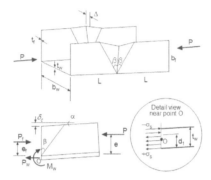

FIGURE 5.8
Plastic collapse mode CF_1.

the moment of the web M_w about the web's midpoint O are

$$P_w = \sigma_s b_w (2d_1 - t_w), \quad M_w = \sigma_s b_w d_1 (t_w - d_1) \qquad (5.17)$$

From the equilibrium condition for the force and the moment about the web's midpoint O,

$$\begin{cases} P = 2P_f - P_w \\ P(\delta_c + e) = 2P_f(e_f + t_w/2) + M_w \end{cases} \qquad (5.18)$$

is obtained, where the deflection δ_c is given by

$$\delta_c = L\alpha, \quad \alpha = \frac{\Delta^2/(2b_f \tan \beta)}{b_f} \qquad (5.19)$$

Therefore, by considering that P_f and e_f are given as functions of Δ and β in Eqs. (5.12) and (5.13), respectively, and assuming, for example, an angle β of the plastic hinge, Eq. (5.18) can be used to determine load P and width d_1 for a given value of Δ. Since the deflection δ_c of the midpoint is similarly obtained as a function of Δ and β from Eq. (5.19), for the collapse mode CF_1, the functional relationship between load P and deflection δ_c of the midpoint is obtained from the above analysis.

5.2.2 Collapse with compressive buckling of the web

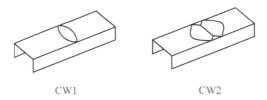

CW1 CW2

FIGURE 5.9
Collapse mode with compressive buckling of the flange.

There are two possible collapse modes depending on the behavior of the plastic hinge that appears on the web [155], as shown in Fig. 5.9. Here, collapse mode CW_1 is discussed as an example.

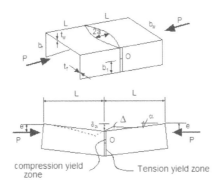

compression yield
zone ⎯ Tension yield zone

FIGURE 5.10
Plastic collapse mode CW_1.

As shown in Fig. 5.10, in collapse mode CW_1, a plastic curved hinge appears as a flip-disc on the web due to compression [155]; the force corresponding to this is assumed to be P_w. Due to bending, the flange experiences both plastic compression yield (the region of width $b_f - b_1$ above point O in the figure) and plastic tensile yield (the region of width b_1 below point O in the figure); the corresponding forces are assumed to be P_1 and P_2, respectively. P_w is obtained from Eq. (5.16) as a function of Δ and a, whereas P_1 and P_2 are obtained as a function of b_1 as follows:

$$\begin{cases} P_1 &= 2\sigma_s(b_f - b_1)t_f \\ P_2 &= 2\sigma_s b_1 t_f \end{cases} \tag{5.20}$$

From the equilibrium condition for the force and the moment about point O,

$$
\begin{cases}
P &= P_w + P_1 - P_2 \\
P(\delta_c + b_f - b_1 - e) &= P_w(b_f - b_1) + P_1(b_f - b_1)/2 + P_2 b_1/2
\end{cases}
\tag{5.21}
$$

is obtained, where deflection δ_c is obtained as a function of Δ, a and b_1:

$$
\delta_c = L\alpha, \qquad \alpha = \frac{\Delta^2/(2a)}{b_f - b_1}
\tag{5.22}
$$

Therefore, for a given value of Δ, the two unknown parameters b_1 and P can be obtained from the balance equation (Eq. (5.21)) by assuming, for example, $a/b = 0.2$. Furthermore, the deflection δ_c of the midpoint can be obtained as a function of Δ from Eq. (5.22), and, as a result, for the collapse mode of CW_1, the functional relationship between the load P and the deflection δ_c of the midpoint is obtained from the above analysis.

5.3   Bending collapse due to buckling

In a thin-walled member with an open cross section subjected to bending moment, collapse begins with buckling of thin-walled plates constituting the member. However, the collapse mode depends on whether the end part of the flange is on the compression side. Here, the bending collapse of U-shaped and V-shaped beams shall be reviewed as typical examples of beams with an open cross section.

5.3.1   Bending collapse with compressive stress applied to the flange

When compressive stress due to bending is applied to the free edge of the flange, collapse begins with buckling of the flange. The three collapse modes under axial compression (Fig. 5.7) can also be used to explain the behavior of the plastic hinge in the bending collapse. Both mode CF_1 and mode CF_2 in Fig. 5.7 are also frequently observed in experiments as well as in FEM numerical simulations. As an example, Fig. 5.11 shows the bending collapse behavior of a U-shaped beam numerically analyzed by FEM. Based on Fig. 5.11, the bending collapse of beams is analyzed by the collapse mechanism shown in Fig. 5.12. For the rotation angle θ of the member, $CDGH$ and $BEFI$ rotate by an angle $\theta_2 = \theta/2$ around hinges CC' and BB', respectively, and accordingly ABC rotates by an angle γ_{BC} around hinge BC.

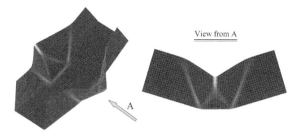

FIGURE 5.11
Bending collapse behavior and contour map of equivalent plastic strain in a U-shaped beam numerically analyzed by FEM.

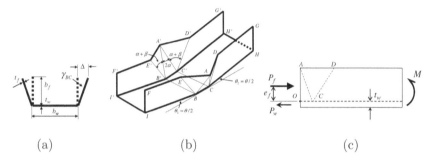

(a)   (b)   (c)

FIGURE 5.12
Bending collapse mechanism of a U-shaped beam: (a) deformation of the central cross section in transverse direction; (b) plastic hinge; (c) force distribution in the cross section of length direction.

From the equilibrium condition for the force and the moment about point O,

$$\begin{cases} 0 = 2P_f - P_w \\ M = 2P_f e_f + M_w \end{cases} \tag{5.23}$$

is obtained, where P_w and M_w are given by

$$P_w = \sigma_s b_w (2d_1 - t_w) , \quad M_w = \sigma_s b_w \left[\frac{t_w^2}{2} - (t_w - d_1)^2 \right] \tag{5.24}$$

By replacing k_{1m} in Eqs. (5.12) and (5.13) with k_{3m}

$$k_{3m} = \sec^2 \alpha + \sec^2 \beta$$

P_f and e_f can be obtained as

$$
P_f = \frac{\sigma_s t_f b_f}{2} \left[\sqrt{\left\{\left(\frac{2\Delta}{k_{3m} t_f}\right)^2 + 1\right\}} - \frac{2\Delta}{k_{3m} t_f} \right.
$$
$$
\left. + \frac{k_{3m} t_f}{2\Delta} \log\left(\sqrt{\left\{\left(\frac{2\Delta}{k_{3m} t_f}\right)^2 + 1\right\}} + \frac{2\Delta}{k_{3m} t_f}\right) \right]
\tag{5.25}
$$

$$
P_f e_f = \frac{\sigma_s t_f^3 b_f^2 k_{3m}^2}{12\Delta^2} \left[\left\{\left(\frac{2\Delta}{k_{3m} t_f}\right)^2 + 1\right\}^{3/2} - 1 - \left(\frac{2\Delta}{k_{3m} t_f}\right)^3 \right]
\tag{5.26}
$$

Here Δ is given by

$$
\Delta = b_f \cdot \gamma_{BC}
\tag{5.27}
$$

Further, the rotation angle γ_{BC} of triangle ABC is given by

$$
\gamma_{BC} = \cos^{-1}\left(\frac{1 - \tan\alpha\tan\beta(1 - \cos\theta_2) - \tan\alpha\sin\theta_2}{\cos\theta_2 + \tan\beta\sin\theta_2}\right)
\tag{5.28}
$$

from the condition that the length of segment AD does not change during the bending collapse: $|AD| = b_f(\tan\alpha + \tan\beta)$.

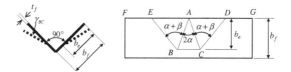

FIGURE 5.13
Bending collapse mechanism of a V-shaped beam.

The bending collapse of a V-shaped beam can be analyzed similarly as shown in Fig. 5.13. However, plastic hinges AC and CD occupy a limited region b_e of the respective flanges rather than the entire width of the flanges. Therefore, Eq. (5.23) can be rewritten as

$$
\begin{cases}
0 = P_f - P_w \\
M = \sqrt{2}P_f e_f + M_w
\end{cases}
\tag{5.29}
$$

where P_w and M_w are given by

$$
\begin{aligned}
P_w &= \sigma_s t_w \left[2d_1 - (b_f - b_e)\right] \\
M_w &= \sqrt{2}\sigma_s t_w \left[\frac{(b_f - b_e)^2}{2} - (b_f - b_e - d_1)^2\right]
\end{aligned}
\tag{5.30}
$$

and P_f, e_f, and γ_{BC} are given by Eqs. (5.25)–(5.27) except that b_e is substituted for b_f. Furthermore, depending on the V-shaped geometry, the angle γ_{BC} is obtained by solving the following equation:

$$
\sin(\gamma_{BC} + \pi/4) + [\cos\theta_2 + \sqrt{2}\tan\beta\sin\theta_2]\cos(\gamma_{BC} + \pi/4)
$$
$$
= \sqrt{2}\Big[1 + \tan\alpha\tan\beta(\cos\theta_2 - 1)\Big] - \tan\alpha\sin\theta_2
\tag{5.31}
$$

In addition, the moment of rotation $M(\theta)$ of the beam can also be obtained from the balance between the strain energy of rotation around the plastic hinge and the work performed by the bending moment. It is seen from Figs. 5.12 and 5.13 that the increment of energy of rotation about the plastic hinge, which corresponds to the rotary angle increment $d\theta$ of the beam, is composed of the following components:

(1) Energy increment dU_1 for rotation about hinges CD, BE, $C'D'$ and $B'E'$

$$
dU_1 = (\sigma_s t_f^2)\frac{b}{\cos\beta} \times d\gamma_{CD}
$$

(2) Energy increment dU_2 for rotation about hinges AC, AB, $A'C'$ and $A'B'$

$$
dU_2 = (\sigma_s t_f^2)\frac{b}{\cos\alpha} \times d\gamma_{AC}
$$

(3) Energy increment dU_3 for rotation about hinges BC and $B'C'$

$$
dU_3 = (\sigma_s t_f^2)b\tan\alpha \times d\gamma_{BC}
$$

In these equations, $b = b_f$ for a U-shaped beam and $b = b_e$ for a V-shaped beam.

(4) For a U-shaped beam, energy increment dU_4 for rotation about hinges CC' and BB'
$$
dU_4 = (\sigma_s t_w^2)b_w \times d\theta_2
$$

For a V-shaped beam, energy increment dU_4 due to tensile plastic strain in the region $b_f - b_e$ of the flange

$$
dU_4 = \sqrt{2}\sigma_s t_f(b_f - b_e)^2 \times d\theta_2
$$

Here, rotation angles γ_{CD} and γ_{AC} around the respective hinges can be obtained as functions of α, β, and θ.

Therefore, $M(\theta)$ is given by

$$
M(\theta) = \frac{dU_1 + dU_2 + dU_3 + dU_4}{d\theta}
\tag{5.32}
$$

In the above analysis, angles α and β are determined by minimizing the strain energy of bending collapse. The analysis gives the values $\alpha \cong 13°$ and $\beta \cong 33°$ for both U-shaped and V-shaped beams. These results agree with the deformation behavior obtained from FEM numerical analysis, as well as with experimental results of Pastor and Roure [161]. With respect to the relationship between moment M and rotation angle θ in the bending collapse of U-shaped and V-shaped beams, Fig. 5.14 compares the results of FEM numerical analysis and the above-mentioned theoretical analyses: the analytic equations Eqs. (5.23)–(5.28) and Eqs. (5.29)–(5.31) deduced from crushing collapse mechanism of a thin plate and the analytic equation Eq. (5.32) deduced from the energy balance for the rotation around the hinge. These analyses are shown by "Th1" and "Th2" in the figure. It is seen from the figures (a) and (b) that, if the plate is as thick as $t/b = 4/50$, the analysis based on collapse mechanism of a thin plate is less accurate than the analysis based on energy balance.

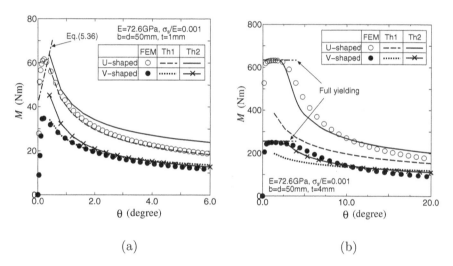

(a)   (b)

FIGURE 5.14
Relationship between moment M and rotation angle θ in bending collapse of U-shaped beam ($b_f = b_w = b$, $t_f = t_w = t$) and V-shaped beam ($b_f = b$, $t_f = t$): (a) $t = 1$ mm; (b) $t = 4$ mm.

Fok et al. [75] conducted a detailed analysis of the bending deformation of U-shaped beams after the elastic buckling of the flange and before the formation of a plastic hinge. On the basis of a study by Rhodes et al. [174], Fok et al. formulated the out-of-plane deformation Δ due to buckling as follows:

$$\Delta = A \sin\left(\frac{\pi x}{e b_f}\right) \frac{y}{b_f} \tag{5.33}$$

where

$$A^2 = \left(\frac{eb_f}{\pi}\right)^2 \frac{\varepsilon_m(20 - 5\alpha)}{3} - \left\{c_1 + c_2\left(\frac{e}{\pi}\right)^2\right\}t_f^2 \tag{5.34}$$

$$c_1 = \frac{5}{9(1 - \nu^2)}, \quad c_2 = \frac{10}{3(1 + \nu)}, \quad \alpha = \frac{b_w + 2b_f}{b_w + b_f} \tag{5.35}$$

and ε_m is the compressive strain at the free end of the flange, and e is the ratio of the wavelength of buckling and the flange width b_f.

From the out-of-plane deformation given by Eq. (5.33) for a member of $t_w = t_f = t$, Fok et al. [75] showed the relationship between bending moment M and rotation angle θ:

$$\theta = L\frac{M - M_{cr}}{EI_r} \tag{5.36}$$

where

$$\begin{cases} M_{cr} = Etb_f^2 \dfrac{\left(\dfrac{t}{b_f}\right)^2 \left(\dfrac{c_1\pi^2}{e^2} + c_2\right)\left(\dfrac{b_w}{8} + \dfrac{b_f}{12}\right)}{\left(b_w + \dfrac{8}{9}b_f\right)} \\[2em] I_r = tb_f^3 \dfrac{\left(\dfrac{b_w}{24} + \dfrac{b_f}{108}\right)}{\left(b_w + \dfrac{8}{9}b_f\right)} \end{cases} \tag{5.37}$$

Fig. 5.14(a) shows the analytic results from Eq. (5.36) for the deformation process of the U-shaped beam after the elastic buckling of the flange and before the formation of a plastic hinge.

5.3.2 Bending collapse with compressive stress applied to the web

• **Bending collapse of U-shaped beams**

When a bending compressive stress is applied to the web, the collapse starts with the initial buckle in the web. Fig. 5.15(a) shows the typical result of FEM analysis on the bending collapse behavior of a U-shaped beam, which occurs when a compressive stress is applied to the web. Further, Fig. 5.15(b) shows the bending collapse behavior of a rectangular tube. As the comparison between Fig. 5.15(a) and (b) shows, the bending collapse behavior of a U-shaped beam with a compressed web closely resembles that of a rectangular tube. Therefore, the mechanism of the bending collapse of a rectangular tube proposed by Kecman [111] (see Section 4.2.4 in Chapter 4) can be applied to the bending collapse of a U-shaped beam, with a suitable correction. Fig. 5.16 shows the bending collapse model of a U-shaped beam based on Kecman's model (Fig. 4.41). Fig. 5.16 differs from Fig. 4.41 in two respects:

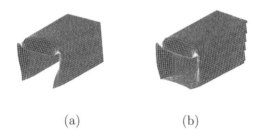

(a)   (b)

FIGURE 5.15
Comparison of collapse behaviors between U-shaped beam and rectangular
tube as numerically analyzed by FEM: (a) U-shaped beam; (b) rectangular
tube.

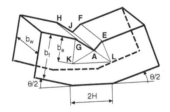

FIGURE 5.16
Theoretical model of the bending collapse of U-shaped beams.

(1) In contrast to rectangular tubes, there is no plastic hinge analogous to
KN and LM (see Fig. 4.41), because there is no base plate in the U-shaped
cross section.

(2) A tensile bending stress occurs in the base plate of a rectangular tube
because the total force in the axial direction should be zero in pure bending. In contrast, in the U-shaped cross section, a tensile bending stress
occurs in the lower part of both sides, and hinges such as KG and LE only
partially affect the side plates.

With respect to the relationship between bending moment M and rotation
angle θ during pure bending deformation of U-shaped beam with its web
compressed, Fig. 5.17 compares the results obtained from FEM numerical
analysis and the revised Kecman's theoretical formula mentioned above. Note
that $b_e/b_f = 0.5$ was assumed in the theoretical analysis.

• **Bending collapse of V-shaped beams**
    When a compressive bending stress is applied to the corner joint of a V-
shaped beam, usually local buckling does not occur, whereas distortion occurs
in the corner joints of the flanges during bending. As shown in Fig. 5.18, the

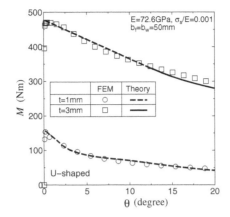

FIGURE 5.17
Relationship between bending moment M and rotation angle θ in the pure bending deformation of the U-shaped beam with compressed bending stress applied to the web.

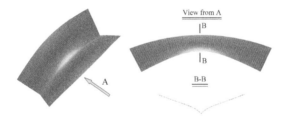

FIGURE 5.18
Behavior of deformation due to bending collapse of a V-shaped beam.

bending deformation behavior of a V-shaped beam is characterized by the corner joint opening and the angle increasing as the deformation proceeds. The moment supported by the beam decreases due to flattening of the cross section.

With respect to four-point bending of a thick V-shaped beam (Fig. 5.19), Yu and Teh [229] proposed a bending collapse mechanism as shown in Fig. 5.20: beam flattening was modeled by flanges that remain straight and corner joints that open. As shown in Fig. 5.20(a), the plastic deformation concentrates on the narrow regions of length λ_p around the points to which the load is applied. The deformation behavior in this region is shown in Fig. 5.20(b), and the cross section D-D in Fig. 5.20(b) is shown in Fig. 5.20(c). As seen in the deformation of cross section $D - D$ in Fig. 5.20(c), the corner

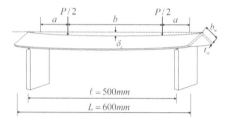

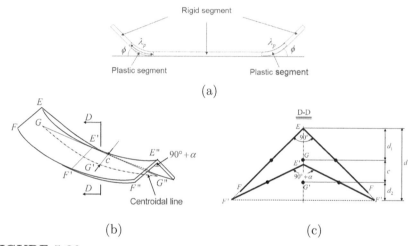

FIGURE 5.19
Four-point symmetric bending of V-shaped beam.

FIGURE 5.20
Bending collapse mechanism of V-shaped beams.

joint opens as the deformation proceeds, and the angle between the flanges
increases from $\pi/2$ before deformation to $\pi/2 + \alpha$ after deformation.

The strain energy for bending a V-shaped beam is the sum of energy U_1
for the overall bending and strain energy U_2 for the change of the flange joint
angle.

$$U = U_1 + U_2 \tag{5.38}$$

Assuming that the material obeys the bilinear hardening rule and assuming
yield stress σ_s and strain-hardening coefficient E_t, the strain energy U_1 for
the overall bending is calculated by

$$
\begin{aligned}
U_1 &= \left[M_{pV} \left(1 - \frac{1}{3}\alpha \right) k + \frac{1}{2} E_t I \left(1 - \frac{2}{3}\alpha \right) k^2 \right] \lambda_p \\
&= M_{pV} \left[1 - \frac{1}{3}\alpha \right] \phi + \frac{1}{2} E_t I \left[1 - \frac{2}{3}\alpha \right] \frac{\phi^2}{\lambda_p}
\end{aligned}
\tag{5.39}
$$

and energy U_2 associated with the change of the angle of the flange joint is calculated by

$$U_2 = M_0 \alpha|_{ave} \lambda_p \cong \frac{1}{6} \sigma_s t^2 \alpha \lambda_p \qquad (5.40)$$

Here, $\alpha|_{ave} = (2/3)\alpha$ is the average angular distortion of the plastic segment and

$$M_{pV} = (\sqrt{2}/4)\sigma_s t b_f^2, \qquad I = t b_f^3/12 \qquad (5.41)$$

where b_f is the width of the flange and k is the bending curvature as given by

$$k = \frac{\phi}{\lambda_p} \qquad (5.42)$$

The unknown factor λ_p is determined by using the condition

$$\frac{\partial U}{\partial \lambda_p} = 0 \qquad (5.43)$$

From Eq. (5.43), the equation for determining λ_p is obtained as follows:

$$\frac{1}{12}\frac{t}{b_f}\left(\frac{\lambda_p}{b_f}\right)^4 + \left(\frac{2\sqrt{2}t}{b_f\phi} - \frac{\sqrt{2}\phi}{36}\right)\left(\frac{\lambda_p}{b_f}\right)^3 - \left(1 + \frac{\phi^2}{216}\frac{E_t}{\sigma_s}\right)\left(\frac{\lambda_p}{b_f}\right)^2 = \frac{3E_t}{2\sigma_s} \qquad (5.44)$$

The bending moment M due to the collapse of a V-shaped beam is determined using

$$M = \frac{dU}{d\phi} \qquad (5.45)$$

For determining M using Eq. (5.45), it is necessary to first determine the relationship between the rotation angle ϕ and the angular distortion α of the corner joint representing the flattening of the cross section. For this purpose, Yu and Teh [229] assume that the flange moves from EF (before deformation) to $E'F'$ (after deformation) as flattening proceeds, and that point F' is on an extension of EF, as shown in Fig. 5.20(c). This allows one to obtain the relationship between rotation angle ϕ of the beam and angular distortion α of the corner joint. That is, distance c between midpoint G of the flange before deformation and midpoint G' of the flange after deformation is given by

$$\begin{aligned} c = d - d_1 - d_2 &= b_f \sin\left(\frac{\pi}{4} + \frac{\alpha}{2}\right) - \frac{b_f}{2\sqrt{2}} - \frac{b_f}{2}\cos\left(\frac{\pi}{4} + \frac{\alpha}{2}\right) \\ &= \frac{b_f}{2\sqrt{2}}\left(3\sin\frac{\alpha}{2} + \cos\frac{\alpha}{2} - 1\right) \end{aligned} \qquad (5.46)$$

In particular, if α is small,

$$2\sqrt{2}\frac{c}{b_f} = \frac{3}{2}\alpha - \frac{1}{8}\alpha^2 \qquad (5.47)$$

On the other hand, as shown in subfigure (b), the median line GG'' becomes circular arc-shaped after deformation; based on the distance between midpoint G' and the median line before deformation (line segment GG'' in the figure), c is given by

$$c = \frac{1}{8}\lambda_p^2 k = \frac{1}{8}\lambda_p \phi \qquad (5.48)$$

From Eqs. (5.47) and (5.48), the relationship between angular distortion α of the corner joint and ϕ is

$$\alpha = \frac{1}{3\sqrt{2}}\frac{\lambda_p^2 k}{b_f} + \frac{1}{12}\alpha^2 \cong \frac{1}{3\sqrt{2}}\frac{\lambda_p^2 k}{b_f} + \frac{1}{216}\frac{\lambda_p^4 k^2}{b_f^2} \qquad (5.49)$$

Fig. 5.21 shows the relationship between force P and deflection δ_c at the midpoint of the beam as determined both from experimental results ($\bullet$ in the figure) and the above-mentioned theoretical analysis of Yu and Teh [229] (dashed lines in the figure) for four-point bending of V-shaped beams as shown in Fig. 5.19. The figure also shows the results of FEM analysis ($\square$ in the figure) for a similar case and the analytic results (solid lines in the figure) using the method discussed in the next section.

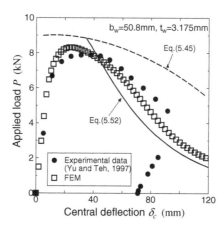

FIGURE 5.21
Relationship between force P and deflection δ_c of a V-shaped beam in four-point bending.

5.4   Bending collapse due to cross-sectional flattening

When the wall of a beam is thick, buckling does not occur during deformation, and bending collapse occurs due to cross-sectional flattening. Here, as an example, the bending collapse of V-shaped and W-shaped beams is reviewed.

The relationship between moment M and rotation angle θ in the bending collapse of a beam can be obtained from the balance equation between the increment dU of strain energy and work due to moment M.

$$dU = M \cdot d\theta \tag{5.50}$$

In a thin beam, the mechanism of collapse is represented with one or more plastic hinges produced in each thin plate constituting the beam. In this case, the correspondence between the rotation angle around the plastic hinge and the rotation angle of the entire beam is obtained from the assumed mechanism, and the increment dU of strain energy corresponding to the increment $d\theta$ of the rotation angle of the beam can be obtained; then, the bending moment M of the beam is determined using

$$M = \frac{dU}{d\theta} \tag{5.51}$$

which is obtained from Eq. (5.50).

On the other hand, in a thick beam, mainly cross-sectional flattening causes the change of the bending moment after the peak load. In this case, it is usually difficult to evaluate the correspondence between cross-sectional flattening and rotation angle. For example, in the above-mentioned analysis model of Yu and Teh [229], in order to obtain the correspondence, it is assumed that point F' is on an extension of EF. However, it is questionable whether such an assumption is applicable to other cases. Therefore, a new technique for determining the relation between the bending moment M and the rotation angle θ is required. In such a case, since the bending moment M corresponding to the flattened cross section and the consumed energy dU from the change in cross-sectional shape can be determined, the relationship between M and θ in the bending collapse can be obtained by

$$d\theta = \frac{dU}{M} \tag{5.52}$$

which is derived also from Eq. (5.50).

5.4.1 Bending collapse of V-shaped beams

Here, bending deformation of a V-shaped beam subjected to a pure bending moment with compressive bending stress applied to the corner joint is considered. As shown in Section 5.3.2, when compressive bending stress is applied to the corner joint of a V-shaped beam, local buckling does not usually occur during bending collapse, and the corner joint opens and the joint angle increases with deformation.

For example, it can be assumed that cross-sectional flattening after opening of a corner joint is approximated by a circular arc-shaped flange of radius ρ as shown in Fig. 5.22. Assuming the central angle $\alpha/2$ of the circular arc, the

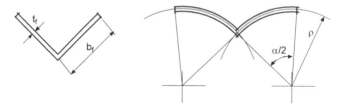

FIGURE 5.22
Cross-sectional flattening model of bending deformation of a V-shaped beam.

cross-sectional shape is described by a single parameter: radius ρ or central angle α.

$$\alpha = \frac{2b_f}{\rho} \tag{5.53}$$

Further, for pure bending the cross-sectional deformation during collapse is concentrated at the central part of the beam—for example, on a region of length λ_p centered at midpoint $x = x_o$; in that region, the flattening parameter α is assumed to change as follows:

$$\alpha = \left[1 - \frac{|x - x_o|}{\lambda_p/2} \right] \alpha_{max} \qquad \text{for } |x - x_o| \leq \lambda_p/2 \tag{5.54}$$

Since flattening parameter α at midpoint $x = x_o$ should be equal to α_{max}, moment M of the beam is given by

$$M = 2\sigma_s t_f \rho^2 \left[2 \sin \left(\frac{\pi}{4} - \frac{\alpha_{max}}{4} \right) - \sin \left(\frac{\pi}{4} - \frac{\alpha_{max}}{2} \right) - \sin \left(\frac{\pi}{4} \right) \right] \tag{5.55}$$

assuming perfectly plastic yielding of the cross section.

The increment of flattening energy in the central region of the beam $x_o - \lambda_p/2 \leq x \leq x_o + \lambda_p/2$, which corresponds to increment $d\alpha_{max}$ of α_{max}, is given by

$$dU = \int_{x_o - \lambda_p/2}^{x_o + \lambda_p/2} M_0 \left[1 - \frac{|x - x_o|}{\lambda_p/2} \right] d\alpha_{max} dx = M_0 \lambda_p d\alpha_{max} \tag{5.56}$$

The α_{max}-dependent relationship between M and $d\theta$ can be evaluated by substituting Eqs. (5.55) and (5.56) into Eq. (5.52). Thus, in order to obtain the relationship between M and θ, it is necessary to determine the initial values of M and θ, namely, the maximum bending moment (from where the beam collapse starts) and the corresponding rotation angle of the beam.

For simplicity, here it is assumed that, as the load reaches a peak, (1) perfectly plastic yielding of the cross section occurs at midpoint $x = x_o$ and

(2) flattening parameter α_{max} at $x = x_o$ in the cross section becomes equal to α_e. Here,

$$\alpha_e = \frac{4b_f \sigma_s}{Et} \tag{5.57}$$

For the beam deformation before the peak load is reached,

$$\theta = \frac{ML}{EI} \tag{5.58}$$

can be used for the relationship between moment M and rotation angle θ of the beam. Here, L is the length of the beam and I is the moment of inertia of area for the neutral axis.

For the case of the pure bending of a V-shaped beam with the corner part compressed, Fig. 5.23 show the relationship between the bending moment M and the rotation angle θ; the figure compares the results from FEM numerical analysis and the above-mentioned theoretical formula (solid line in the figure). Furthermore, the dashed line in the figure shows the analytic results from the method of Yu and Teh [229] discussed in the preceding section.

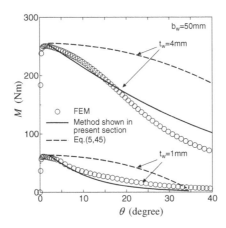

FIGURE 5.23
Change of moment in the bending collapse of a V-shaped beam.

5.4.2 Bending collapse of W-shaped beams

For a beam of thickness t and length L having the waveform cross section shown in Fig. 5.24(b), an analysis is conducted to determine deflection δ_Q of point Q assuming that force P is applied to point Q, which is apart from the left edge with distance a, in the beam simply supported at both ends as shown in Fig. 5.24(a) [31].

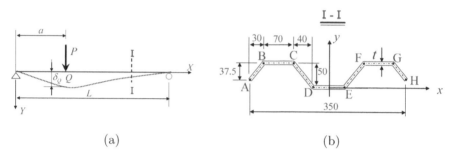

FIGURE 5.24
Geometry and loading conditions for a W-shaped beam [31]: (a) loading condition; (b) cross-sectional view.

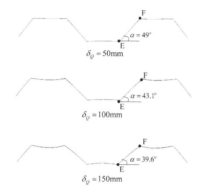

FIGURE 5.25
Cross-sectional deformation behavior in bending of a beam having a waveform cross section [31].

Fig. 5.25 shows how the cross section of the beam changes as the beam deforms as obtained from FEM analysis. As shown in the change of the angle α between the horizontal line and plane EF in the figure, the degree of cross-sectional flattening increases as deflection δ_Q increases.

Based on Fig. 5.25, after the maximum load, beam collapse due to flattening is modeled as in Fig. 5.26(b). That is, (1) dihedral angle β does not change; (2) slopes AB, CD, EF and GH undergo rotation as a rigid body; and (3) top planes BC and FG and lower plane DE do show not in-plane tensile deformation but circular arc-shaped bending of radius R_1.

It is seen from Fig. 5.26(b) that central angle γ corresponding to the circular arcs in the upper and lower planes is given by

$$\gamma = l_1/R_1 \tag{5.59}$$

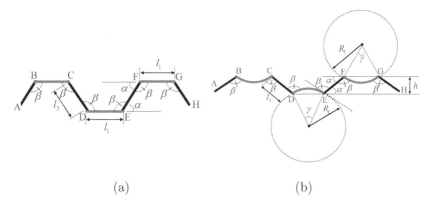

(a)   (b)

FIGURE 5.26
Flattening deformation model [31]: (a) initial shape of cross section; (b) approximation of deformed shape.

and that rotation angle α of the slope as given by

$$\alpha = \pi - \beta - \gamma/2 \qquad (5.60)$$

decreases as angle γ increases with deformation of the upper and lower planes. Therefore, as the deformation proceeds, cross-sectional height h decreases as follows:

$$h = l_3 \sin \alpha \qquad (5.61)$$

As discussed above, after the peak load, the cross-sectional flattening of a beam proceeds along with deformation, whereby the flattened cross section can be described by a single shape parameter: radius R_1 in Fig. 5.26(b).

Assuming radius R_{1a} of the upper and lower circular arcs in the cross section at point $X = a$ where the largest bending moment occurs, the corresponding load P can be obtained as a function of R_{1a} as discussed below. Note that perfectly plastic yielding of the cross section occurs where the maximum moment occurs (the point to which the load is applied); therefore, the maximum moment M corresponding to peak load P is given by

$$M = \int_c \sigma_s |y - y_c| t \, ds \qquad (5.62)$$

where $\int_c \cdots ds$ is the path integral along the center line of the cross section (polygon $ABCDEFGH$ in Fig. 5.24(b)). Further, y_c is the y coordinate of the neutral axis, satisfying

$$\int_c (y - y_c) ds = 0 \qquad (5.63)$$

After M is obtained, P is calculated by

$$P = \frac{L}{a\,(L-a)}M \tag{5.64}$$

To obtain displacement δ_Q corresponding to load P during collapse, Eq. (5.52) is applied so that the increment of displacement $d\delta_Q$ is determined from the increment dU of the beam strain energy by the following equation:

$$P \cdot d\delta_Q = dU \tag{5.65}$$

In the bending of a beam, the external work goes toward increasing the bending strain energy through expansion and contraction deformation of fibers in the axial direction of the beam; this occurs during bending with increasing load. In the collapse process after the maximum load, it is thought that the external work increases the strain energy corresponding to the collapse deformation, namely, the flattening deformation.

Assume the strain energy U for flattening of the entire beam up until the radius of the circular arc in upper and lower planes changes to R_{1a} at point $X = a$. To obtain U, it is necessary to know the change of the flattening parameter along the length: radius R_1. Here, as an approximation, a linear curvature distribution is assumed as follows:

$$\frac{1}{R_1} = \begin{cases} \dfrac{X}{a} \times \dfrac{1}{R_{1a}} & (X \le a) \\[2mm] \dfrac{(L-X)}{(L-a)} \times \dfrac{1}{R_{1a}} & (X \ge a) \end{cases} \tag{5.66}$$

Therefore, the flattening strain energy udX along the short length dX of beam is given as a function of flattening radius R_1:

$$udX = \begin{cases} 3 \times \dfrac{Et^3 l_1}{24R_1^2}dX & R_1 > R_e \\[3mm] 3 \times \left[\dfrac{Et^3 l_1}{24R_e^2}dX + \dfrac{l_1\sigma_s t^2}{4}\left(\dfrac{1}{R_1} - \dfrac{1}{R_e}\right)dX \right] & R_1 < R_e \end{cases} \tag{5.67}$$

where

$$R_e = \frac{Et}{2\sigma_s} \tag{5.68}$$

By substituting Eq. (5.66) into Eq. (5.67), U is obtained:

$$U = \int_0^L udX \tag{5.69}$$

As a result, the flattening strain energy U can be expressed as

$$U = \frac{Et^3 l_1 L}{24R_{1a}^2} \tag{5.70}$$

in the case of $R_e < R_{1a}$ and by

$$
\begin{aligned}
U = & \; Ll_1 \frac{3\sigma_s t^2}{4R_{1a}} \left[\frac{1}{2} - \left(\frac{R_{1a}}{R_e} \right) + \frac{1}{2} \left(\frac{R_{1a}}{R_e} \right)^2 \right] \\
& + Ll_1 \frac{Et^3}{8R_e^2} \left[1 - \frac{2}{3} \left(\frac{R_{1a}}{R_e} \right) \right]
\end{aligned}
\tag{5.71}
$$

in the case of $R_e \geq R_{1a}$.

From the above, load P and strain energy U are respectively obtained from Eqs. (5.62) and (5.64) and Eqs. (5.70) and (5.71) for a given value of the flattening parameter R_{1a} at $X = a$; therefore, the relationship between the load and the increment of displacement can be obtained by using Eq. (5.65).

To obtain the relationship between load and displacement, it is necessary to determine their initial values, namely, the peak load at the very beginning of beam collapse and the corresponding displacement. For simplicity, it is assumed here that when the peak load is reached (1) cross-sectional plastic yield occurs and (2) R_{1a} becomes equal to R_e. In addition, to obtain the relationship between load P and displacement δ_Q before the peak load, the relational formula for the three-point elastic bending is used, which is given by

$$
P = \frac{3EIL}{a^2(L-a)^2} \delta_Q
\tag{5.72}
$$

Here, I is the moment of inertia of area for the neutral axis.

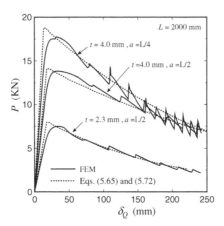

FIGURE 5.27
Comparison between results of FEM and proposed method for $P - \delta_Q$ curve [31].

In Fig. 5.27 the solid lines show the relation between deflection δ_Q and load P obtained from FEM for $t = 2.3$ mm and $a = L/2$, as well as for $t = 4.0$ mm and $a = L/4,\ L/2$. Furthermore, in Fig. 5.27, the dotted lines show the predicted P-δ_Q curves from the proposed theoretical analysis. Before the peak load, the curve is plotted by using Eq. (5.72). On the curve of Eq. (5.72), a point with P equaling the peak load given by Eq. (5.62) with $R_{1a} = R_e$ is taken as the beginning of collapse, from which the curve is plotted by using Eq. (5.65). Fig. 5.27 show good agreement between the FEM analysis and the proposed theoretical analysis.

6

Torsion

Pure torsional load is seldom applied to thin-walled members in use, but both axial and bending deformations are frequently accompanied by torsional load. To understand deformation behavior under such combined loads, it is first necessary to clarify collapse deformation of thin-walled members under torsional loads.

6.1 Shear buckling of plates

Elastic buckling of plates under shear load has long been studied and is discussed in the book by Timoshenko [203].

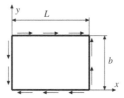

FIGURE 6.1
Plate subjected to shear stress.

In elastic buckling of plates of length L, width b and thickness t under shear stress, as shown in Fig. 6.1, the differential equation for the displacement w in the out-of-plane direction is given by

$$\frac{\partial^4 w}{\partial x^4} + 2\frac{\partial^4 w}{\partial x^2 \partial y^2} + \frac{\partial^4 w}{\partial y^4} = -\frac{2\tau_{xy}t}{D}\frac{\partial^2 w}{\partial x \partial y} \tag{6.1}$$

where

$$D = \frac{Et^3}{12(1-\nu^2)}$$

Similarly to elastic buckling of compressed plates, if the plates are simply supported on all four sides and the displacement w is assumed to be

$$w = \sum_{m=1}^{\infty} \sum_{n=1}^{\infty} A_{mn} \sin \frac{m\pi x}{L} \sin \frac{n\pi y}{b} \tag{6.2}$$

then the buckling stress τ_{buc} is evaluated by using

$$\tau_{buc} = k \frac{\pi^2 D}{b^2 t} \tag{6.3}$$

where the coefficient k depends on the ratio $\beta = L/b$ between the plate length and width. Here,

$$k = 5.35 + 4/\beta^2 \qquad (\beta \geq 1) \tag{6.4}$$

Plates that are clamped on all four sides can be similarly analyzed; if Eq. (6.3) is assumed to give the buckling stress, then the coefficient k is given by

$$k = 8.98 + 5.6/\beta^2 \qquad (\beta \geq 1) \tag{6.5}$$

From Eqs. (6.4) and (6.5), the coefficient is largest for square plates (namely, $L = b$), reaching 9.35 and 14.58 for simply supported and clamped plates, respectively. Buckling stress decreases as the rectangular plate becomes longer, and the coefficient k reaches limits as $L/b \to \infty$ of 5.35 and 8.98 for simply supported and clamped plates, respectively.

Buckling load has also been studied for cases in which shear stress and either compressive or tensile stress were simultaneously applied [19]: buckling shear stress decreased with compressive stress and increased with tensile stress.

For plastic shear buckling of plates, Gerard [79] did a series of tests on long 2024-0 aluminum alloy plates under shear and proposed use of the shear secant modulus as the plasticity-reduction factor for this case.

6.2   Torsion of cylinders

Schwerin [180] and Donnell [65] were the first to study torsional buckling of cylinders. Since then, many researchers, including Timoshenko [202], Kromm [116] and Batdorf [16] have performed more accurate analyses.

Fig. 6.2 shows the relationship between torsional moment M, torsion angle θ, and the deformation behavior for thin-walled cylinders under torsion. The figure shows that as the torsion angle θ increases, the shear stress at the cylinder surface (e.g., shear stress τ_A at point A shown in insert in the figure) first increases and then begins to decrease at about $\theta = 3.5°$, indicating that buckling occurs on the cylinder wall's surface. As buckling occurs, circumferential waves that spiral around the cylinder appear on the surface, and the torsional moment M begins to decrease. However, the decrease in load after buckling is not as fast as in axial compression buckling.

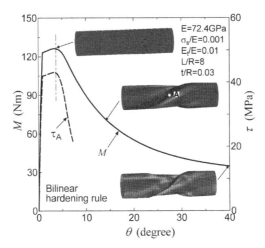

FIGURE 6.2
Relationship between torsional moment and torsion angle of a cylinder subjected to torsional load.

6.2.1 Elastic buckling of cylinders under torsion

When using Donnell's governing equations for the buckling of cylinders [65] (Eq. (1.6) in Section 1.1.1) to analyze elastic buckling of a cylinder of radius R, thickness t and length L, as shown in Fig. 6.3, it is assumed that the stress state before buckling is given by

$$N_{x0} = N_{\theta 0} = 0, \quad N_{x\theta 0} = \tau_{x\theta} t, \quad w_0 = \text{constant} \tag{6.6}$$

Then, differential equations for determining the torsional buckling stress $\tau_{x\theta}$ can be obtained; these equations are composed from the following two equations:

$$D\nabla^4 \hat{w} + \frac{1}{R}\hat{f}_{,xx} - 2\tau_{x\theta} t \hat{w}_{,x\theta} = 0 \tag{6.7a}$$

$$\nabla^4 \hat{f} - \frac{Et}{R}\hat{w}_{,xx} = 0 \tag{6.7b}$$

where the first equation, Eq. (6.7a), expresses the equilibrium of forces in the out-of-plane direction and the second equation, Eq. (6.7b), expresses the compatibility condition. Here,

$$D = \frac{Et^3}{12(1-\nu^2)}$$

holds.

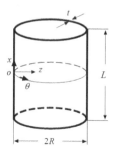

FIGURE 6.3
Geometrical shape of a cylinder subjected to torsional load.

From Eq. (6.7),

$$DV^8\hat{w} + \frac{Et}{R^2}\frac{\partial^4\hat{w}}{\partial x^4} - 2\tau_{x\theta}\frac{t}{R}\nabla^4\left(\frac{\partial^2\hat{w}}{\partial x\partial\theta}\right) = 0 \tag{6.8}$$

is obtained.

Theoretical torsional buckling stress in cylinders is much more difficult to analyze analytically than is axial compression stress. This is because an equation with both odd- and even-order derivatives of the off-plane displacement $\hat{w}$ does not have a separable solution that can be expressed as a product of sine and cosine functions.

To theoretically analyze the torsional buckling stress of a cylinder by satisfying Eq. (6.7) and boundary conditions at both ends of a specified cylinder, there are two approaches, as follows.

(1) From the shape of the spiral rotation of buckling waves on the cylinder surface, as shown in Fig. 6.2, Timoshenko [203] assumed that displacement in the torsional buckling of a cylinder is as follows:

$$\begin{cases} \hat{u} = A\cos\left(\dfrac{m\pi x}{L} - n\theta\right) \\[2mm] \hat{v} = B\cos\left(\dfrac{m\pi x}{L} - n\theta\right) \\[2mm] \hat{w} = C\sin\left(\dfrac{m\pi x}{L} - n\theta\right) \end{cases} \tag{6.9}$$

From Eq. (6.8),

$$\frac{\tau_{x\theta}}{E} = \frac{1}{12(1-\nu^2)}\left(\frac{t}{R}\right)^2\frac{(\bar{m}^2+n^2)^2}{2\bar{m}n} + \frac{\bar{m}^3}{2n(\bar{m}^2+n^2)^2} \tag{6.10}$$

is obtained, where

$$\bar{m} = \frac{m\pi R}{L}$$

For a given set of values of $\tau_{x\theta}/E$, t/R, ν and n, Eq. (6.10) can be considered as an eighth degree function of an unknown quantity $\bar{m}$ and assume that the roots of the equation are $\bar{m}_1, \bar{m}_2, \cdots, \bar{m}_8$; then, the buckling displacements are given by[1]

$$
\begin{cases}
\hat{u} = \displaystyle\sum_{k=1}^{8} A_k \cos\left(\frac{\bar{m}_k x}{R} - n\theta\right) \\[2mm]
\hat{v} = \displaystyle\sum_{k=1}^{8} B_k \cos\left(\frac{\bar{m}_k x}{R} - n\theta\right) \\[2mm]
\hat{w} = \displaystyle\sum_{k=1}^{8} C_k \sin\left(\frac{\bar{m}_k x}{R} - n\theta\right)
\end{cases}
\tag{6.11}
$$

Here, the coefficients A_k, B_k; $k = 1, 2, \cdots, 8$ are functions of C_k, $k = 1, 2, \cdots, 8$, being chosen to satisfy Eq. (6.7). There are thus only eight independent coefficients: C_k, $k = 1, 2, \cdots, 8$. These eight coefficients are chosen to satisfy the eight boundary conditions. For example, if the boundary condition for the simple support of both ends of a cylinder is given by[2]

$$\hat{u} = \hat{v} = \hat{w} = \hat{w}_{xx} = 0 \qquad \text{for } x = 0, L \tag{6.12}$$

then the eight homogeneous linear equations with unknown constants C_k, $k = 1, 2, \cdots, 8$ are obtained by substituting Eq. (6.11) into boundary conditions such as Eq. (6.12). For these homogeneous equations to yield a non-trivial solution, the determinant $\Delta|_{C_k}$ of the coefficients should vanish:

$$\Delta|_{C_k} = 0 \tag{6.13}$$

A cylinder length L satisfying the condition $\Delta|_{C_k} = 0$ (i.e., corresponding to the values given for $\tau_{x\theta}/E$, t/R, ν, and the possible values of n) can be obtained by applying the Galerkin method to Eq. (6.13).

(2) The second method first assumes a function $\hat{w}$ whose boundary conditions are satisfied. For example, if the boundary condition for $\hat{w}$ is given by $\hat{w} = \hat{w}_{,xx} = 0$ at $x = 0, L$, then $\hat{w}$ is assumed to be

$$
\hat{w} = t\left\{\sum_{m=1}^{N} C_{mn} \sin\left(m\pi \frac{x}{L}\right)\right\} \sin(n\theta) + t\left\{\sum_{m=1}^{N} D_{mn} \sin\left(m\pi \frac{x}{L}\right)\right\} \cos(n\theta)
\tag{6.14}
$$

and the compatibility equation (6.7b) is solved exactly for the stress function $\hat{f}$ in terms of the assumed radial displacement $\hat{w}$. The constant of integration for the general solution of $\hat{f}$ is determined from the boundary

---

[1] Although Eq. (6.10) may have complex roots, they always appear in conjugate pairs because all coefficients of the equation are real. Hence, the complex numbers $\hat{u}, \hat{v}, \hat{w}$ can be converted into real numbers.

[2] The boundary conditions are analyzed in detail in [225].

conditions for u and v, such as $\hat{u} = \hat{v} = 0$ at $x = 0, L$. Therefore, exact expressions of $\hat{w}$ and $\hat{f}$ that satisfy both boundary and compatibility conditions can be obtained.

By substituting the obtained values of $\hat{w}$ and $\hat{f}$ into the equilibrium equation, Eq. (6.7a), homogeneous linear equations can be obtained for the coefficients C_{mn} and D_{mn} in Eq. (6.14). To produce non-trivial solutions, the determinant $\Delta|_{C_{mn}D_{mn}}$ of the coefficients of C_{mn} and D_{mn} in these equations should vanish:

$$\Delta|_{C_{mn}D_{mn}} = 0 \qquad\qquad (6.15)$$

From Eq. (6.15), the buckling stress under the given geometrical condition can be determined by using the Galerkin method in the same way as before.

The buckling shear stress τ_{buc} corresponds to the n at which $\tau_{x\theta}$ is minimized. By repeated calculation, curves representing buckling shear stresses as functions of the geometrical dimensions of the structure can be constructed. Fig. 6.4 shows the buckling shear stress τ_{buc} obtained from the above-mentioned analysis based on Donnell's equation. The figure also shows the buckling stresses obtained by the finite element method (FEM) analysis.

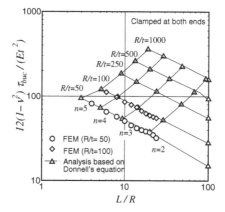

FIGURE 6.4
Values of τ_{buc}.

As shown in Fig. 6.4, the number of circumferential waves n changes mainly with the relative cylinder length L/R. In a sufficiently long cylinder, the buckling mode of $n = 2$ with two circumferential waves corresponds to the smallest torsional buckling load. As the cylinder length decreases, n increases ($n = 3, 4, \cdots$). Table 6.1 shows the relationship between the geometrical shapes of cylinders and the number n of circumferential waves for torsional buckling of the cylinders; these values are as found by FEM analysis.

TABLE 6.1
Relationship between the number n of circumferential waves and the geometrical shape of cylinders as obtained from FEM analysis.

L/R \\ t/R	0.01	0.02	0.03	0.04	0.05	0.06	0.08	0.1
2	8	6	5	5	5	4	4	4
4	6	5	4	4	4	4	3	3
6	5	4	3	3	3	3	3	3
10	4	3	3	3	3	3	2	2
20	3	3	2	2	2	2	2	2

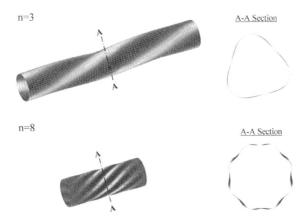

FIGURE 6.5
Buckling of cylinders of different lengths (for $n = 3$ and $n = 8$).

In addition, Fig. 6.5 shows the external appearance of buckling circumferential waves for $n = 3$ and $n = 8$.

In addition, for a cylinder long enough to neglect boundary effects, such as a cylinder satisfying the condition $L/\sqrt{Rt} > 10$ [225], the buckling shear stress τ_{buc} can be approximately obtained by neglecting the boundary effects and using Eq. (6.10). From Eq. (6.10) and for $n = 2$ and $\bar{m} \ll 1$,

$$\tau_{x\theta} \cong \frac{E}{3(1 - \nu^2)} \left(\frac{t}{R}\right)^2 \frac{1}{\bar{m}} + \frac{E}{64}\bar{m}^3 \qquad (6.16)$$

is obtained, where if $\bar{m}$ is chosen to minimize $\tau_{x\theta}$, then the corresponding buckling shear stress τ_{buc} is given by

$$\tau_{buc} = 0.272\frac{E}{(1-\nu^2)^{3/4}}\left(\frac{t}{R}\right)^{3/2} \qquad \text{for } \nu = 0.3, \ \tau_{buc} \cong 0.29E\left(\frac{t}{R}\right)^{3/2}$$

(6.17)

As many studies have shown, when $n = 2$, Donnell's approach gives a solution about 10% larger than that produced by exact analysis [74]; this accuracy is not good enough. Note that the error for Donnell's solution is about 2–4% when $n = 3$, 4 and very small when $n \geq 5$.

When the cylinder is simply supported at both ends, the buckling shear stress τ_{buc} depends on the cylinder length and is classified into three categories. Analytical results such as the following are often used for these categories.

(1) The buckling shear stress τ_{buc} of a very short cylinder is equal to the buckling stress of an elongated plate and is given, from Eq. (6.4) in the preceding section if it is assumed that $\beta \to \infty$, by

$$\tau_{buc} = 5.35\frac{\pi^2 D}{L^2 t}$$

(6.18)

(2) For a moderately long cylinder whose length satisfies $(L/R) \leq 8.7(R/t)^{0.5}$, the buckling shear stress τ_{buc} is evaluated by using

$$\tau_{buc} = \frac{0.85\pi^2 E}{12(1-\nu^2)^{5/8}}\left(\frac{t}{R}\right)^{5/4}\left(\frac{R}{L}\right)^{1/2}$$

$$\text{for } \nu = 0.3, \ \tau_{buc} \cong 0.75E\left(\frac{t}{R}\right)^{5/4}\left(\frac{R}{L}\right)^{1/2}$$

(6.19)

In some cases, Eq. (6.19) is multiplied by a coefficient C_τ given as

$$C_\tau = \sqrt{1 + 42\left(\frac{R}{L}\right)^3\left(\frac{t}{R}\right)^{1.5}}$$

so that the limit at infinity (i.e., as $L \to 0$) of τ_{buc} is equal to the value in Eq. (6.18).

(3) For a longer cylinder whose length satisfies $(L/R) > 8.7(R/t)^{0.5}$, the buckling shear stress τ_{buc} is evaluated by using

$$\tau_{buc} = \frac{E}{3\sqrt{2}(1-\nu^2)^{3/4}}\left(\frac{t}{R}\right)^{3/2} \qquad \text{for } \nu = 0.3, \ \tau_{buc} \cong 0.25E\left(\frac{t}{R}\right)^{3/2}$$

(6.20)

Euler buckling (i.e., $n = 1$) may occur in a very long cylinder. The shear stress $\tau|_{Euler}$ of Euler buckling due to torsion is evaluated by using

$$\tau|_{Euler} = \pi E\frac{R}{L}$$

(6.21)

where $\tau|_{Euler}$ decreases as the cylinder length L increases.

For the buckling shear stress τ_{buc} and the number of circumferential waves n, by introducing non-dimensional parameters defined as

$$k_s = \frac{tL^2}{\pi^2 D}\tau_{buc}\,, \quad \beta = \frac{L}{\pi R}n$$

these parameters can be expressed as functions of the parameter Z and the Poisson's ratio ν:

$$k_s = f_{k_s}(Z,\nu)\,, \quad \beta = f_\beta(Z,\nu)$$

where

$$Z = \sqrt{1-\nu^2}\,\frac{L^2}{Rt}$$

Thus, the influence of the geometrical shape of cylinders on buckling stress and on the number of circumferential waves can be expressed by a single parameter Z [225].

Fig. 6.6 shows the values of k_s and β obtained from theoretical analysis and FEM simulation.

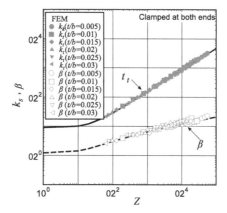

FIGURE 6.6
Values of k_s and β.

Many researches have been made on the post-buckling behavior of circular tubes under torsion (for example [127, 157]). Yamaki [223] conducted precise experimental studies on the behavior after elastic buckling of circular tubes under torsion. Also, based on the Donnell's basic equations, analytic solutions for the post-buckling behavior have been obtained by Yamaki [224], which were found to be in reasonable agreement with experimental results. Fig. 6.7, which is obtained from the analytic solutions given by Zhang and Han [231], shows curves of shear stress versus twisting angle for circular tubes with various lengths. It is found from Fig. 6.7 that the curves slope downward immediately after buckling for $L/R \geq 2$, and the greater the ratio of

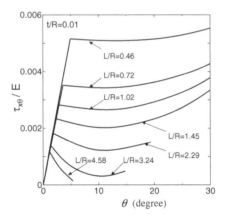

FIGURE 6.7
Shear stress versus twisting angle for circular tubes with various lengths.

tube length to radius L/R is, the lower the post-buckling path becomes. In general, the post-buckling equilibrium paths of circular tubes subjected to torsion are unstable and the relatively shorter tubes have higher post-buckling equilibrium paths.

The buckling of cylinders due to the application of both compressive and torsional load has also been studied. In elastic buckling, the interaction between axial compression and torsion in the simply supported case is nearly linear [18].

6.2.2   Plastic buckling of cylinders under torsion

Plastic buckling occurs if the wall of a cylinder is thick. Similarly to the case of elastic buckling of a twisted cylinder, the plastic buckling stress of a twisted cylinder depends on cylinder length: plastic buckling stress increases as the cylinder length decreases, as shown in Fig. 6.8. Fig. 6.8(a) shows relations between shear stress and rotation angle for circular tubes with the same ratio of thickness to radius ($t/R = 0.04$) and various lengths, assuming a material that obeys the bilinear hardening rule, a yield stress of $\sigma_s/E = 0.001$, and a strain hardening coefficient of $E_t/E = 0.01$. As shown in the figure, the shorter the tube is, the higher the plastic buckling stress becomes, and for long tubes the effect of the tube length on the plastic buckling stress becomes very small. If a cylinder becomes to some extent long ($L/R > 12$ in the figure), the plastic buckling stress will become almost constant. Fig. 6.8(b) compares deformation behaviors just after buckling through contours of strain in the tube surface for tubes analyzed in Fig. 6.8(a) with $L/R = 4, 12, 20$ and 32. It is seen from the figure that after buckling the most deformed region covers the

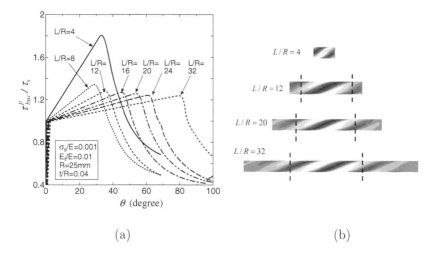

(a)   (b)

FIGURE 6.8
Torsional plastic buckling stress increases as the cylinder length decreases: (a) relations between shear stress and rotation angle; (b) contours of strain just after buckling.

full length of the tube for the relatively short tube (see sample of $L/R = 4$), and it occupies only a part of the full length for long tubes such as $L/R \geq 12$.

Many methods have been proposed for analyzing the plastic buckling of twisted cylinders; for additional details refer to, for example, [122, 158, 142].

Gerard [80] proposed an analytical technique for determining the plastic buckling stress τ_{buc}^p of a cylinder under torsion based on the J_2 deformation theory, in which, similarly to the case of plastic buckling stress in the axial crushing of a cylinder, τ_{buc}^p is evaluated by using the corresponding elastic buckling stress multiplied by a plasticity reduction factor η_p as follows:

$$\tau_{buc}^p = \eta_p \tau_{buc}^e \qquad (6.22)$$

Here, η_p is given as

$$\eta_p = \left(\frac{1 - \nu^2}{1 - \nu_p^2} \right)^{3/4} E_s/E \qquad (6.23)$$

where E_s is the secant modulus of a stress–strain curve, and ν_p is the Poisson's ratio for plastic deformation, which is obtained from the elastic Poisson's ratio ν by

$$\nu_p = \frac{1}{2} - \left(\frac{1}{2} - \nu \right) \frac{E_s}{E} \qquad (6.24)$$

Moreover, Rhodes [171] proposed the following formula for τ_{buc}^p based on the assumption that the strain at plastic buckling was equal to that at elastic buckling:

$$\tau_{buc}^p = f_\tau\left(\tau_{buc}^e/G\right) \tag{6.25}$$

where $f_\tau(\gamma)$ is the shear stress corresponding to the shear strain γ.

Furthermore, it is known that, as Timoshenko showed, the influence of plastic yield on the local bending deformation associated with the creation of buckling folds can be evaluated simply by replacing Young's modulus with

$$E_r = \frac{4EE_t}{(\sqrt{E} + \sqrt{E_t})^2} \tag{6.26}$$

in the formula for elastic buckling stress [203]. Therefore, the plastic buckling stress τ_{buc}^p can also be evaluated using a formula for the elastic buckling stress τ_{buc}^e by replacing E with E_r.

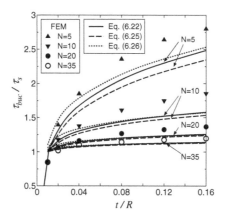

FIGURE 6.9
Torsional plastic buckling stress.

Fig. 6.9 and Table 6.2 show the torsional plastic buckling stress for cylinders with different thicknesses, as obtained by FEM simulation for a material whose stress–strain relation obeys the following equation:

$$\varepsilon = \frac{\sigma}{E}\left[1 + \frac{3}{7}\left(\frac{\sigma}{\sigma_s}\right)^{N-1}\right] \tag{6.27}$$

Fig. 6.9 and Table 6.2 also show the prediction values obtained from Eqs. (6.22), (6.25) and (6.26). It is seen from Fig. 6.9 and Table 6.2 that although the theoretical prediction of the plastic buckling stress of a cylinder under torsion is qualitatively in agreement with the analysis by FEM, the prediction accuracy is low and prediction is difficult at present.

TABLE 6.2
Torsional plastic buckling stress τ_{buc}/τ_s for cylinders with different thicknesses made of a material whose stress–strain relation obeys Eq. (6.27) (Numeric values in [ ] are the relative error of τ_{buc} obtained from Eq. (6.22), (6.25) or (6.26) to τ_{buc} obtained by FEM).

N	t/R	FEM	Eq. (6.22)		Eq. (6.25)		Eq. (6.26)	
	0.04	1.85	1.68	-9.4%	1.59	-14.0%	1.79	-3.3%
5	0.08	2.36	2.06	-12.7%	1.95	-17.3%	2.13	-9.4%
	0.12	2.64	2.30	-12.9%	2.17	-17.6%	2.36	-10.6%
	0.16	2.80	2.48	-11.3%	2.35	-16.2%	2.53	-9.6%
	0.04	1.38	1.30	-5.3%	1.27	-7.8%	1.34	-2.8%
10	0.08	1.61	1.44	-10.5%	1.40	-13.0%	1.45	-9.7%
	0.12	1.75	1.52	-13.0%	1.48	-15.4%	1.52	-12.9%
	0.16	1.85	1.58	-14.8%	1.54	-17.2%	1.57	-15.1%
	0.04	1.16	1.14	-1.6%	1.13	-2.9%	1.16	-0.4%
20	0.08	1.27	1.20	-5.2%	1.18	-6.6%	1.20	-5.0%
	0.12	1.33	1.23	-6.9%	1.22	-8.2%	1.23	-7.1%
	0.16	1.37	1.26	-8.0%	1.24	-9.4%	1.25	-8.5%
	0.04	1.11	1.08	-2.3%	1.07	-3.1%	1.09	-1.6%
35	0.08	1.14	1.11	-2.9%	1.10	-3.7%	1.11	-2.8%
	0.12	1.18	1.13	-4.2%	1.12	-5.0%	1.13	-4.3%
	0.16	1.20	1.14	-4.8%	1.13	-5.6%	1.14	-5.1%

After plastic buckling, the collapse deformation concentrates on a narrower region. The deformation behaviors in the collapse process after buckling are shown in Fig. 6.10 by taking the case of $L/R = 12$ in Fig. 6.8 as an example. A collapse section usually appears as waveform with circumferential wave number of $n = 2$ in many cases, as seen from a sample of $E_t/E = 0.01$ shown in Fig. 6.11. If the tube material has a large value of the strain hardening coefficient, the circumferential wave number becomes $n = 3$, as seen from a sample of $E_t/E = 0.1$ shown in Fig. 6.11.

6.3 Torsion of square tubes

Many reports have been published on the collapse behavior of thin-walled square tubes under torsional loads (e.g., [153, 135, 178, 55]). Murray (1984) investigated in detail maximum torsional load applied to thin-walled square tubes and proposed an evaluation method that uses either the maximum torsional load due to elastic buckling or the full-section yield load due to plastic yield stress under the assumption of rigid perfectly plastic materials. Chen and Wierzbicki [55] theoretically investigated the maximum torsional load by

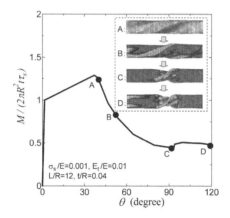

FIGURE 6.10
Moment response and deformation behavior for a circular tube of $\sigma_s/E = 0.001$, $E_t/E = 0.01$, $L/R = 12$, $t/R = 0.04$.

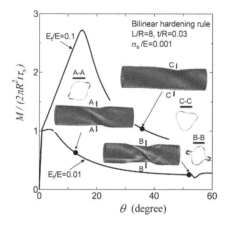

FIGURE 6.11
Comparison of moment response and deformation behavior between tubes with different strain hardening coefficients $E_t/E = 0.01$ and $E_t/E = 0.1$.

FEM simulation based on the concave structure in the cross section of a square tube created during large torsional plastic deformation.

In a square tube subjected to torsional load, torsional collapse depends on both the wall thickness of the tube and the strain-hardening coefficient of the material. There are two distinct collapse modes, and which occurs depends on the factors controlling the maximum moment M_{col} in the square tube [40].

The first collapse mode occurs when the limit stress is reached at the tube wall due to either buckling or the full-section plastic yield, and the second, caused by cross-sectional flattening, occurs when the wall of the square tube is thick and the strain hardening of the material is large. These collapse modes will be discussed below.

6.3.1 Elastic buckling

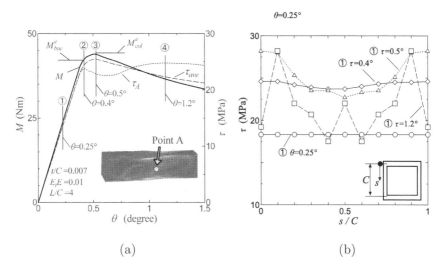

FIGURE 6.12
Collapse behavior due to elastic buckling [40]: (a) variation of torsional moment and shear stress, depending on the torsion angle θ; (b) distribution of shear stress within the cross section.

Fig. 6.12 shows the variation of torsional moment M and shear stress in a twisted square tube as an example of collapse by elastic buckling. To highlight the relation between buckling and the timing of the occurrence of the maximum torsional moment, the figure (a) also shows the variation of shear stress τ_A at the midpoint A of the wall surface of the square tube, where buckling wrinkles are created (see the deformation diagram in Fig. 6.12(a)) and the variation of average shear stress τ_{ave} in the cross section through point A of the square tube. With the shear stress τ_A as shown, the occurrence of buckling in the wall surface of a square tube can be determined by examining decline in the shear stress τ_A as the torsion is increased. From the fact that the shear stress τ_A shown in Fig. 6.12 peaks earlier than torsional moment does, it is clear that the torsional moment continuously increases when the shear stress at point A on the wall surface falls from the buckling of the surface of the square tube. Here, the torsional moment at buckling is defined as buckling load M_{buc}^e; the maximum torsional moment is defined as collapse load M_{col}^e. Torsional moment does not have a peak at buckling and continues to increase after buckling. This pattern resembles the load increase after buckling in the axial crushing of a thin-walled plate, and can also be understood from the shear stress change after buckling on the cross section of a square tube passing through point A. Fig. 6.12(b) shows the distribution of shear stress along the cross section passing through point A (coordinate s in Fig. 6.12(b)) during the respective deformation stages (before, during and after buckling), which are marked by ①, ②, ③, ④ in Fig. 6.12(a). Although the stress decreases in the middle part of the tube side plate as buckling occurs, the stress increases near both edges owing to the constraint imposed by the corner. Thus, both τ_{ave} and the torsional moment increase as shown in Fig. 6.12(a). Note that the torsional moment has a peak M_{col}^e because, in addition to the decrease of stress in the middle part, the increase of stress at the corner reaches a limit, which is shown in Fig. 6.12(b).

Torsional buckling of a square tube is caused by shear stress in the four plates of the tube and can thus be analyzed as a buckling problem for a plate subjected to shear stress. Elastic shear buckling stress in a plate of thickness t, width b and length L is given by Eq. (6.3). In applying Eq. (6.3) to a square tube under torsion, the plate length L should correspond to the region straddled by the buckling wrinkle rather than the full length of the actual plate (here, the length L of the square tube). With respect to the region of the torsional buckling wrinkles of the square tube, Mahendran and Murray [135] experimentally determined that the inclined angle of the buckling wrinkles was about 30°. Note that this fact can also be inferred from Fig. 6.13, which shows the FEM-simulated deformation behavior of a square tube.

FIGURE 6.13
Appearance of the buckling corrugation of a square tube [40].

Letting $L = b/\tan 30°$ and putting this into Eq. (6.4), k is given by

$$k = 5.35 + (2\tan 30°)^2 \cong 6.68 \tag{6.28}$$

and therefore the elastic buckling shear stress of a thin-walled square tube subjected to torsional load can be evaluated by using

$$\tau_{buc}^e \cong 6.68 \frac{\pi^2 E}{12\left(1 - \nu^2\right)} \left(\frac{t}{C}\right)^2 \tag{6.29}$$

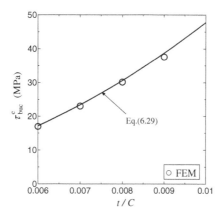

FIGURE 6.14
Elastic buckling stress in a square tube under torsion.

Fig. 6.14 compares the elastic buckling stress τ_{buc}^e predicted by Eq. (6.29) with the result from FEM analysis; good agreement is seen between them.

As shown in Fig. 6.12(a), the torsional moment in a square tube increases even after buckling of the wall of the tube. This phenomenon is similar to that observed in axial crushing. In axial crushing of a plate, the crushing stress σ_{col} can be interpreted by Karman's idea of effective width [211]. Various formulas for evaluation of the collapse stress σ_{col} were proposed, using a parameter λ called the slenderness of the plate and defined by the yield stress σ_s and the elastic buckling stress σ_{buc}^e of the plate as follows:

$$\lambda = \sqrt{\frac{\sigma_s}{\sigma_{buc}^e}} \tag{6.30}$$

For example, Karman proposed

$$\frac{\sigma_{col}}{\sigma_s} = \frac{1}{\lambda} = \sqrt{\frac{\sigma_{buc}^e}{\sigma_s}} \tag{6.31}$$

and Winter [221] proposed

$$\frac{\sigma_{col}}{\sigma_s} = \frac{1}{\lambda}\left(1 - 0.25\frac{1}{\lambda}\right) = \sqrt{\frac{\sigma_{buc}^e}{\sigma_s}}\left(1 - 0.25\sqrt{\frac{\sigma_{buc}^e}{\sigma_s}}\right) \tag{6.32}$$

By applying Eqs. (6.31) and (6.32) to the evaluation of the collapse moment after elastic buckling of a square tube under torsion, the torsional collapse moment can be evaluated by using

$$\frac{M_{col}}{M_s} = \sqrt{\frac{\tau_{buc}^e}{\tau_s}} \tag{6.33}$$

and

$$\frac{M_{col}}{M_s} = \sqrt{\frac{\tau^e_{buc}}{\tau_s}} \left(1 - 0.25\sqrt{\frac{\tau^e_{buc}}{\tau_s}} \right) \tag{6.34}$$

respectively [40]. Here,

$$M_s = 2C^2 t \tau_s, \qquad \tau_s = \sigma_s \big/ \sqrt{3} \tag{6.35}$$

Fig. 6.15 shows the values given by Eqs. (6.33) and (6.34), the results of FEM analysis, and the experimental results of Mahendran and Murray [135]; approximate agreement is seen among them. The figure also shows the collapse moment M_{col} as obtained from the collapse shear stresses τ_{col} of a square tube under torsion by the methods proposed by Basler [15], Hoglund [93], Herzog [92] and Richtlinie [175].

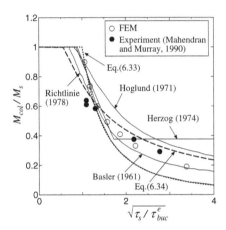

FIGURE 6.15
Torsional collapse moment of square tubes [40].

6.3.2   Plastic buckling

As the relative thickness t/C of a square tube becomes moderately thick, plastic buckling will occur because the torsional shear stress will reach and exceed the yield stress before buckling develops in the tube. Fig. 6.16 shows an example of torsional collapse caused by plastic buckling. The occurrence of buckling in the tube wall can be understood from the decline in the shear stress τ_A as the torsion is increased.

    In order to accurately predict the maximum torsional moment, an evaluation method for the torsional plastic buckling shear stress that takes the strain

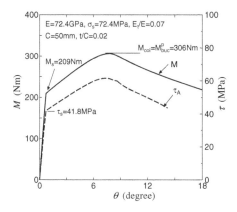

FIGURE 6.16
Torsional collapse due to plastic buckling $(t/C=0.014$ and $E_t/E=0.01)$.

hardening of the material into consideration is required. It is presently difficult to establish an effective evaluation system although many approximate
equations have been proposed for plastic buckling stress.

A possible approximation approach is the use of Stowell's formula [194] on
the ratio $\eta_p|_{axi}$ between plastic and elastic buckling stresses:

$$\eta_p|_{axi} = \frac{E_s}{2E}\left(1 + \sqrt{\frac{1}{4} + \frac{3E_t}{4E_s}}\right) \qquad (6.36)$$

for a plate under axial compression. That is, although torsional buckling is
evaluated by shear stress, the compressive principle stress plays an important
role. Therefore, it is considered that by assuming the ratio $\eta_p|_{tor}$ between
plastic and elastic buckling stresses for a plate under torsion is equal to the
ratio $\eta_p|_{axi}$ (i.e. $\eta_p|_{tor} \cong \eta_p|_{axi}$), the plastic buckling stress of a square tube
under torsion can be approximately evaluated by

$$\tau^p_{buc} = \eta_p|_{tor} \times \tau^e_{buc}, \quad \eta_p|_{tor} \cong \eta_p|_{axi} \qquad (6.37)$$

Furthermore, it is also possible to apply approaches based on Gerard's
method of Eq. (6.22) and Rhodes's method of Eq. (6.25), shown in Subsection
6.2.2 for evaluation of the plastic buckling stress of a square tube under torsion.

Fig. 6.17 and Table 6.3 show the torsional plastic buckling stresses for a
square tube of a material whose stress–strain relation obeys Eq. (6.27) obtained from FEM simulation and from evaluation of Eq. (6.37), Eq. (6.22)
and Eq. (6.25), respectively. All of the results obtained from various analyses of Eq. (6.37), Eq. (6.22) and Eq. (6.25) are almost the same and are
approximately in agreement with the numerical analysis result of FEM.

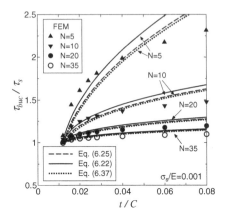

FIGURE 6.17
Values of plastic buckling stress.

TABLE 6.3
Torsional plastic buckling stress τ_{buc}/τ_s for square tubes with different thicknesses made of a material whose stress–strain relation obeys Eq. (6.27). (Numeric values in [ ] are the relative error of τ_{buc} obtained from Eq. (6.22), (6.25) or (6.37) to τ_{buc} obtained by FEM.)

N	t/C	FEM	Eq. (6.22)		Eq. (6.25)		Eq. (6.37)	
5	0.02	1.61	1.48	-7.9%	1.41	-12.4%	1.38	-14.0%
	0.04	1.99	2.08	4.9%	1.97	-0.8%	1.94	-2.2%
	0.06	2.18	2.48	14.2%	2.35	7.9%	2.32	6.5%
	0.08	2.31	2.80	21.0%	2.65	14.4%	2.61	12.9%
10	0.02	1.25	1.23	-1.4%	1.20	-4.0%	1.18	-5.6%
	0.04	1.37	1.45	5.8%	1.41	2.9%	1.39	1.7%
	0.06	1.43	1.58	10.3%	1.54	7.2%	1.52	6.1%
	0.08	1.48	1.68	13.5%	1.63	10.3%	1.61	9.2%
20	0.02	1.10	1.11	0.9%	1.10	-0.4%	1.08	-1.5%
	0.04	1.15	1.21	4.7%	1.19	3.2%	1.18	2.5%
	0.06	1.18	1.26	6.9%	1.24	5.4%	1.23	4.7%
	0.08	1.20	1.30	8.4%	1.28	6.8%	1.27	6.2%
35	0.02	1.05	1.06	1.4%	1.05	0.6%	1.05	0.0%
	0.04	1.07	1.11	3.6%	1.10	2.7%	1.10	2.3%
	0.06	1.09	1.14	4.7%	1.13	3.9%	1.13	3.5%
	0.08	1.10	1.16	5.5%	1.15	4.7%	1.15	4.3%

6.3.3 Cross-sectional flattening

It is known that cross-sectional flattening is observed in torsional deformation of a square tube also, as shown in Fig. 6.18. Therefore, cases where the torsional moment begins to descend may also be observed during torsional deformation due to cross-sectional flattening, even if torsional shear stress does not fall.

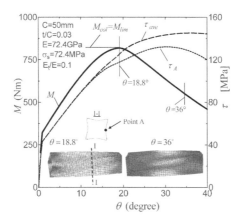

FIGURE 6.18
Relations of θ and M, τ_A, τ_{ave} and deformed shapes when $t/b=0.03$ and $E_h/E=0.1$ [40].

Fig. 6.18 shows the variation of torsional moment M, shear stress τ_A at point A (the point of the center of the cross section I-I in Fig. 6.18) and average shear stress τ_{ave} on the cross section I-I for a square tube with $E_t/E = 0.1$ and $t/C = 0.03$ subjected to torsional load. As the figure shows, the shear stress τ_A and the average shear stress τ_{ave} continuously increase, even after the torsional moment reaches a peak; that is, when both the relative thickness t/C and the strain-hardening coefficient E_t/E of a square tube are large, buckling does not occur despite occurrence of a maximum moment.

The reason that torsional moment shows a peak is cross-sectional flattening. Fig. 6.18 shows a view of the deformed shape of the tube and cross section at the peak ($\theta=18.8°$) also, where a large cross-sectional flattening is observed. The moment around the axis of a square tube becomes small due to flattening deformation in the cross section. Furthermore, unlike torsional collapse caused by elastic buckling or plastic buckling, local collapse does not occur and the flattening deformation in the cross section extends over the entire length of the square tube, just as shown in the view of the deformed shape in Fig. 6.18, where $\theta=36°$.

Therefore, as shown by Eq. (4.41) for the bending deformation of tubes, **the maximum moment** M_{col} in the torsional deformation of a tube equals

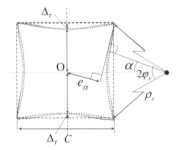

FIGURE 6.19
Flattening shape of cross section and approximation model for a square tube under torsion [40].

one of either **the critical torsional moment** M_{cri} caused by the buckling at the wall plate due to the shear stress or **the limit torsional moment** M_{lim} due to flattening.

In collapse caused by flattening of cross sections, it is necessary to know the flattening quantity to predict the maximum torsional moment. As an example, assuming $\theta = 18.8°$ (as in Fig. 6.18), Fig. 6.19 shows the cross section of a square tube obtained by FEM simulation as a thin line, and the cross section before deformation as a broken curve. As the figure shows, the flattening deformation of a cross section is almost the same from right to left and from top to bottom and is largest in the middle. Here, the flattening quantity is denoted by Δ_τ. Fig. 6.20 shows the relationship between the relative flattening quantity Δ_τ/C and the torsion angle θ for $E_t/E = 0.07$ and 0.1, $t/b = 0.04, 0.05$ and 0.06. The relative flattening quantity Δ_τ/C is independent of the strain-hardening characteristics of the material and can be expressed as a function of torsion angle θ and relative thickness t/C, as shown in the approximation formula Eq. (6.38). The value obtained by Eq. (6.38) is also shown in Fig. 6.20.

$$\frac{\Delta_\tau}{C} = 0.0033 \left(\frac{\theta C^2}{tL} \right)^2 \tag{6.38}$$

The variation of the moment can be calculated by using the flattening quantity Δ_τ obtained during torsion changes, as shown in Fig. 6.20.

The dotted curve in Fig. 6.19 approximates the flattening surface of the wall by a circular arc of radius ρ_s and central angle $2\phi_s$. The radius ρ_s and central angle $2\phi_s$ can be obtained from the following simultaneous equations:

$$\begin{cases} 2\rho_s\phi_s = C \\ \rho_s (1 - \cos\phi_s) = \Delta_\tau \end{cases} \tag{6.39}$$

as a function of flattening quantity Δ_τ. Therefore, the torsional moment M

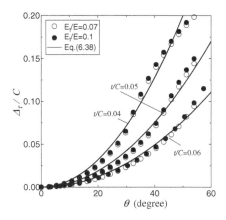

FIGURE 6.20
Relationship between flattening quantity and torsion angle for square tubes
[40].

due to shear stress τ in the cross section is given by

$$M = 4 \int_{-\phi_s}^{\phi_s} \tau \times e_\alpha \times (t\rho_s d\alpha) \tag{6.40}$$

where e_α is distance from the central point O of the tube to the line tangent
to the circular arc at the point at angle α (see Fig. 6.19) and is obtained by
using

$$e_\alpha = (\rho_s + d_\tau)\cos\alpha - \rho_s \tag{6.41}$$

$$d_\tau = \rho_s \sin\phi_s - \Delta_\tau \tag{6.42}$$

Furthermore, if the shear strain γ is assumed to be constant in the cross
section, it is given by

$$\gamma = d_\tau \frac{\theta}{L} \tag{6.43}$$

from the deformation in the central part of the wall surface. Thus, the shear
stress τ is given by

$$\tau = \begin{cases} G\gamma & (\gamma < \tau_s/G) \\ \tau_s + G_t(\gamma - \tau_s/G) & (\gamma < \tau_s/G) \end{cases} \tag{6.44}$$

where

$$G_t = E_t/3 \tag{6.45}$$

Fig. 6.21 shows the values obtained from Eq. (6.40) and the results of
FEM analysis. In deriving the torsional moment from Eq. (6.40), the flattening quantity Δ_τ is required; Fig. 6.21 shows the two types of moment from

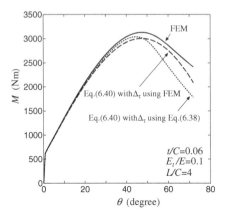

FIGURE 6.21
Comparison of the torsional moment of a square tube calculated from the
flattening quantity Δ_τ and the results of FEM analysis [40].

Eq. (6.40): In one type, Δ_τ is obtained from Eq. (6.38), and in another type,
Δ_τ is directly obtained from FEM analysis. The torsional moments obtained
from Eq. (6.40) and FEM analysis are in good agreement when the flattening quantity Δ_τ obtained from FEM analysis is used. Therefore, the torsional
moment can be predicted by approximating the flattening shape of the cross
section with a circular arc without the need to accurately reproduce the cross-
sectional shape of the tube under torsion.

6.3.4 Torsional collapse of square tube

Fig. 6.22 shows the relation between torsional moment and rotation angle as
analyzed by FEM numerical simulation of a square tube under torsion with
both ends of the tube fixed in the axial direction. The torsional moment, after
reaching an ultimate point, dropped significantly with increasing torsional
rotation due to the plastic sectional collapse of the tube, and, appeared to
reach constant asymptotic value at large rotations. The deformed shapes are
also shown in the figure. Chen and Wierzbicki [55] developed a simple torsional
collapse model, which captured the basic feature shown in Fig. 6.22. In the
model, the torsion of a tube is divided into three phases: the pre-buckling,
post-buckling, and collapse spreading phases. In the pre-buckling phase, all
sections rotate as rigid bodies without sectional deformation. The walls of the
tube become spiral surfaces. After buckling, walls collapse inward, as shown in
the figure. This relieves the membrane strains and reduces the load-carrying
capacity of the tube. The sectional distortion is the largest at the mid-section
of the tube and it decreases to zero at the two ends. At a certain twisting

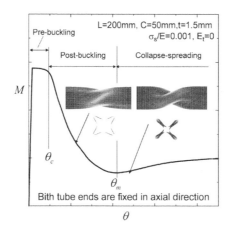

FIGURE 6.22
Three different deformation phases in torsion of a square tube.

rotation angle (transitional twisting rotation θ_m), internal touching occurs at the mid-section of the tube, as shown in the figure. From this point on, the torsion enters a new phase called the **collapse spreading phase**, and the most-collapsed section will spread toward the two ends from the mid-section. The dimensional torsional moments in the three phases are given by Chen and Wierzbicki [55] as follows:

$$\tau_M = 0.58 + 0.05r^2\theta^2 \quad (0 \le \theta \le \theta_c) \tag{6.46}$$

for the pre-buckling phase,

$$\tau_M = 0.58 - 0.21r^{-0.22}\theta^{0.34} \quad (\theta_c \le \theta \le \theta_m) \tag{6.47}$$

for the post-buckling phase, and

$$\tau_M = 0.58 - 0.21r^{-0.22}\theta_m^{0.56}\theta^{-0.22} \quad (\theta \ge \theta_m) \tag{6.48}$$

for the collapse spreading phase. Here,

$$\theta_c = \frac{2.2(1+\nu)}{(1-\nu^2)^{0.75}}\left(\frac{2t}{C}\right)^{1.5}\frac{L}{C}, \quad \theta_m = \pi/2, \quad r = C/L \tag{6.49}$$

and

$$\tau_M = \frac{M}{2\sigma_0 C^2 t} \tag{6.50}$$

The moment responses obtained from FEM numerical simulations for square tubes under torsion with both ends of the tube fixed in the axial direction are shown in Fig. 6.23, together with the analytical solutions derived above.

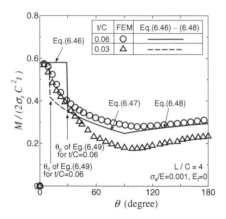

FIGURE 6.23

Comparison of the torsional moment of a square tube calculated from Eqs. (6.46)–(6.48) and the results of FEM analysis.

Bibliography

[1] E.F. Abdewi, S. Sulaiman, and A.M.S. Hamouda. Effect of geometry on the crushing behaviour of laminated corrugated composite tubes. *Materials Processing Technology*, 172:394–399, 2006.

[2] W. Abramowicz. The effective crushing distance in axially compressed thin-walled metal columns. *Int. J. Impact Engng.*, 1:309–317, 1983.

[3] W. Abramowicz and N. Jones. Dynamic axial crushing of circular tubes. *Int. J. Impact Engng.*, 2(3):263–281, 1984.

[4] W. Abramowicz and N. Jones. Dynamic axial crushing of square tubes. *Int. J. Impact Engng.*, 2(2):179–208, 1984.

[5] W. Abramowicz and N. Jones. Dynamic progressive buckling of circular and square tubes. *Int. J. Impact Engng.*, 4(4):243–270, 1986.

[6] W. Abramowicz and N. Jones. Transition from initial global bending to progressive buckling of tubes loaded statically and dynamically. *International Journal of Impact Engineering*, 19:415–437, 1997.

[7] W. Abramowicz and T. Wierzbicki. Axial crushing of multiconer sheet metal columns. *Journal of Applied Mechanics*, 56:113–120, 1989.

[8] C.S. Ades. Bending strength of tubing in the plastic range. *Journal of Aerospace Science*, 24:605–610, 1957.

[9] E.L. Aksel'rad and F.A. Emmerling. Collapse load of elastic tubes under bending. *Isr. Journal of Technology*, 22:89–94, 1984.

[10] J.M. Alexander. An approximate analysis of the collapse of thin cylindrical shells under axial loading. *Quart. J. Mech. Appl. Math.*, 13(1):10–15, 1960.

[11] American Iron and Steel Institute. *North American specification for the design of cold-formed steel structural members (NAS)*. Washington, DC, 2001. Draft Edition November 9, 2001; 1st Printing February 2002.

[12] A. Andronicou and A.C. Walker. A plastic collapse mechanism for cylinders under uniaxial end compression. *J. Const. Steel Res.*, 1:23–34, 1981.

309

[13] M.R. Bambach. Local buckling and post-local buckling redistribution of stress in slender plates and sections. *Thin Walled Structures*, 44:1118–1128, 2006.

[14] M.R. Bambach. Design of uniformly compressed edge-stiffened flanges and sections that contain them. *Thin-Walled Structures*, 47:277–294, 2009.

[15] K. Basler. Strength of plate girders in shear. *JDStructDDiv., Proc-DASCE*, (ST7):151–197, 1961.

[16] S.B. Batdorf. A simplified method of elastic stability analysis for thin cylindrical shells. Technical Report NACA Rep. 874, 1947.

[17] S.B. Batdorf. Theories of plastic buckling. *Journal of the Aeronautical Sciences*, 16(7):405–408, 1949.

[18] S.B. Batdorf, M. Stein, and M. Schildcrout. Critical combinations of torsion and direct axial stress for thin-walled cylinders. Technical Report NACA TN-1345, 1947.

[19] S.B. Batdorff and M. Stein. Critical combinations of shear and direct stress for simply supported rectangular flat plates. Technical Report NACA TN-1223, 1947.

[20] S.C. Batterman. Plastic buckling of axially compressed cylindrical shells. *AIAA Journal*, 3(2):316–325, 1965.

[21] S.C. Batterman and L.H.N. Lee. Effect of modes on plastic buckling of compressed cylindrical shells. *AIAA Journal*, 4(12):2255–2257, 1966.

[22] P.P. Bijlaard. On the plastic stability of thin plates and shells. *Proc. Koninkl. Ned. Akad, Wetenschap*, 50:765–775, 1947.

[23] P.P. Bijlaard. Theory and tests on the plastic stability of plates and shells. *Journal of the Aeronautical Sciences*, 16(9):529–541, 1949.

[24] R.S. Birch and N. Jones. Dynamic and static axial crushing of axially stiffened cylindrical shells. *Thin-Walled Structures*, 9:29–60, 1990.

[25] J.G. Bouwkamp and R.M. Stephen. Large diameter pipe under combined loading. *Transp. eng. J. ASCE*, 99(TE3):521–536, 1973.

[26] L.G. Brazier. On the flexure of thin cylindrical shells and other "thin" sections. *Proceedings of the Royal Society of London, Series A*, 116:104–114, 1927.

[27] E.J. Brunelle. Buckling of transversely isotropic mindlin plates. *AIAA Journal*, 9:1018–1022, 1971.

[28] E.J. Brunelle and S.R. Robertson. Initially stressed mindlin plates. *AIAA Journal*, 12:1036–1045, 1974.

[29] D. Bushnell. Elastic-plastic bending and buckling of pipes and elbows. *Comp. Struct.*, 13:241–248, 1981.

[30] J.F. Carney and R.J. Sazinski. Portable energy absorbing system for highway service vehicles. *Transportation Engineering Journal*, TE4:407–421, 1978.

[31] D.H. Chen. The collapse mechanism of corrugated cross section beams subjected to three-point bending. *Thin-Walled Structures*, 51:82–86, 2012.

[32] D.H. Chen, T. Fujita, and K. Ushijima. The initial peak load in axial impact of thin-walled circular tubes with consideration of strain rate effect of material. *Transactions of JSME*, A75:1476–1483, 2009.

[33] D.H. Chen, K. Hattori, and S. Ozaki. Axial crushing characteristics of circular tubes with radial corrugation. *Journal of Computational Science and Technology*, 3(2):437–448, 2009.

[34] D.H. Chen, K. Hattori, and S. Ozaki. Crushing characteristics of square tubes with radial corrugation. *Transactions of JSME*, A75:142–149, 2009.

[35] D.H. Chen and K. Masuda. Prediction of maximum moment of rectangular tubes subjected to pure bending. *Journal of Environment and Engineering*, 6(3):554–566, 2011.

[36] D.H. Chen, K. Masuda, and S. Ozaki. Study on elastoplastic pure bending collapse of cylindrical tubes. *Transactions of JSME*, A74:520–527, 2008.

[37] D.H. Chen, K. Masuda, and S. Tsunoda. Axial crushing characteristics of thin-walled conic absorber with corrugated surface. *Transactions of JSME*, A77:261–270, 2011.

[38] D.H. Chen, K. Masuda, K. Ushijima, and S. Ozaki. Deformation modes for axial crushing of cylindrical tubes considering the edge effect. *Journal of Computational Science and Technology*, 3(1):339–350, 2009.

[39] D.H. Chen, T. Masuzawaand, and S. Ozaki. Axial- and ring-stiffened circular tubes under axial compression. *Materials*, 57(7):696–703, 2008.

[40] D.H. Chen, M. Okamoto, and K. Masuda. Torsional collapse of thin-walled square tubes. *Transactions of JSME*, A76:1170–1177, 2010.

[41] D.H. Chen and S. Ozaki. Circumferential strain concentration in axial crushing of cylindrical and square tubes with corrugated surfaces. *Thin-Walled Structures*, 47(5):547–554, 2009.

[42] D.H. Chen and S. Ozaki. Numerical study of axially crushed cylindrical tubes with corrugated surfaces. *Thin-Walled Structures*, 47(11):1387–1396, 2009.

[43] D.H. Chen and S. Ozaki. Theoretical analysis of axial crushing of cylindrical tubes with corrugated surfaces. *Journal of Computational Science and Technology*, 3(1):327–338, 2009.

[44] D.H. Chen and S. Ozaki. Axial collapse behavior of plate. *Thin-Walled Structures*, 48:77–88, 2010.

[45] D.H. Chen, H. Sakaizawa, and S. Ozaki. Crushing behaviour of hexagonal thin-walled tube with partition plates (1st report: analysis of deformation mode). *Transactions of JSME*, A72:1978–1984, 2006.

[46] D.H. Chen, H. Sakaizawa, and S. Ozaki. Crushing behaviour of hexagonal thin-walled tube with partition plates (2nd report: Analysis of compressive stress). *Transactions of JSME*, A72:1985–1991, 2006.

[47] D.H. Chen and Y. Shimizu. Axially crushed square tubes with corrugated surface. *Transactions of JSME*, 72A(723):1668–1675, 2006.

[48] D.H. Chen, D. Tanaka, and S. Ozaki. Telescopic deformation of stepped circular tube subjected to axial crushing. *Journal of Computational Science and Technology*, 3(1):351–362, 2009.

[49] D.H. Chen and K. Ushijima. Evaluation of first peak stress based on buckling theory for axial collapse of circular cylindrical shell. *Transactions of Society of Automotive Engineers of Japan*, 37(5):19–24, 2006.

[50] D.H. Chen and K. Ushijima. Telescopic deformation of stepped circular tube subjected to oblique load. *Transactions of JSME*, 76A(769):1178–1185, 2010.

[51] D.H. Chen and K. Ushijima. Eccentricity in the progressive crushing of circular tubes. *Journal of Solid Mechanics and Materials Engineering*, 5(5):219–229, 2011.

[52] D.H. Chen and K. Ushijima. Estimation of the initial peak load for circular tubes subjected to axial impact. *Thin-Walled Structures*, 49(7):889–898, 2011.

[53] D.H. Chen and K. Ushijima. Evaluation of quasi-static axial crushing characteristics of laterally grooved square tube. *Journal of Solid Mechanics and Materials Engineering*, 5(3):151–163, 2011.

[54] D.H. Chen, S. Yoshida, and S. Ozaki. Deformation behavior for axial crushing of three-fold point corner. *Journal of Computational Science and Technology*, 3(2):426–436, 2009.

[55] W. Chen and T. Wierzbicki. Torsional collapse of thin-walled prismatic columns. *Thin-Walled Structures*, 36:181–196, 2000.

[56] W. Chen, T. Wierzbicki, and S. Santosa. Bending collapse of thin-walled beams with ultralight filler: numerical simulation and weight optimization. *Acta Mechanica Journal*, 153:183–206, 2002.

[57] J. Christoffersen and J.W. Hutchinson. A class of phenomenological corner theories of plasticity. *Journal of the Mechanics and Physics of Solids*, 27:465–487, 1979.

[58] Cold-Formed Steel Structures. *Australian/New Zealand Standard AS/NZS 4600*. Printed in Australia, 2005.

[59] E. Corona and S. Vaze. Buckling of elastic-plastic square tubes under bending. *International Journal of Mech. Sci.*, 38:753–775, 1996.

[60] H.L. Cox. Buckling of thin plates in compression. Technical Report 1554, British ARC Reports and Memoranda, 1934.

[61] G.H. Daneshi and S.J. Hosseinipour. Elastic-plastic theory for initial buckling load of thin-walled grooved tubes under axial compression. *J. Materials Processing Tech.*, 125-126:826–832, 2002.

[62] W.R. Dean. On the theory of elastic stability. *Proceedings of the Royal Society of London*, 107(Series A):734–759, 1925.

[63] W.J. Dewalt and W.B. Herbein. Energy absorption by compression of aluminium tubes. Technical Report No. 12-72-23, Alcoa Research Laboratories, Alcoa Center, Pennsylvania, 1972.

[64] B.P. DiPaolo and J.G. Tom. A study on an axial crush configuration response of thin-wall, steel box components: The quasi-static experiments. *International Journal of Solids and Structures*, 43:7752–7775, 2006.

[65] L.H. Donnell. Stability of thin-walled tubes under torsion. Technical Report NACA Rep. 479, 1933.

[66] M. Elchalakani, X.L. Zhao, and R.H. Grzebieta. Plastic mechanism analysis of circular tubes under pure bending. *International Journal of Mechanical Sciences*, 44:1117–1143, 2002.

[67] A.M. Elgalai, E. Mahdi, A.M.S. Hamouda, and B.S. Sahari. Crushing response of composite corrugated tubes to quasi-static axial loading. *Composite Structures*, 66:665–671, 2004.

[68] Y. Estrin, A.V. Dyskin, A.J. Kanel-Belov, and E. Pasternak. Materials with novel architectonics: Assemblies of interlocked elements. *IUTAM Symposium on Analytical and Computational Fracture Mechanics of Non-Homogeneous Materials*, pages 51–55, 2002.

[69] O. Fabian. Collapse of cylindrical, elastic tubes under combined bending, pressure and axial loads. *International Journal of Solids and Structures*, 13:1257–1270, 1977.

[70] O. Fabian. Elastic-plastic collapse of long tubes under combined bending and pressure load. *Ocean Engng*, 8:295–330, 1981.

[71] G.L. Farley. Effects of specimen geometry on the energy absorption capability of composite materials. *Journal of Composite Materials*, 20:390–400, 1986.

[72] D. Faulkner. A review of effective plating for use in the analysis of stiffened plating in bending and compression. *Journal of Ship Research*, 19:1–17, 1975.

[73] W. Flügge. Die Stabilitat der Kreiszylinderschale. *Ingenieur-Archiv*, 3:463–506, 1932.

[74] W. Flügge. *Stresses in Shells*. Springer Verlag, Berlin, 2nd edition, 1973.

[75] W.C. Fok, G. Lu, and L.K. Seah. A simplified approach to buckling of plain c channels under pure bending. *Proceedings of the Institution of Mechanical Engineers, Part C: Journal of Mechanical Engineering Science*, 207:255–262, 1993.

[76] European Committe for Standardization. *Eurocode 3, Design of steel strength - General Rules - Supplementary Rules for Cold-Formed Thin Gauge Members and Sheeting*. 1996.

[77] D.A. Galib and A. Limam. Experimental and numerical investigation of static and dynamic axial crushing of circular aluminum tubes. *Thin-Walled Structures*, 42:1103–1137, 2004.

[78] S. Gellin. The plastic buckling of long cylindrical shells under pure bending. *International Journal of Solids and Structures*, 16:397–407, 1980.

[79] G. Gerard. Critical shear stress of plates above the proportional limit. *Jour. Appl. Mech.*, 15:7–12, 1948.

[80] G. Gerard. Compressive and torsional buckling of thin-wall cylinders in yield region. *Nat. Adv. Comm. Aeronaut., Technical Note No. 3726*, 1956.

[81] G. Gerard. On the role of initial imperfection in plastic buckling of cylinders under axial compression. *Journal of the Aeronautical Sciences*, 29:744–745, 1962.

[82] G. Gerard and H. Becker. Handbook of structural stability, part i-buckling of flat plates. Technical Report NACA TN 3781, 1957.

[83] T.L. Gerber. Plastic deformation of piping due to pipe-whip loading. *ASME paper*, 74-NE-1, 1974.

[84] T.R. Graves Smith and S. Sridharan. Elastic collapse of thin-walled column. In J. Rhodes and A.C. Walker, editors, *Thin-walled Structures*. Granada Publishers, London, 1980.

[85] R.H. Grzebieta. An alternative method for determining the behaviour of round stocky tubes subjected to an axial crush load. *Thin-Walled Structures*, 9(4):61–89, 1990.

[86] R.H. Grzebieta and N.W. Murray. Rigid-plastic collapse behavior of an axially crushed stocky tube. *Proc. ASME Winter Annual Meeting, AMD-Vol.*, 105:10–15, 1989.

[87] F. Guarracino. On the analysis of cylindrical tubes under flexure: theoretical formulations, experimental data and finite element analyses. *Thin-Walled Structures*, 41:127–147, 2003.

[88] S.R. Guillow, G. Lu, and R.H. Grzebieta. Quasi-static axial compression of thin-walled circular aluminium tubes. *Int. J. Mech. Sci.*, 43:2103–2123, 2001.

[89] J.J. Harrigan, S.R. Reid, and C. Peng. Inertia effects in impact energy absorbing materials and structures. *Int J Impact Eng*, 22:955–279, 1999.

[90] S.W. Hasan and G.J. Hancock. Plastic bending test of cold-formed rectangular hollow sections. *J. Australian Inst. Steel Constr.*, 23(4):2–19, 1989.

[91] R.J. Hayduk and T. Wierzbicki. Extensional collapse modes of structural members. In *Proc. Symp. Advances and Trends in Structural and Solid Mechanics*, pages 405–434, 1982.

[92] M. Herzog. Die traglast unversteifter und versteifter, dunnwandiger blechtrager unter reinem schub und schub mit biegung nach versuchen,. *Bauingenieur*, 49:382–389, 1974.

[93] T. Höglund. Livets verkningssatt och barformaga hos tunnavaggig i-balk, divdbuilding statics and structual engineering. In *Royal Inst-DTech., Bulletin No. 93, Stockholm*, 1971.

[94] S.J. Hosseinipour and G.H. Daneshi. Energy absorbtion and mean crushing load of thin-walled grooved tubes under axial compression. *Thin-Walled Structures*, 41:31–46, 2003.

[95] X. Huang and G. Lu. Axisymmetric progressive crushing of circular tubes. *Int. J. Crashworthiness*, 8:87–95, 2003.

[96] J.W. Hutchinson. Plastic buckling. *Advances in Applied Mechanics*, 14:67–144, 1974.

[97] T. Inoue. Analysis of plastic buckling of rectangular steel plates supported along their four edges. *Int. J. Solids Structures*, 31:219–230, 1994.

[98] Ø. Jensen, M. Langseth, and O.S. Hopperstad. Experimental investigations on the behaviour of short to long square aluminium tubes subjected to axial loading. *International Journal of Impact Engineering*, 30:973–1003, 2004.

[99] W. Johnson. *Impact Strength of Materials*. Edward Arnold, London, 1972.

[100] W. Johnson, P.D. Soden, and S.T.S. Al-Hassani. Inextensional collapse of thin-walled tubes under axial compression. *J. Strain Analysis*, 12(4):317–330, 1977.

[101] N. Jones. *Structural Impact*. Cambridge University Press, Cambridge, UK, 1989.

[102] N. Jones and E.A. Papageorgiou. Dynamic and plastic buckling of stringer stiffened cylindrical shells. *Int. J. Mech. Sci.*, 24:1–20, 1982.

[103] N. Jones and T. Wierzbicki. *Structural Crashworthiness and Failure*. Elsevier Applied Science, London, 1993.

[104] G.T. Ju and S. Kyriakides. Bifurcation and localization instabilities in cylindrical shells under bending-ii: Predictions. *International Journal of Solids and Structures*, 29:1143–1171, 1992.

[105] D. Karagiozova, M. Alves, and N. Jones. Inertia effects in axisymmetrically deformed cylindrical shells under axial impact. *International Journal of Impact Engineering*, 24(10):1083–1115, 2000.

[106] D. Karagiozova and N. Jones. Dynamic elastic-plastic buckling phenomena in a rod due to axial impact. *Int J Impact Engng*, 18:919–947, 1996.

[107] D. Karagiozova and N. Jones. Dynamic elastic-plastic buckling of circular cylindrical shells under axial impact. *Int J Solids Struct*, 37(14):2005–2034, 2000.

[108] D. Karagiozova and N. Jones. Dynamic effects on buckling and energy absorption of cylindrical shells under axial impact. *Thin-Walled Structures*, 39(7):583–610, 2001.

[109] D. Karagiozova and N. Jones. Influence of stress waves on the dynamic progressive and dynamic plastic buckling of cylindrical shells. *International Journal of Solids and Structures*, 38:6723–6749, 2001.

[110] G.N. Karam. On the ovalisation in bending of nylon and plastic tubes. *International Journal of Pressure Vessels and Piping*, 58:147–149, 1994.

[111] D. Kecman. Bending collapse of rectangular and square section tubes. *International Journal of Mechanical Sciences*, 25:623–636, 1983.

[112] T.H. Kim and S.R. Reid. Bending collapse of thin-walled rectangular section columns. *Computers and Structures*, 79:1897–1911, 2001.

[113] W.T. Koiter. The effect width of flat plates for various longitudinal edge conditions at loads far beyoun buckling load. *National Luchtvaart-Laboratorium (Netherlands), Report S*, 287, 1943.

[114] W.T. Koiter. On the Stability of Elastic Equilibrium. PhD Thesis, Delft University (in Dutch), 1945.

[115] W.T. Koiter. The effect of axisymmetric imperfections on the buckling of cylindrical shells under axial compression. *Proc. Kon. Ned. Akad. Wet.*, 66-B:265–279, 1963.

[116] A. Kromm. Die stabilitätsgrenze der kreiszylinderschale bei beanspruchung durch schub-und längskräfte. In *Jahrbuch 1942 der deutschen Lufifahrtorschung*, pages 602–616, 1942.

[117] S. Kyriakides and P.K. Shaw. Response and stability of elasto-plastic circular pipes under combined bending and external pressur. *International Journal of Solids and Structures*, 18:957–973, 1982.

[118] S. Kyriakides and P.K. Shaw. Inelastic buckling of tubes under cyclic bending. *Journal of Pressure Vessel Technology*, 109:169–178, 1987.

[119] M. Langseth and O.S. Hopperstad. Static and dynamic axial crushing of square thin-walled aluminium extrusions. *Int J Impact Eng*, 18:949–968, 1996.

[120] M. Langseth, O.S. Hopperstad, and T. Berstad. Crashworthiness of aluminum extrusions: validation of numerical simulation, effect of mass ratio and impact velocity. *Int J Impact Eng*, 22:829–854, 1999.

[121] L.H.N. Lee. Inelastic buckling of initially imperfect cylindrical shells subjected to axial compression. *Journal of Aeronautical Science*, 29:87–95, 1962.

[122] L.H.N. Lee and C.S. Ades. Plastic torsional buckling strength of cylinders including the effect of imperfections. *Journal of Aerospace Sciences*, 24:241–248, 1957.

[123] S. Li and S.R. Reid. Relationship between the elastic buckling of square tubes and rectangular plates. *ASME J. of Applied Mechanics*, 57:969–973, 1990.

[124] S. Li and S.R. Reid. The plastic buckling of axially compressed square tubes. *ASME J. of Applied Mechanics*, 59:276–=282, 1992.

[125] A. Libai and C.W. Bert. A mixed variational principle and its application to the nonlinear bending problem of orthotropic tubes - ii. application to nonlinear bending of circular cylindrical tubes. *International Journal of Solids and Structures*, 31, 1994.

[126] W.E. Lilly. The economic design of columns. *Transactions of the Institution of Civil Engineers of Ireland (Dublin)*, 33:67–93, 1906.

[127] T.T. Loo. Effects of large deflections and imperfections on the elastic buckling of cylinders under torsion and axial compression. In *Proc Second US Natl Congr Appl Mech*, pages 345–357, 1954.

[128] R. Lorenz. Achsensymmetrische verzerrungen in dünnwandigen hohlzylinder. *Zeitschrift des Vereines Deutscher Ingenieure*, 52:1706–1713, 1908.

[129] E.E. Lundquist. Strength tests of thin-walled duralumin cylinders in compression. Technical Report NACA No. 473, 1933.

[130] M.A. Macaulay and R.G. Redwood. Small scale model railway coaches under impact. *The Engineer*, 25:1041–1046, 1964.

[131] M. Madhavan and J.S. Davidson. Elastic buckling of i-beam flanges subjected to a linearly varying stress distribution. *Journal of Constructional Steel Research*, 63:1373–1383, 2007.

[132] C.L. Magee and P.H. Thornton. Design consideration in energy absorption by structural collapse. *Society of Automotive Engineers (SAE)*, (780434), 1978.

[133] E. Mahdi, B.B. Sahari, A.M.S. Hamouda, and Y.A. Khalid. An experimental investigation into crushing behaviour of filament-wound laminated cone-cone intersection composite shell. *Comp. Struct.*, 51:211–219, 2001.

[134] E. Mahdi, B.B. Sahari, A.M.S. Hamouda, and Y.A. Khalid. Crushing behaviour of cone-cylinder-cone composite system. *Mech. Comp. Struct.*, 2:99–117, 2002.

[135] M. Mahendran and N.W. Murray. Ultimate load behavior of box-columns under combining loading of axial compression and torsion. *Thin-Walled Structures*, 9:91–120, 1990.

[136] A.G. Mamalis and W. Johnson. The quasi-static crumpling of thin-walled circular cylinders and frusta under axial compression. *J. Mech. Sci.*, 25:713–732, 1983.

[137] A.G. Mamalis, D.E. Manolakas, A.K. Baldoukas, and G.L. Viegelahn. Deformation characteristics of crashworthy thin walled steel tubes subjected to bending. *Proceedings of Institute of Mechanical Engineers. Journal of Mechanical Science*, 203:411–417, 1989.

[138] A.G. Mamalis, D.E. Manolakos, and A.K. Baldoukas. Energy dissipation and associated failure modes when axially loading polygonal thin-walled cylinders. *Thin Walled Structures*, 12:17–34, 1991.

[139] A.G. Mamalis, D.E. Manolakos, G.L. Viegelahn, N.M. Vaxevanidis, and W. Johnson. The inextensional collapse of grooved thin-walled cylinders of pvc under axial loading. *Int. J. Impact Engng.*, 4(1):41–56, 1986.

[140] A.G. Mamalis, G.L. Viegelahn, D.E. Manolakos, and W. Johnson. Experimental investigation into the axial plastic collapse of steel thin-walled grooved tubes. *Int. J. Impact Engng.*, 4(2):117–126, 1986.

[141] R. Mao and G. Lu. Nonlinear analisis of cross-ply thick cylindrical shells under axial compression. *Int. J. Solids Structures*, 35:2151–2171, 1998.

[142] R. Mao and G. Lu. Plastic buckling of circular cylindrical shells under combined in-plane loads. *Int. J. Solids Structures*, 38:741–757, 2001.

[143] K. Marguerre. The apparent width of the plate in compression. Technical Report NACA TA, No. 833, 1937.

[144] K. Masuda and D.H. Chen. Study on role of partition plates in square tube subjected to pure bending. *Transactions of JSME*, A75:580–587, 2009.

[145] K. Masuda and D.H. Chen. Maximum moment of rectangular tubes subjected to pure bending. *Transactions of JSME*, A78:1340–1347, 2012.

[146] K. Masuda, D.H. Chen, and S. Ozaki. Study on pure bending collapse of square tubes in consideration of work-hardening effect. *Transactions of JSME*, A75:13–20, 2009.

[147] A.G. McFarland. Hexagonal cell structures under post-buckling axial load. *AIAA Journal*, 1:1380–1385, 1963.

[148] Q. Meng, S.T.S. Al-Hassani, and P.D. Soden. Axial crushing of square tubes. *Int. J. Mech. Sci.*, Special Issue for Structural Crashworthiness Conf., 25:747–773, 1983.

[149] R. Milligan, G. Gerand, C. Lakshmikantham, and H. Becker. General instability of orthotropic stiffened cylinders under axial compression. *AIAA Journal*, 4(11):1906–1913, 1966.

[150] R.D. Mindlin. Influence of rotatory inertia and shear on flexural motion of isotropic elastic plates. *J. Appl. Mech.*, 18:31–38, 1951.

[151] S. Morita, S. Haruyama, and D.H. Chen. Study of axially crushed cylindrical tubes with corrugated surface based on experiment and fem. *Transactions of JSME*, A76(762):215–222, 2010.

[152] D. Munz and C. Mattheck. Cross-sectional flattening of pipes subjected to bending. *International Journal of Pressure Vessels and Piping*, 10:421–429, 1982.

[153] N.W. Murray. *Introduction to the theory of thin-walled structure*. Oxford Press, London, 1984.

[154] N.W. Murray and P. Bilston. Local buckling of thin-walled pipes being bent in plastic range. *Thin-Walled Structures*, 14:411–434, 1992.

[155] N.W Murray and P.S. Khoo. Some basic plastic mechanisms in the local buckling of thin-walled steel structures. *Int. J. Mech. Sci.*, 23(12):703–713, 1981.

[156] R. Narayanan and S.L. Chan. Effective widths of plates under uniformly varying edge displacements. *Int. J. Mech. Sci.*, 28(6):393–409, 1986.

[157] W.A. Nash. Buckling of initially imperfect cylindrical shells subject to torsion. *J Appl Mech*, 24:125–130, 1957.

[158] K.W. Neale. Bifurcation in an elastic plastic cylindrical shell under torsion. *ASME J Appl Mech*, 40:826–828, 1973.

[159] E. Ore and D. Durban. Elastoplastic buckling of axially compressed circular cylindrical shells. *Int. J. Mech. Sci.*, 34:727–742, 1992.

[160] J.K. Paik. Some recent advances in the concepts of plate-effectiveness evaluation. *Thin-Walled Structures*, 46:1035–1046, 2008.

[161] M.M. Pastor and F. Roure. Open cross-section beams under pure bending: I. Experimental investigations. *Thin-Walled Structures*, 46:476–483, 2008.

[162] F. Paulsen and T. Welo. Cross-sectional deformations of rectangular hollow sections in bending: Part ii: analytical models. *International Journal of Mechanical Sciences*, 43:131–152, 2001.

[163] N.K. Prinja and N.R. Chitkara. Post-collapse cross-sectional flattening of thick pipes in plastic bending. *Nuclear Engineering and Design*, 83:113–121, 1984.

[164] A. Pugsley. On the crampling of thin tubular struts. *Quart. J. Mech. Appl. Math.*, 32(1):1–7, 1979.

[165] A. Pugsley and M. Macaulay. The large-scale crampling of thin cylindrical columns. *Quart. J. Mech. Appl. Math.*, 13:1–9, 1960.

[166] F.G. Rammerstorfer. *Leichtbau Repetitorium*. Oldenbourg Verlag, Wien, München, 1992.

[167] B.D. Reddy. An experimental study of the plastic buckling of circular cylinders in pure bending. *International Journal of Solids and Structures*, 15:669–683, 1979.

[168] B.D. Reddy. Plastic buckling of a cylindrical shell in pure bending. *International Journal of Mech. Sci.*, 21:671–679, 1979.

[169] E. Reissner. The effect of transverse shear deformation on the bending of elastic plates. *J. Appl. Mech.*, 12:69–77, 1945.

[170] E. Reissner. On finite bending of pressurized tubes. *Journal of Applied Mechanics*, 26:386–392, 1959.

[171] J. Rhodes. Manual of crashworthiness engineering—Vol. iv: Ultimate strength of thin-walled components. Technical report, Center for Transportation Studies, Massachusetts Institute of tchnology, Cambridge, MA, 1989.

[172] J. Rhodes. Buckling of thin plates and members - and early work on rectangular tubes. *Thin-Walled Structures*, 40:87–108, 2002.

[173] J. Rhodes and J.M. Harvey. Effects of eccentricity of load or compression on the buckling and post-buckling behaviour of flat plates. *Int. J. Mech. Sci.*, 13:867–879, 1971.

[174] J. Rhodes, J.M. Harvey, and W.C. Fok. The load-carrying capacity of initially imperfect eccentrically loaded plates. *Int. J. Mech. Sci.*, 17:161–175, 1975.

[175] D. Richtlinie. *Beulsicherheitsnachweise fur platten*. Deutsche Ausschuss fur stahlbeton, 1978.

[176] A. Robertson. The strength of tubular struts. *Proceedings of the Royal Society of London*, 121, Series A:558–585, 1928.

[177] A. Rusch and J. Lindner. Application of level 1 interaction formulae to class 4 sections. *Thin-Walled Structures*, 42:279–293, 2004.

[178] S. Santosa and T. T. Wierzbicki. Effect of an ultralight metal filler on the torsional crushing behavior of thin-walled prismatic columns. Impact and Crushworthiness Laboratory Technical Report 5, Massachusetts Institute of Technology, 1997.

[179] L. Schuman and G. Back. Strength of rectangular plates under edge compression. Technical Report NACA TR 356, 1930.

[180] E. Schwerin. Die torsionsstabilität des dünuwandigen rohres. *ZAMM*, 5:235–243, 1925.

[181] P. Seide and V.I. Weingarten. On the buckling of circular cylindrical shells under pure bending. *Transactions of the ASME*, 28:112–116, 1961.

[182] M. Seitzberger, F.G. Rammerstorfer, H.P. Gradinger, H.P. Degischer, M. Blaimschein, and C. Walch. Experimental studies on the quasi-crushing of steel columns filled with aluminium foam. *International Journal of Solids and Structures*, 37:4125–4147, 2000.

[183] I. Sigalas, M. Kumosa, and D. Hull. Trigger mechanisms in energy-absorbing glass cloth/epoxy tubes. *Comp. Sci. Technol.*, 40:265–287, 1991.

[184] A.A. Singace. Axial crushing analysis of tubes deforming in the multi-lobe mode. *Int. J. Mech. Sci.*, 41:865–890, 1999.

[185] A.A. Singace and H. El-Sobky. Behaviour of axially crushed corrugated tubes. *International Journal of Mechanical Sciences*, 39:249–268, 1997.

[186] A.A. Singace, H. El-Sobky, and M. Petsios. Influence of end constrains on the collapse of axially impacted frusta. *Thin Walled Structures*, 39:415–428, 2001.

[187] A.A. Singace, H. Elsobky, and T.Y. Reddy. On the eccentricity factor in the progressive crushing of tubes. *Int. J. Solids Structures*, 32(24):3589–3602, 1995.

[188] J. Singer. Buckling of integrally stiffened cylindrical shells—A review of experiment and theory. In *In Contributions to the Theory of Aircraft Structures*, pages 325–357. Delft University Press, 1972.

[189] B. Skocezen and J. Skizypek. Application of the equivalent column concept to the stability of axially compressed bellows. *International Journal of Mechanical Sciences*, 34:901–916, 1992.

[190] L.H. Sobel and S.Z. Newman. Plastic buckling of cylindrical shells under axial compression. *Trans. ASME J. Pressure Vessel Technology*, 102:40–44, 1980.

[191] R.V. Southwell. On the general theory of elastic stability. *Philosophical Transactions ot the Royal Society of London*, 213, Series A:187–244, 1914.

[192] P.K. Stangl and S.A. Meguid. Experimental and theoretical evaluation of a novel shock absorber for an electrically powered vehicle. *Int. J. Impact Engng*, 11:41–59, 1991.

[193] W.B. Stephens, J.H. Strarnes, and B.O. Almorth. Collapse of long cylindrical shells under combined bending and pressure loads. *American Institute of Aeronautics and Astronautics Journal*, 13:20–25, 1975.

[194] E.Z. Stowell. A unified theory of the plastic buckling of columns and plates. Technical Report NACA TN 1556, 1948.

[195] W.J. Stronge and T.X. Yu. *Dynamic Models for Structural Plasticity.* Springer-Verlag, 1993.

[196] L.L. Tam and C.R. Calladine. Inertia and strain-rate effects in a simple plate-structure under impact loading. *Int J Impact Engng*, 11:349–377, 1991.

[197] B.F. Tatting, Z. Gurdal, and V.V. Vasiliev. The brazier effect for finite length composite cylinders under bending. *International Journal of Solids and Structures*, 34:1419–1440, 1997.

[198] P.H. Thornton, H.F. Mahmood, and C.L. Magee. Energy absorption by structural collapse. In *Structural Crashworthiness, (Jones N. and Wierzbicki T. eds.)*, pages 96–117. Butterworths, London, 1983.

[199] S. Timoshenko. Einige stabilittsprobleme der elastizitätstheorie. *Zeitschrift für Mathematik und Physik*, 58:337–385, 1910.

[200] S. Timoshenko. Bending stresses in curved tubes of rectangular cross-section. *Trans. ASME*, 45:135, 1923.

[201] S. Timoshenko. *History of Strength of Materials.* McGraw-Hill Book Company, New York/ Toronto/ London, 1953.

[202] S.P. Timoshenko. *Theory of Elastic Stability.* McGraw-Hill, 1936.

[203] S.P. Timoshenko and J.M. Gere. *Theory of Elastic Stability.* McGraw-Hill, 2nd edition, 1996.

[204] P. Tugoce and J. Schroeder. Plastic deformation and stability of pipes exposed to external couples. *International Journal of Solids and Structures*, 15:643–658, 1979.

[205] S. Ueda. Moment-rotation relationship considering flattening of pipe due to pipe whip loading. *Nuclear Engineering and Design*, 85:251–259, 1985.

[206] K. Ushijima and D.H. Chen. Evaluation of energy absorption capacity for thin-walled tapered tube. *Transactions of Society of Automotive Engineers of Japan*, 39(3):77–82, 2008.

[207] K. Ushijima, D.H. Chen, K. Masuda, and S. Haruyama. Estimation of average compressive load for tubes under axial loading in consideration of strain hardening. In *Advanced Studies in Mechanical Engineering, Proceedings of JSSM2006, Korea*, pages 224–227, 2006.

[208] K. Ushijima, S. Haruyama, and D.H. Chen. Evaluation of first peak stress in axial collapse of circular cylindrical shell. *Transactions of JSME*, A70(700):1695–1702, 2004.

[209] K. Ushijima, S. Haruyama, K. Fujita, and D.H. Chen. Study on axially crushed cylindrical tubes with grooved surface. *Transactions of JSME*, A71(707):1015–1022, 2005.

[210] K. Ushijima, S. Haruyama, H. Hanawa, and D.H. Chen. Strain concentration for cylindrical tubes subjected to axial compression. *Transactions of JSME*, A71(707):1023–1029, 2005.

[211] T. von Karman, E.E. Sechler, and L.H. Donnell. The strength of thin plates in compression. *ASME Applied Mechanics Transactions*, 54:53–57, 1932.

[212] X. Wang and F.G. Rammerstorfer. Determination of effective breadth and effective width of stiffened plates by finite strip analysis. *Thin-Walled Structures*, 26:261–286, 1996.

[213] M.D. White, N. Jones, and W. Abramowicz. A theoretical analysis for the quasi-static axial crushing of top-hat and double-hat thin walled sections. *Int. J. Mech. Sci.*, 41:209–233, 1999.

[214] T. Wierzbicki. Crushing analysis of metal honeycombs. *Int. J. Impact Engng.*, 1(2):157–174, 1983.

[215] T. Wierzbicki. Optimum design of integrafed front panel against crash. Technical report, Ford Motor Company, Vehicle Component Dept., 1983.

[216] T. Wierzbicki and W. Abramowicz. On the crushing mechanics of thin-walled structures. *Journal of Applied Mechanics*, 50:727–734, 1983.

[217] T. Wierzbicki, S.U. Bhat, and W. Abramowicz. Alexander revisited - a two folding elements of progressive crushing of tubes. *Int. J. Solids Structures*, 29(24):3269–3288, 1992.

[218] T. Wierzbicki, L. Recke, W. Abramowicz, T. Gholami, and J. Huang. Stress profilest in thin-walled prismatic columns subjected to crush loading. ii. bending. *Comput. Struct.*, 51:625–641, 1994.

[219] J.C. Wilhoit and J.E. Merwin. Critical plastic buckling parameter for tubing in bending under axial tension. In *5th Ann. Offshore Tech. Conf. Houston, paper, OTC. 1874*, 1973.

[220] W.M. Wilson and E.D. Olson. Tests of cylindrical shells. *University of Illinois Engineering Experiment Station, Bulletin*, (331), 1941.

[221] G. Winter. *Strength of Thin Steel Compression Flenges. Reprint No.32.*
Ithaca, NY, USA: Cornell University Engineering Experimental Station,
1947.

[222] T. Yamada, Y. Takayama, D. Abe, and T. Nishimura. Analysis of buckling behavior of aluminum structures for crash absorbing. In *SAE International World Congress, 2004-01-1614*, pages 10–15, 2004.

[223] N. Yamaki. Experiments on the postbuckling behavior of circular cylindrical shells under torsion. In *Buckling of structures (Budiansky B,
editor)*, pages 312–330. New York: Berlin-Heidelberg, 1976.

[224] N. Yamaki. Postbuckling behavior of circular cylindrical shells under
torsion. *Ingenieur Archiv*, 45(2):79–89, 1976.

[225] N. Yamaki. *Elastic Stability of Cylindrical Shell.* North-Holland, Ams-
terdam, 1984.

[226] Y. Yoshida, X. Wan, M. Takahashi, and T. Hosokawa. Bending energy
absorption of extruded aluminum beams. *JSAE Review*, 18:385–392,
1997.

[227] C. Yu and B.W. Schafer. Effect of longitudinal stress gradient on the
ultimate strength of thin plates. *Thin-walled Structures*, 44:787–799,
2006.

[228] T.X. Yu, S.R. Reid, and B. Wang. Hardening softening behaviour of
tubular cantilever beams. *International Journal of Mech. Sci.*, 35:1021–
1033, 1993.

[229] T.X. Yu and L.S. Teh. Large plastic deformation of beams of angle-
section under symmetric bending. *International Journal of Mechanical
Sciences*, 39:829–839, 1997.

[230] L.C. Zhang and T.X. Yu. An investigation of the brazier effect of a
cylindrical tube under pure elastic-plastic bending. *Int. J. Pres. Ves. &
Piping*, 30:77–86, 1987.

[231] X. Zhang and Q. Han. Buckling and postbuckling behaviors of imperfect
cylindrical shells subjected to torsion. *Thin-Walled Structures*, 45:1035–
1043, 2007.

[232] X.L. Zhao and G.J. Hancock. Experimental verification of the theory
of plastic moment capacity of an inclined yield line under axial force.
Thin-Walled Structures, 15:209–233, 1993.

Index